BRINGING

BIOLOGY TO LIFE

BRINGING BIOLOGY TO LIFE

An Introduction to the Philosophy of Biology

MAHESH ANANTH

broadview press

BROADVIEW PRESS— www.broadviewpress.com
Peterborough, Ontario, Canada

Founded in 1985, Broadview Press remains a wholly independent publishing house. Broadview's focus is on academic publishing; our titles are accessible to university and college students as well as scholars and general readers. With over 600 titles in print, Broadview has become a leading international publisher in the humanities, with world-wide distribution. Broadview is committed to environmentally responsible publishing and fair business practices.

The interior of this book is printed on 100% recycled paper.

© 2018 Mahesh Ananth

All rights reserved. No part of this book may be reproduced, kept in an information storage and retrieval system, or transmitted in any form or by any means, electronic or mechanical, including photocopying, recording, or otherwise, except as expressly permitted by the applicable copyright laws or through written permission from the publisher.

Library and Archives Canada Cataloguing in Publication

Ananth, Mahesh, 1969-, author
    Bringing biology to life : an introduction to the philosophy of biology / Mahesh Ananth.

Includes bibliographical references and index.
ISBN 978-1-55111-990-8 (softcover)

    1. Biology—Philosophy. I. Title.

QH331.A53 2018   570.1   C2018-902360-0

Broadview Press handles its own distribution in North America
PO Box 1243, Peterborough, Ontario K9J 7H5, Canada
555 Riverwalk Parkway, Tonawanda, NY 14150, USA
Tel: (705) 743-8990; Fax: (705) 743-8353
email: customerservice@broadviewpress.com

Distribution is handled by Eurospan Group in the UK, Europe, Central Asia, Middle East, Africa, India, Southeast Asia, Central America, South America, and the Caribbean. Distribution is handled by Footprint Books in Australia and New Zealand.

Canada

Broadview Press acknowledges the financial support of the Government of Canada for our publishing activities.

Edited by Robert M. Martin
Book design by Michel Vrana

PRINTED IN CANADA

For Michael Bradie and David J. Depew

CONTENTS

ACKNOWLEDGMENTS xi

INTRODUCTION xiii

CHAPTER 1 | ARISTOTLE: THE FIRST PHILOSOPHER OF BIOLOGY

I. Giving Aristotle His Due   1
II. Standing on Shoulders: Influences on Aristotle's Philosophy of Biology   3
III. Aristotle's Philosophy of Science   5
IV. From Philosophy of Science to Philosophy of Biology   7
V. The Four Causes   12
VI. Biological Function/Teleology   15
VII. Objections to Aristotle's Teleology   17
VIII. Conclusion   33
Chapter Review: Discussion and Questions   34

CHAPTER 2 | CHARLES DARWIN: THE GRANDEUR OF A PHILOSOPHICAL NATURALIST

I. Introduction: The Newton of a Blade of Grass?   39
II. Life and Times: Darwin's Transformation into a Naturalist   41
III. A Sketch of Darwin's Argument in the *Origin*   43
IV. The Origin of Darwin's Philosophy of Science   45
V. Natural Selection: Probably True or True to the Highest Degree Probable?   49
VI. A Necessary Deviation: The Divine Challenge to Natural Design   52
VII. Darwin's "One Long Argument"   53
VIII. Taking Stock: Assessing Darwin's Argument   65
IX. Two Lingering Worries   76
X. Conclusion and Darwin's Influence on the Philosophy of Biology   78
Chapter Review: Discussion and Questions   79

CONTENTS

CHAPTER 3 | THE UNIT(S) OF SELECTION

I. Introduction: Historical Reminder and the Keeper of Genes   85
II. Laying Out the Unit(s) of Selection Debate   88
III. Setting the Stage: Darwin on Individual and Group Selection   91
IV. The Genetic Turn: The Nonsense of Individual and Group Selection?   93
V. Gould's Visibility and Kaleidoscope Argument:
A Further Reply to Dawkins   100
VI. Genic Pluralism as the Solution?   103
VII. What about Group Selection?   103
VIII. Taking Natural Selection Seriously:
Multi-Level Selection to the Rescue?   107
IX. A Snowball's Chance? The Developmental Systems Alternative   110
X. Conclusion   119
Chapter Review: Discussion and Questions   120

CHAPTER 4 | BIOLOGICAL FUNCTION

I. Introduction: A Fascination with Knives   125
II. Function Talk and Talk of Teleology   127
III. Basic Criteria for Functional Analysis   129
IV. Systemic Functionalism   130
V. Varieties of Evolutionary Functionalism   139
VI. Conclusion   156
Chapter Review: Discussion and Questions   157

CHAPTER 5 | THE SPECIES DEBATE

I. Introduction: A Confusing Day at the Zoo   161
II. Who Cares? Why Care?   165
III. Just a Bunch of Individuals? The Nominalist Challenge   166
IV. Criteria for Species Ascription   171
V. Organizing the Species Debate   172
VI. Engaging the Species Debate: Essentialism and Natural Kind   174
Phenetic Species Concept   179
Biological Species Concept   182
Species as Individual Lineages   185
Phylogenetic-Cladistic Species Concept   192
Relational Biosimilarity Species Concept   198
VII. A Plea for Pluralism and Concluding Remarks   204
Chapter Review: Discussion and Questions   205

CHAPTER 6 | NATURALIZING THE MORAL SENSE: EVOLUTION AND ETHICS

I. Introduction   211
II. The Three Faces of Morality   214
III. Ethics Projects and Evolutionary Ethics Projects   215
IV. Faces of Evolution: Evolutionary Altruism and Human Morality   217
V. Objections to Naturalizing Ethics   220
VI. The Evolution of the Moral Sense: Contemporary Evolutionary Ethics   232
VII. How Should We Proceed? Kitcher to the Rescue?   246
VIII. Conclusion   247
Chapter Review: Discussion and Questions   247

CHAPTER 7 | EVOLUTIONARY PSYCHOLOGY

I. Introduction: Just What the Prophet Ordered!   253
II. Principles of SEP   255
  1. Mental Modules   256
  2. The Environment of Evolutionary Adaptedness   266
  3. Is Mother Goose in the House of Biology?   271
III. Conclusion   273
Chapter Review: Discussion and Questions   273

CHAPTER 8 | OF GNATS AND MEN: EVOLUTION, RELIGION, AND INTELLIGENT DESIGN

I. Science and Religion: A Troubled Romance   277
II. Design, Design, Design: Paley's Challenge   279
III. Darwin's Dismay and Proposed Resolution   280
IV. Paley's New Defenders: The Rise of the Intelligent Design Camp   282
V. The Pink Demarcation Elephant in the Room   293
VI. Conclusion   302
Chapter Review: Discussion and Questions   302

Index   309

ACKNOWLEDGMENTS

This book is the product of the work that I have done with both my undergraduate and graduate students over the years. I am fortunate to have had such wonderful "guinea pigs" in the process of putting this manuscript together. I hope that they can see and appreciate their contributions throughout the text. I know that this project could not have come to fruition without their insightful feedback and willingness to be guided through material that was quite recondite with respect to most of their educational backgrounds.

I pass along a deep thanks to Ben Dixon, Catherine Herbert-Annis, and Michael Scheessele. These scholar-teachers provided detailed comments on earlier drafts of each of the chapters—and did so with much joy. Their demands for content clarification and reminders of my target audience moved me to re-write many sections of the text. Their steadfastness to the support of this project has not been lost on me in any way. I look forward to returning the favor to each of them on each of their future scholarly endeavors. I also owe a tremendous debt of gratitude to Fred D. Miller, who has been a consummate scholar-teacher archetype. His guidance in helping me to understand Aristotle's philosophical works has been a source of enormous inspiration. Early versions of the Aristotle chapter owe much to his feedback. Additionally, over the years, I have bounced-off ideas related to this book to my colleagues, Louise Collins, Matthew Shockey, Warren Shrader, and Lyle Zynda. I thank each of them for their professional support.

I also am very thankful to the editorial team at Broadview Press. Specifically, Stephen Latta has guided this stop-and-go project with much patience, excellent content and scope feedback, and skillful prodding. I feel very fortunate to have worked with him. Additionally, both the blind reviewer, Bob Martin, and Tara Lowes offered detailed suggestions on improving the content of the text. Not only did they catch a great many gaffs on my part, but they provided a philosophical luster that was lacking.

My wife, Rekha Murthy, has been my loving companion and support system. I could not fathom this life's journey without her and my boys, Kathan, Rohan, and Millan. Academics, in their pursuit of scholarly projects, require freedom from not only academic teaching and service requirements, but also from a range of family-oriented chores. This usually places additional burdens on other family members and my family is no exception. Fortunately, during my absences, Rekha has managed our family needs with a loving flair and caring handle that is unique to her. There are not enough heartfelt thanks to extend to her ...

Finally, my interests in both the history of biology and the philosophy of biology have and continue to be inspired by my remarkable teachers, David J. Depew and Michael Bradie. David introduced me—with much passion and panache—to the works of Aristotle and Darwin during my undergraduate days and was kind enough to read and offer detailed comments on numerous parts of this text and, over the years, on a great deal of my scholarship. Passing me along to Michael for my graduate studies proved to be a great exchange by David in terms of locating a deeply dedicated teacher-scholar molder. Michael not only supported my philosophy of biology interests and provided feedback on many of my research projects (including this one!), but also taught me how to engage this area of study with a kind of analytic care I could not have imagined—a kind of attention to detail that I learned to include in my own teaching pedagogy. I feel very fortunate to have studied with such fine mentors whom I now call friends. I can only hope that the content of this text meets with their approval, especially since this work is dedicated with an enduring gratitude to both of them.

M.A.

INTRODUCTION

DISCIPLINES AND SUB-DISCIPLINES: LOCATING THE PHILOSOPHY OF BIOLOGY

One of the core disciplines within *philosophy* is the *philosophy of science*. One of the core sub-disciplines within the philosophy of science is the *philosophy of biology*—the topic of this text. As a way of understanding the philosophy of biology as a sub-discipline, a brief account of philosophy and philosophy of science will prove useful. Although there is no consensus amongst philosophers on how to define either of these areas of study, philosophy can be viewed, maybe conservatively, as a kind of critical thinking (with special attention to logical structure and rational argumentation) about the procedures and assumptions of other disciplines. It encompasses a preoccupation with understanding issues as clearly and fairly as possible with the aid of its logical toolkit. Such fixation on precision and (to the extent that it is possible) unbiased curiosity includes engaging in complex and frequently unusual lines of reasoning that could very well result in a rejection of one's own considered view, further endorsement of one's own view, or even the suspension of judgment regarding one's view about the issue at hand. On this view of philosophy, critical thinking requires employing one's intellectual resources in defense of plausible answers to questions that are tentative and yet still worthy of believing. Indeed, philosophical understanding is acquired *in the journey* of critical exploration of the many arguments surrounding an issue as one moves to embrace tentatively a particular position on the issue.

Philosophy of science is concerned with a critical examination of the core concepts, reasoning, logical structure, and explanations found within the field of science. Put another way, philosophy of science is an examination of the nature of science and the relevant historical, epistemological (i.e., knowledge), metaphysical, moral, and political implications of the discipline of science. For example, when physicists talk about elementary objects such as strings, are they claiming that such entities *actually exist*? What sort of *logical reasoning* informs their claims about the existence of strings? If strings do exist, how do they *know* that such entities exist? If such entities exist, what does this tell us about related macroscopic entities and subatomic particles? Does the existence of strings have any implications for human *morality* or *political* organizations and practices? These are the sorts of questions that are explored within the philosophy of science.

A similar comparison can be made in terms of history and sociology and their respective science-related sub-discipline. If we view the study of history as a critical analysis of the causal factors relevant to the chronological development of ideas, people, places, and events, then the history of science can be understood as the study of the causal factors pertinent to the development of various fields and aspects of science that includes empirical, theoretical, and practical knowledge about the natural world. Likewise, if we understand sociology as the systematic study of human social behavior within the context of various kinds of social institutions, then the sociology of science can be viewed as the critical study of the social structure and dynamics applicable to science. This sociological research will include the comparative study of the social structure of various institutions of science and their influence on other social institutions and how scientific knowledge correspondingly affects human activities and vice versa.

Keeping in mind the idea of disciplines and sub-disciplines, one way of understanding contemporary philosophy and philosophy of science is to view them, in part, as second-order disciplines. What this means is that philosophy investigates the primary discussions of other fields. For example, the philosophy of law is a critical exploration of issues that encompass the conceptual and practical nuances of the province of law. This evaluation, in part, includes an analytical examination of the core concepts, reasoning, and logical structure that captures the distinctiveness of primary specialties (like law). At its best, philosophy—as a second-order discipline—provides analytic precision and insights that might not be on the radar of industrious and occupied practitioners of primary disciplines.

The philosophy of biology, as a sub-discipline of the philosophy of science, is just this sort of second-order arena. Its practitioners investigate the core concepts, reasoning, and logical structure of arguments and explanations that are the primary domain of biology. More specifically, the philosophy of biology is primarily, though not exclusively, organized around conceptual and practical issues pertaining to evolutionary biology. In a sense, the philosophy of biology is the philosophy of evolution. Nowadays there are new areas of inquiry that are receiving attention from a philosophy of biology point of view. For example, the topics of adaptation, development, units of selection, biological function, knowledge, metaphysics, species, history of biology, and ethics are all traditional topics within the philosophy of biology that are currently under re-examination by philosophers. Moreover, areas of inquiry like the human genome project, race, environment, health, culture, emotions, language, free will, biotechnology, and psychology are fast becoming the "hot" loci within the philosophy of biology.

Before turning to the synopsis of the contents of this book, a point about philosophical analysis is warranted. As mentioned above, philosophers have a tendency to formalize arguments in a way that is a bit foreign to those who have had very little exposure to philosophy. This book abides by this tradition, but aims to provide the kind of hand-holding that makes the formal arguments easily accessible—even to beginning students in philosophy. This is accomplished in most of the chapters with a primary argument that governs most of the topic under discussion. Then, subsequent arguments within the chapter are connected back to the primary argument if it is relevant to do so. This may seem hard-going at the start, but the diligent student (with the help of the instructor) will come to grasp and appreciate this approach as the text unfolds. As a form of guidance, each chapter ends with its own set of questions designed to reinforce understanding of the reading material. With this in mind, there is a suggestion that should be taken seriously: *students should do themselves a favor and answer these questions as part of their reading assignments*. Anything unclear in the readings and questions can be directly addressed in class to ensure comprehension and retention of the material.

NAVIGATING THIS TEXT:

A MAP AND CHAPTER SUMMARIES

At an introductory level, this text will provide a guided tour into the philosophy of biology. This tour will canvass three broad areas: (1) the history of biological thought, (2) traditional philosophy of biology debates, (3) and more recent philosophy of biology issues pertaining to humans. Although an exhaustive coverage of each of these areas is not possible, the hope is that this exposure will not only motivate students to continue to explore the topics examined here and other related topics, but also to realize that there is ample room for them to offer insightful critical comments of their own. Indeed, the genesis, organization, and the many arguments of this book owe a great deal to the discerning undergraduates with whom I have had the pleasure to work over the years.

This philosophy of biology tour commences with the history of biology. Chapter 1 appropriately begins with the first philosopher of biology in Western philosophy, Aristotle. Although most students have likely encountered Aristotle by way of his ethical treatises, his biological writings are every bit as impressive in terms of his study of animal life and the theoretical/logical framework he offers. The chapter describes both the medical and chemical theories that influenced his thinking and the prominence he gives to logical structure, biological function, and causation in making sense of animal features and behaviors. Additionally, the chapter dispels many erroneous criticisms of Aristotle's account of causation, revealing just how impressive a thinker he is in the area of biology. Indeed, even though some of his viewpoints are difficult to support from a contemporary perspective, the chapter makes clear why even the mighty Charles Darwin held Aristotle in such high esteem.

Skipping a great deal of the history of biology, Chapter 2 turns to the remarkable work of Charles Darwin and his overall defense of evolution by natural selection in his famous *Origin of Species*. It is in this chapter that students are exposed to a formal argument reconstruction of the primary argument (of the *Origin of Species* and of this chapter) and then the subsequent arguments in defense of the overarching opening argument. Students are given the chance to see how Darwin systematically delineates how life plausibly came into existence and developed, and the role natural selection must have played in this. Crucial to this account is Darwin's commitment to (1) geologic time, (2) a continuously changing Earth, (3) the analogy of artificial selection/breeding, (4) and the role of resource scarcity and its influence on population dynamics. The chapter concludes with a reminder of the momentous influence Darwin's thinking came to have not only within the narrow bounds of biology, but also across a great many areas of inquiry. In fact, it was Darwin's efforts that paved the way for abandoning nearly two thousand years of Aristotelian biology and setting the table for what would become modern biology and philosophy of biology. This stands as one of the great intellectual feats in human history.

Chapter 3 represents the first of three chapters focusing on traditional philosophy of biology issues, all of which were directly or indirectly addressed by both Aristotle and Darwin. The first of these issues discussed in this chapter is the "Unit(s) of Selection" debate. The primary concern focuses on the question "Over what does natural selection range?" Scholars have argued that one of the following is favored by natural selection over the other units: (1) individual organisms, (2) groups, (3) genes, (4) multiple levels, or (5) the developmental system. The central arguments for and against each of these five alternatives are put on display for the reader after a little bit of "genetics 101" is explained. The hope is that the reader will come away with an understanding of what questions are deemed most relevant by defenders of these different alternative units of selection and the arguments for why some questions are given pride of place over other questions.

Chapter 4 examines the concept of biological function. It may seem straightforward to say that the performance of a specific task by something captures what it means for that something to have a function. For example, a triangular solid wooden wedge has the function to keep doors open because that is the task it is designed to perform. Similarly, the function of the heart is to pump blood because that is the primary task that the heart performs. Once Darwinian lenses are flipped on, however, a number of vexing philosophical concerns creep to the forefront, rendering the artifact analogy dubious. After providing criteria for what counts as a cogent concept of biological function, some of the conceptual worries surrounding the biological function debate are addressed in this chapter by way of the following four alternative theories: (1) Systemic Functionalism, (2) Etiological Evolutionary Functionalism, (3) Propensity Functionalism, and (4) Mixed Evolutionary Functionalism. It is argued that the mixed account of (4) may well be worth endorsing in the light of some damaging objections to (1)–(3).

Chapter 5 gives special attention to the species debate. Biologists, as far back as Aristotle, have tried to carve-up the natural world into neat categories by relying on a set of criteria related to organism appearance, reproduction, and population dynamics. Yet, the natural world appears to be immune to the somewhat arbitrary "biological slicing" accounts of species such that even Darwin insisted that the pursuit of a univocal species concept is futile. Darwin's concerns are not without merit and this chapter explores the difficulties of submitting a persuasive species concept, even in the light of the advances in genetics and field biology. Specifically, the following theories will be discussed: (i) species as natural kinds, (ii) pheneticism, (iii) biological species concept, (iv) species as individuals, (v) phylogenetic-cladistic species concept, and (vi) relational biosimilarity species concept. With some reservation, this analysis reveals that (vi) does an admirable job of contending with both the biological and philosophical demands that constitute the species debate.

The final three chapters investigate attempts to incorporate evolutionary biology into the human condition. Chapter 6 examines to what extent evolution can assist in making sense of human morality. If humans are in possession of a "moral sense" (a faculty for determining what is morally right and wrong), then it is worth exploring the origin of this feature. Restricting the domain of discourse to naturalistic accounts, it is worth determining whether evolution has anything interesting to contribute to this moral sense faculty. To this end, upon sketching the various senses of morality (everyday morality, moral theory, and metaethics), the possible evolutionary causes (kin selection, group selection, reciprocal altruism, combination accounts), and the common objections that should be avoided (is-ought fallacy, objection from agency, and objection to progressive evolution), a number of contemporary evolutionary ethical theories are explained and evaluated. The result of this analysis is that many of the contemporary attempts at an evolutionary ethic fall short of making sense of human morality. Still, if we accept that humans are a product of the tree of life, then it is reasonable to think that the set of moral capacities, which comprises the moral sense faculty and contributes to modern human morality, is the product of hominid evolution.

At the end of the *Origin of Species*, Darwin speculated that human psychology would be transformed for the better by coming under the umbrella of evolutionary biology. The last thirty years appear to vindicate Darwin's pronouncement, as the field of evolutionary psychology has emerged as a distinct area of inquiry regarding the nature of human behavior and mind. Chapter 7 investigates this newer area of psychology. After making a distinction between Weak Evolutionary Psychology (WEP) and Strong Evolutionary Psychology (SEP), the chapter focuses on the ambitious efforts of defenders of SEP. While WEP merely reminds researchers not to ignore possible insights offered by the field of evolutionary biology, SEP boldly advances the belief in an "innate" universal human mind made up of distinct evolved modules, which are crucial to making sense of the "surface" cultural-psychological variation we observe in the world today. The overall *prima facie* conclusion defended in this chapter is that, although SEP might be championed by some scholars, it should leave the rest of us with very little about which to cheer—even prescient Darwin might bid defenders of SEP to continue their efforts with a bit more restraint!

Chapter 8 concludes the philosophy of biology tour and the human-centered topics with an examination of the seemingly never-ending battle between advocates of evolution and stalwarts of religion. In a sense, this final chapter is an expansion of a similar discussion raised in the second chapter of this book. Basically, one way to make sense of Darwin's *Origin of Species* is to see it as an argument against the view that a divine-designer is the proximate cause of the natural world. Darwin goes to great lengths to reveal the absurdity of thinking that a God-like being could be the creator of the sundry and odd life forms that did and do roam this planet. Rather, argues Darwin, it is much more likely that the process of natural selection makes best sense of life as we know it on this planet. Similarly, in an updated rendition of this argument, contemporary "intelligent design" loyalists advocate that a God-like being is the only rational explanation of the complexities of life. As a way of entering this highly contentious debate, this chapter will focus on two of the intelligent design camp's prominent defenders, Phillip Johnson and Michael Behe, and the critical rejoinder put forth against their arguments. This analysis will show that there is very little of the Intelligent Design project that is worth salvaging by way of Johnson's and Behe's efforts.

The depth and content of topics covered and style of presentation in this text are, in part, the product of the work I have done with students over the years. As a result, I am sure that there will be concerns regarding omitted topics, scope and content of material covered, and style of exposition. I am quite sympathetic with these concerns, but I hope that the quality and variety of topics covered still suffice to motivate both instructors and students to move beyond what is offered here. I hope, at the very least, to have brought the philosophy of biology to life.

Chapter 1

ARISTOTLE
The First Philosopher of Biology

For in all natural things there is something marvelous.... For what is not haphazard but rather for the sake of something is in fact present most of all in the works of nature; the end for the sake of which each animal has been constituted or comes to be takes the place of the good. If someone has considered the study of the other animals to lack value, he ought to think the same thing about himself....

—*Aristotle*, Parts of Animals
(Book I.5, 645^{a}16–17 and 645^{a}24–28; trans. Lennox, 2001b)

I. GIVING ARISTOTLE HIS DUE

Born in 384 BCE in the northern Greek township of Stagira, Aristotle eventually found his way to Athens and Plato's prestigious Academy around 367 BCE. He would study in this center for higher learning with his teacher and mentor, Plato, for about the next 20 years. Due to political upheaval and warring factions, however, Aristotle left and returned to Athens on numerous occasions during his philosophical development.[1] Fortunately, as part of his traveling adventures, Aristotle took up the practice of what we might now call field and experimental biology. He studied, dissected, and experimented on many marine and land organisms; as well, he studied and learned of a great variety of flowers and plants.[2] The following statement does much justice to Aristotle's accomplishments in the field of what we now call biology: "Though Aristotle's work in zoology was not without errors, it was the grandest

---

1 See Shields (2007) and Anagnostopoulos (2009) for more on Aristotle's life.
2 Aristotle was Alexander the Great's personal teacher. This relationship allowed Aristotle to gain access to a great many life forms because, as part of his territorial conquests, Alexander would ship to Aristotle organisms, fauna, and flora from these local areas—especially Asia. See David Salter (2001), p. 137.

biological synthesis of the time, and remained the ultimate authority for many centuries after his death."[3]

In the light of this range of his biological studies, it is no stretch of the imagination to claim that Aristotle's philosophical views may have been influenced by his biological studies and vice-versa.[4] To the extent that this is so, Aristotle not only studied animals as a kind of "bench biologist," but he also brought a theoretical perspective to his interactions with, observations of, and reflections on biological systems. The upshot is that, because of both his close engagement with plant and animal life and his ability to put forth a theoretical framework with respect to this engagement, Aristotle can legitimately be thought of as both a biologist and a philosopher of biology—indeed, he can likely be considered one of the first of each of these in the history of Western thought.[5]

The work of Aristotle that has come down to us is a staggering collection of essays and lecture notes including complex discussions concerning history, logic, psychology, physics, rhetoric, aesthetics, politics, biology, ethics, and much more. As remarkable as the scope of Aristotle's areas of research is, his biological works alone are quite impressive. Specifically, Aristotle's *History of Animals*, *Parts of Animals*, and *Generation of Animals* (amongst other biological works) reveal a systematic and careful empirical scientist, whom Grant fittingly labels "the founder of biology as a discipline" (Grant, 2007, p. 33). Indeed, as Jonathan Barnes notes with equal admiration, Aristotle's "scientific fame rests primarily on his work in zoology and biology: his study on animals laid the foundations of the biological sciences; and they were not superseded until two millennia after his death" (Barnes, 2000, p. 16).

Comments from the very person whose ideas aided in its supersession, Charles Darwin, are worth our attention. In a letter of gratitude (presented in part below) to William Ogle (1882/1987),[6] Darwin expresses both his praise and awe of Aristotle's accomplishments in the field of biology:

> You must let me thank you for the pleasure which the Introduction to the Aristotle book has given me. I have rarely read anything which has interested me more; though I have not read as yet more than a quarter of the book proper. From quotations which I had seen I had a high notion of Aristotle's merits, but I had not the most remote notion what a wonderful man he was. Linnaeus and Cuvier have been my two gods, though in different ways, but they are mere school-boys to old Aristotle.... I am glad that you have explained in so probable a manner some of the grossest mistakes attributed to him.—I never realized before reading your book to what an enormous summation of labour we owe even our common knowledge. (Darwin, 1882)[7]

---

3   http://www.ucmp.berkeley.edu/history/aristotle.html.
4   For such a defense, see Marjorie Grene (1963) and Anthony Preus (1975).
5   That Aristotle should be considered an experimental biologist is persuasively defended by Cosans (1998).
6   Ogle is an Ancient Greek Philosopher who translated Aristotle's *Parts of Animals* and presented it to Darwin.
7   The entire letter and an analysis of it can be found in Gotthelf (1999).

Darwin seems quite pleased to have some common misconceptions about Aristotle's views clarified by Ogle. Indeed, not only does Darwin concede that he had been paying homage to the wrong "gods" of his day, but he also goes on to acknowledge that the very conceptual frameworks used to make sense of even mundane aspects of human life owe much to Aristotle's toil. It is no surprise, then, to learn that Darwin thought Aristotle "was one of the greatest, if not the greatest observer, that ever lived (Darwin, 1879)."[8] Again, it is not casually or mere hero worshipping that Aristotle is being given the appellations of "first biologist" and "first philosopher of biology."

Why such high praise from Darwin and others? This opening chapter attempts to answer this question by providing a glimpse into Aristotle's philosophy of biology. To this end, first, a brief overview of those individuals who influenced Aristotle's thinking about biology will be explained. Second, Aristotle's view of nature, that is, his philosophy of science will be made clear. Third, with his philosophy of science in hand, Aristotle's philosophy of biology will be explained; specifically, the logical structure that Aristotle employs to make sense of biological systems will be put forth, along with his four-fold causal account. Fourth, from a modern perspective, a critical look at Aristotle's four-fold causal account will be presented. Finally, given the enormous importance of *teleology/final causation* in Aristotle's philosophy in general and in his reflections on biology more specifically, five major objections to Aristotle's notion of final cause will be examined to determine how seriously final causation should be taken. Whether or not one is entirely persuaded by this final section, the reader should get an exposure to (1) some of Aristotle's more general philosophical views, (2) his keen skills as a bench biologist, and (3) his main biological works (*History of Animals*, *Parts of Animals*, and *Generation of Animals*). The overall goal of this chapter is to show that Aristotle's philosophy of biology reveals a remarkable degree of sophistication, revealing that Darwin and others were right on target with their praise.

II. STANDING ON SHOULDERS:
INFLUENCES ON ARISTOTLE'S PHILOSOPHY OF BIOLOGY

Aristotle was the son of a doctor. It should not be surprising, then, that he was influenced by ideas associated with the field of medicine. Indeed, he was eager to employ analogies from the field of medicine to address philosophical problems (e.g., *Metaphysics*, IV.2, 1003^{a}35–^{b}4 and *Metaphysics*, IV.2, 1003^{b}10–15). This reveals that he had more than a casual interest in the history of medicine and its corresponding influence on making sense of nature. As Van Der Eijk stresses, "Aristotle and his followers were well aware of earlier and contemporary medical thought ... and readily acknowledged the extent to which doctors contributed to the study of nature. This attitude was reflected in the reception of medical ideas in their own research and in the interest they took in the historical development of medicine" (2005, p. 14). The

---

8   This quotation is taken from Gotthelf (1999).

main practitioners of this field of medicine, namely the Hippocratics, held a basic view of biology that stands as part of the foundation of Aristotle's biology.

Basically, the Hippocratics believed that there are two sets of opposites—hot/cold and wet/dry—which could be combined in four ways to produce the following natural elements:

(1) earth by way of combining dry and cold,
(2) air by combining hot and wet,
(3) fire by combining hot and dry, and
(4) water by combining wet and cold.[9]

It is the combining of these four elements, argue some of the Hippocratics, that produce the following four basic elements (humors) of the body that determine its health status:

1. black bile (earth)
2. yellow bile (air)
3. blood (fire)
4. phlegm (water)

So long as these four elements were in proper balance (a sort of equilibrium or homeostasis), the body was considered to be in a state of health. In contrast, a diseased body was thought of as an imbalance or disproportionate mixture of 1–4. This view of a proper chemical mixture of the body can be considered the Hippocratic homeostasis philosophy of medicine.

Aristotle embraced much of the above account of the basic chemistry of the body.[10] He did not, however, accept entirely the Hippocratic scientific methodology. That is, there was tension amongst the Hippocratics on how to proceed with respect to caring for patients. Some argued that a theoretical framework about the body was mostly irrelevant with respect to diagnosis and treatment because each patient's needs were unique; a kind of medical casuistry (focus only on the details of a given case) or strict empiricist perspective was defended by these medical Hippocratics.[11] In contrast, other Hippocratics insisted that some kind of theoretical framework

---

9   This is a broad generalization and does not necessarily reflect the view of all the medical practitioners that fall under the title of "Hippocratics." See Edelstein (1967), Lloyd (1979), Grene and Depew (2004), and Boylan (2006) for some of the subtle differences amongst the Hippocratics.
10 For an insightful discussion of how Aristotle comes to embrace hot/cold and wet/dry as the foundation of his chemistry, see ch. 17 of Solmsen (1960).
11 A good example is Diocles, who may have been a student of Aristotle's. He argued against abstract allegiance to causal explanation and insisted that "one should give credence to the [things] that have been well grasped on the basis of experience over a long time." This quotation is from Van Der Eijk (2005), p. 78. Note, however, that Van Der Eijk argues that Diocles' considered views about causal explanations may be more similar to the views of both Aristotle and Theophrastus. See Van Der Eijk (2005), pp. 92–100.

or set of (causal) "laws" about the body is necessary in order to diagnosis and treat accurately the variety of ailments of patients.[12]

As we will see, Aristotle's philosophy of science embraces a little from both of the above camps. At a glance, for now, along with accepting the homeostasis philosophy of medicine of the Hippocratics, Aristotle is also wedded to a four-fold causal framework (see below); yet he insisted on careful empirical studies in order to locate the specific details associated with each of his causes. The result is that Aristotle defends a **mixed causal-empiricist methodology** in which he is willing to modify his theoretical views in the light of new empirical data. Aristotle's philosophy of science, for the sake of this chapter, refers to this causal-empiricist methodology/methodological framework and his account of soul (see next section). It is this methodology that is clearly established in his summary judgment about bees:

> This then appears to be the state of affairs with regard to the generation of bees, so far as theory can take us, supplemented by what are thought to be the facts about their behaviour. But the facts have not been sufficiently ascertained; and if at any future time they are ascertained, then credence must be given to the direct evidence of the senses more than to theories—and to theories too provided that the results which they show agree with what is observed. (*Generation of Animals*, III.10, 760^{b}28–34; trans. Peck, 1942)

Aristotle cautiously gives weight to empirical observation over theory, if one must choose between the two. He notes, however, that theory must be given its due as well if it aligns with observation. In this sense, Aristotle is wedded to both observation and theory when trying to make sense of the natural world and biological systems in particular.

III. ARISTOTLE'S PHILOSOPHY OF SCIENCE

For Aristotle, the study of living things is housed within the broader category of nature or natural scientific philosophy—that is, physics. His study of biological entities, in fact, is the culmination of his overarching "distinctly organized investigation of the natural world" (Falcon, 2005, p. 2). He tells us, "For those things are natural which, by a continuous movement originated from an internal principle, arrive at some end" (*Physics*, II.8, 199^{b}15–17; trans. Hardie and Gaye, 1984). This internal principle, Aristotle tells us, is soul (*Parts of Animals*, I.1, 641^{a}25–32; trans. Lennox, 2001b).[13]

In terms of animals, Aristotle makes clear that the nature of such beings requires taking into consideration that which constitutes both the source of movement and the end to which that movement is directed. In the case of organisms, this movement and

---

12 Hippocrates appears to have held this position.
13 For more on how Aristotle understood the nature of organisms in his biological works, see Berryman (2007) and Henry (2009).

end-state is the soul. For Aristotle, then, 'nature' refers to both the soul and matter of organisms. It is because the soul is both the source of motion and the ultimate end-state for the sake of which movement occurs, Aristotle thinks, that the soul is of far more importance than matter in terms of studying nature; indeed, he makes it clear that soul gives matter its substantial nature rather than the other way around.[14]

The reason for giving this importance to soul is that Aristotle understood 'science' and 'physics' to refer to those things that grow, develop, and locate themselves (i.e., move themselves) in a non-accidental and non-coincidental way (see Grene and Depew, 2004, p. 5 and Depew, 2008). Aristotle needed some organizing principle that could account for the fact that natural systems develop and change, always or for the most part, in their characteristically consistent manner. For example, acorns—always or for the most part—develop toward becoming oak trees.[15] This is no random or accidental event. To the contrary, natural processes show a very high degree of precision and organization with respect to their unfolding and movements. Thus, since Aristotle did not think that soul-less matter could self-organize, he thought that some internal organizing principle, i.e., soul, was the source of such organic organization. Aristotle drives home the point as follows:

> [T]hat there is no science of the accidental is obvious; for all science is either of that which is always or of that which is for the most part. For how else is one to learn or teach another? The thing must be determined as occurring either always or for the most part, e.g. that honey-water is useful for a patient in a fever is true for the most part. (*Metaphysics*, VI.2, 1027^{a}20–24; trans. Ross, 1984)[16]

---

14   Aristotle's theory of soul cannot be explicated fully here. For an attempt to make sense of Aristotle's theory of soul from a contemporary philosophy of mind perspective, see Miller (1999).

15   It is worth pointing out that Aristotle is using 'always or for the most part' in terms of a living thing's species-specific natural developmental unfolding. Of course, as it turns out, most acorns do not actualize their species-specific end-state due to a mix of biological and environmental factors, but this is more indicative of the sort of reproduction under consideration. For instance, some species reproduce in large numbers and little parental investment with a few offspring surviving various sorts of less than stable environmental stressors (known as *r reproductive strategy*; e.g., salmon, coral, insects, and bacteria), while other species reproduce in small numbers offering high-level parental investment with these small numbers of offspring frequently able to navigate life's challenges (known as *k reproductive strategy*; e.g., elephants). As it turns out, oak trees first produce about 50,000 acorns only after 20 years of growth and then produce even more of these high energy-rich seeds every few years (with only a few of them actually germinating); that is, oak trees employ the *r reproductive strategy*. So, 'always or for the most part' with respect to the oak tree-acorn example suggests that the internal developmental unfolding of most of their seeds would produce oak trees if environmental conditions permitted. It is the environmental and development patterns of oak trees that accounts for the infrequency of acorn germination and not so much about the internal capability of the acorn. So, Aristotle's use of 'always or for the most part' is not greatly disturbed by the acorn-oak tree example. For more on this issue and related debates, see Lennox (2001c) and Sober and Gottlieb (2017).

16   Note that I am ignoring aspects of Aristotle's discussion—namely his different senses of necessity—to make the analysis manageable. See Hankinson (1995) for an insightful analysis.

It is clear that Aristotle rejects the possibility of a scientific account of random events. The very randomness, he reasons, precludes being able to understand and teach about the nature of such events. Thus, since Aristotle understands the practice of science to be a knowledge-gathering endeavor, he thinks that the lack of understanding with respect to random events renders a science of random events impossible.[17]

Now, it may seem odd that Aristotle employs honey-water as part of his account of what is natural; for how does honey-water possess an internal principle of change? He addresses this, as part of his analysis of the distinction between "natural entities" and "artifacts," in his *Physics* (Aristotle, *Physics*, II.1, 192^{b}6–19 and 192^{b}29–34; trans. Hardie and Gaye, 1984). To summarize, the reason Aristotle gives honey-water as an example is because water behaves—always or for the most part—in the same manner as a result of its natural constitution. For example, given a certain temperature, water will always or for the most part be in a solid, liquid, or gaseous state. Aristotle likely means that, when combined, water and honey have predictable changes and effects on a certain class of sickly patients. It follows, then, that honey-water will aid—always or for the most part—in the health of a fever-ridden patient by reducing fever, increasing energy by way of sugars, and increasing hydration. The same is true, thinks Aristotle, for the changes that earth, fire, air, animals, and plants undergo.[18]

In contrast, artifacts, like beds, may include natural entities in them, but their ability to change is the product of external forces (e.g., the bed maker) and not the result of some internal constitution to change. Indeed, as Aristotle says above, it is a mere coincidence that wood/earth is present in a given bed—again, there is no internal principle or force that produces an original or second generation wood/earth-made bed from within the wood/earth itself or the first generation bed itself. In contrast, liquid water can produce gas states or solid states (and vice-versa) as a result of its internal features reacting to ambient temperature conditions. Thus, Aristotle insists that all and only those entities whose changes originate from a non-accidental internal organizing principle, force, or impetus are worthy of scientific—that is, knowledge gathering—pursuit.

IV. FROM PHILOSOPHY OF SCIENCE TO PHILOSOPHY OF BIOLOGY

Given the scope of Aristotle's natural philosophy, how does he propose to make sense of the various biological systems in the world? More formally, what is Aristotle's methodological framework (i.e., his philosophy of science) for making sense of living beings? The scholarship with respect to this question reveals no definitive consensus, but recent work on Aristotle's biology suggests that the organization of his biological

---

17   One can only speculate about what Aristotle would have thought about the nature of quantum phenomena, which are thought by many contemporary physicists to be fundamentally random moving entities. Given his own resistance to the atomists of his day, he might hold out for a more unifying theory, like string theory. Still, given his empiricist leanings, it is unclear.
18   For more on Aristotle's views on nature and motion, see Ch. 2 of Lang (1998).

works follows the pattern he lays out in his logical works.[19] From this perspective, Aristotle first attempts to categorize and gather factual information about organisms, in his *History of Animals*, following his scientific method in both his *Prior Analytics* and *Posterior Analytics*. A hint at Aristotle's penchant for fact gathering and categorizing can be found in his empirical judgments concerning apes with respect to humans:

> Some animals dualize in their nature with man and the quadrupeds, e.g., the ape, the monkey, and the baboon.... Apes are hairy on the back, in virtue of their being quadrupeds, and hairy on the fronts, in virtue of being man-like (that, as I mentioned before, is a point in which the arrangement in man is the reverse of that in the quadrupeds), except that the hair is coarse; hence apes are covered with thick hair both back and front. Its face shows many resemblances to that of man: it has similar nostrils and ears, and teeth, both front and molars, like man's. Furthermore, whereas other quadrupeds have no lashes on one of the two eyelids, the ape has them on both, though they are extremely fine, especially the lower ones, and very small indeed. This is unique, for the other quadrupeds lack these lower eyelashes. (*History of Animals*, II.8, 502^{a}17–34; trans. Peck, 1965)

Notice that Aristotle is only explaining the similarities and differences between apes and humans based on a comparison between the class of quadrupeds (organisms that walk on four feet) and the class of humans. He observes that, although apes belong in the category of quadrupeds, their overall nature overlaps with that of humans. To reinforce this comparison, he goes so far as to explain that apes are the only quadrupeds that have eyelashes that are comparable to humans. Importantly, it is clear that Aristotle is acutely aware of shared characteristics and different ones (e.g., differences in thickness and coarseness of hair and similarities with respect to facial features) amongst species and is sensitive to describe biological systems accordingly (comparative anatomy).

Moreover, and this is the crucial point for now, Aristotle does not shy away from gathering the "facts" about organisms before pronouncing judgments about them. It is the gathering of facts about some phenomenon that moves one in the philosophical and scientific knowledge-acquisition direction that represents an intellectual interest of the natural world that is beyond mere curiosity or wonder.[20] As Kosso stresses, "A

---

19   See Lennox (2006, 2001a, and 2001b) for further references and commentary related to both Aristotle's *Parts of Animals* and the technical issues surrounding Aristotle's overall philosophy of biology.

20   This is not to deny the importance of wonder. As Aristotle famously says, "For it is owing to their wonder that men both now begin and at first began to philosophize ..." (*Metaphysics*, I.2 982^{b}11–13; trans. Ross, 1984). Yet, as part of moving beyond the initial perplexity and ignorance that accompanies wonder, Aristotle offers the following summary judgment: "Evidently we have to acquire knowledge of the original causes (for we say we know each thing only when we think we recognize its first cause)" (*Metaphysics*, I.2 983^{a}24–25; trans. Ross, 1984).

scientific observation is more than a casual glance. It is not just a wide-eyed, gullible, seeing-is-believing look at nature. It must be more knowledgeable observation that considers the circumstances" (Kosso, 1998, p. 19). It is no accident, then, that Aristotle-as-scientist painstakingly describes parts and behaviors of organisms in his biological treatises.

With respect to categorizing, Aristotle notes that apes belong to the group of quadrupeds as distinct from humans who are bipedal. Although he does not make the point explicit, it is clear that the way these two species move helps to make sense of why the thickness and coarseness of their respective body hairs on their chests and backs are the way they are. Furthermore, given his use of 'dualize,' Aristotle is cognizant of the fact that the natural world is not carved up into easily observed distinct kinds of organisms with corresponding distinct physical characteristics. As he illustrates with location and type of bodily hair, many physiological differences between organisms can be understood as differences in degree, as opposed to strict differences in kind. It is reasonable, then, to infer that Aristotle allows for overlapping characteristics amongst different species and sub-species as part of his attempt to categorize biological systems.

Second, as part of offering what he thinks to be a complete scientific analysis, Aristotle attempts to determine the causes of various features of organisms in his *Parts of Animals*. Upon providing a fairly robust conceptual framework of how organisms can be differentiated in his *History of Animals*, Aristotle offers the following summary judgment:

> What has just been said has been stated thus by way of outline, so as to give a foretaste of the matters and subjects which we have to examine; detailed statements will follow later; our object being to determine first of all the differences that exist and the actual facts in the case of all of them. Having done this *we must attempt to discover the causes*. And, after all this is the natural method of procedure—to do this only after we have before us the ascertained facts about each item, for this will give us a clear indication of the subjects with which our exposition is to be concerned and the principles upon which it must be based. (*History of Animals*, I.6 491^{a}7–14; my italics; trans. Peck, 1965)

As he makes clear in the above quotation, Aristotle reminds the reader of the importance of both gathering the facts and categorizing each type of organism. Then, he proceeds to the next stage—what he thinks is a natural transition (presumably for rational beings)—of studying organisms, namely, determining the causal forces at play that influence the presence of both the core differences between organisms and their similarities (as the case may be) and the incidental features of organisms. Thus, Aristotle's approach to studying biological systems is as follows:

> (1) begin with a specific question/concern about an observed phenomenon by way of some initial (albeit potentially labor intensive) fact gathering,

(2) offer logical structure/categorization of the observed phenomenon, and

(3) provide an answer to the specific question/concern by way of a causal analysis.

Each of these will be explained in turn by way of a contrived example. Then, we will apply the lessons of the contrived example to one of Aristotle's own examples.

(1) *Specific question about an observed phenomenon*

Imagine that you are at the park on a cold and windy November afternoon. Much to your surprise, you notice a number of squirrels lifting their tails over their backs. You find this behavior peculiar. After observing this same behavior on subsequent trips to the park (i.e., you have established that this behavior is a fact and not some delusion you were having in the bitter cold weather) during various times of the year, you ask yourself: **why do squirrels lift their tails over their backs during the winter?**

This question satisfies Aristotle's first requirement; that is, you have located a phenomenon and formulated a question/concern about it.

(2) *Logical structure/categorization of the observed phenomenon*

At this point, you start off by making clear that having a tail follows being a squirrel, that tails are distinguished in different ways, and that having a concaved almost-cobra-like-hood-shaped fluffy tail is a feature that belongs to squirrels in the windy and cold winter season. You come up with the following argument:

P1. Tail-raising belongs, always or for the most part, to the class of thermal regulation behavior.

P2. Thermal regulation behavior belongs, always or for the most part, to every species of squirrel.

C1. So, tail-raising belongs, always or for the most part, to every species of squirrel.

With the syllogistic argument in place, you have satisfied Aristotle's second requirement of providing a formal structure (of the observed phenomenon) upon which to reflect. In fact, Aristotle employs the syllogistic argument structure and points out the following: "It results, therefore, that in all our searches we seek either if there is a middle term or what the middle term is. For the middle term is the [causal] explanation, and in all cases that is sought" (*Posterior Analytics*, II.2, 90a5–7; trans. Barnes, 1984; bracketed addition mine). Indeed, with respect to syllogistic arguments, you remind yourself that the subject term is 'tail-raising,' the predicate term is 'species of squirrel,' and the middle term is 'thermal regulation behavior.'[21]

(3) *Causal analysis*

Examining your argument, you realize that 'Thermal regulation behavior belongs to every species of squirrel' masks a basic causal claim. In this case, you determine that tail-raising *is caused by* the need to secure thermal regulation. At this point, the material biological mechanisms (i.e., the various physical ways the body goes about generating and maintaining body heat) that are causally associated with how the body secures and maintains heat would be needed to complete the causal account of tail-raising.

In a simplified way, these three requirements capture Aristotle's methodology. To see this method in play with respect to biology, Aristotle offers his own biological example. He asks the question: why it is the case that some trees shed their leaves at certain times of the year?[22] This question includes Aristotle's first requirement of having an observed phenomenon so as to secure a domain of discourse. Second, some kind of formal/logical structure is needed in order to organize clearly the concern at hand with respect to the observed phenomenon. Aristotle recommends the following syllogistic argument in which there is a subject class, a predicate class, and a middle class:

P1. The process of shedding leaves belongs, always or for the most part, to everything which undergoes solidification of sap at the seed connection.

P2. Solidification of sap at the seed connection belongs, always or for the most part, to every broad-leaved plant.

C1. So, the process of shedding leaves belongs, always or for the most part, to every broad-leaved plant.[23]

---

21   I will assume that instructors using this text will assist students who are not familiar with the logical structure discussed here.
22   In fact, Aristotle uses this example in his *Posterior Analytics* to illustrate how one goes about any sort of specialized knowledge-seeking (i.e., scientific) endeavor. See Aristotle, *Posterior Analytics*, II.2, 90a7–91.
23   Aristotle's own account of the shedding of leaves can be found in his *Posterior Analytics*, II.16–17, 99a21–29. This argument reconstruction is a slightly modified version of the one offered by Lennox (2001a, p. 12). For a recent treatment of the interplay between Aristotle's philosophy of science and his philosophy of biology, see Leunissen (2010).

Hidden in **P2** in the above argument is the role of causation. Basically, for Aristotle, one can infer that the middle class—'Solidification of sap at the seed connection'—provides the causal account with respect to the predicate class—'broad-leaved plant.' So, the reason why the process of shedding leaves belongs to everything which undergoes solidification of sap at the seed connection is because solidification of sap at the seed connection (at a certain time of the year) *causes* the leaves to detach.

In order to appreciate the role causation plays in Aristotle's philosophy of biology, the next section will introduce his famous "four causes" and how they are designed to make sense of biological systems. Moreover, this discussion will introduce the notion of *biological function* and *teleology*, which prove to be concepts that are still being debated today in our post-Darwinian world.

V. THE FOUR CAUSES

The importance of causation for Aristotle cannot be overstated. He is quite explicit about causation and its role with respect to knowledge. Knowledge of X requires a causal account of X (*Physics*, II.3, 194^{b}17–23; trans. Hardie and Gaye, 1984). Specifically, Aristotle argues that there are four causes that must be made clear in order to be able to claim that one has acquired knowledge of something (*Physics*, II. 7, 198^{a}22–27; trans. Hardie and Gaye, 1984). For illustrative purposes, let us suppose that a shoe is the phenomenon to be understood within Aristotle's four causes. The explanation unfolds as follows.

> (1) *Material Cause*: The material from which X is made, or comes into existence, or into which it perishes. So, the material cause of a shoe would be, for example, the leather that is needed for its construction.
>
> (2) *Formal Cause*: The essence or structure of X that is necessary for X to perform its task. Thus, the essence or structure of a shoe is similar to that of a foot so as to assist in various sorts of ambulatory movements.[24]
>
> (3) *Efficient Cause*: The agent Y that is responsible for the action or movement that brought X into existence or took X out of existence. Therefore, the shoemaker is the agent that brings the shoe into being.
>
> (4) *Final Cause*: The purpose, function or goal for which X exists. Hence, the purpose of a shoe is to protect the foot and/or facilitate ambulatory movements.

---

24 There is no easy rendering of Aristotle's conception of formal causation. 'Essence' or 'structure' is not entirely satisfactory, but this will have to do for our purposes. Johnson recommends (among numerous other translations) "what it is for something to be" or "that which it is to be something." For further discussion, see Johnson (2008, pp. 46–49). In a different context, Grene and Depew (2004, p. 21, fn. 15) suggest "What-it-is-that-makes-something-the-thing-it-is."

With the four-fold causal account in place, Aristotle's discussion of leaf falling can now be made transparent. As a reminder, Aristotle is clear that causal accounts must be given, not only for things that come into existence, but also for those things that perish. So, the shoe is an entity that comes into existence, but the leaf falling phenomenon is an example of an entity that perishes. In order to be able to claim that one knows the nature of leaf falling, one would need to make sense of this biological process within his four-fold causal approach as follows:

(1) *Material Cause*: The material cause of leaf falling is the material nature of deciduous leaves. Aristotle hints at an answer when he speaks of the solidification of sap at the point where the leaf meets the branch. From this perspective, much like leather is the material cause of shoes, the appropriate mixture of earth, air, and water that make up leaves constitute the material nature of leaves such that they can easily fall off as a result of sap solidification during the winter season.

(2) *Formal Cause*: In this case, the essence, structure, and or shape of leaves must be of a sort that ensures that leaf falling occurs. Fortunately, Aristotle tells us that the leaves must be broad. This is a fuzzy term, but it would allow for a range of structures or essences of deciduous leaves (think of the various shapes of maple leaves). So, like the formal cause of shoes is the footed-shape structure, the formal cause of leaf falling is broadness-shape of leaf.

(3) *Efficient Cause*: The efficient cause of leaf falling would be responsible for "making" these leaves fall or "doing" something to the leaves. So, much like the shoemaker is the efficient cause of the shoe, the efficient cause of leaf falling would be the solidification of sap at the seed connection. Restated, the hardening of the sap where the end of the leaf meets the branch causes the leaf to fall off at that juncture.[25]

(4) *Final Cause*: The final cause of leaf falling is "that for the sake of which" leaf falling occurs. Basically, in much the same way shoes are for the sake of facilitating ambulatory movements, leaves fall for the sake of energy and water conservation.[26]

---

25 Our contemporary scientific efficient causes of leaf falling are the enzymes called **pectinases** and **cellulases**. See the following website for further details: http://www.botany.uwc.ac.za/ecotree/leaves/decidu.htm.

26 In order to contend with harsh winters, deciduous trees go into a state of hibernation, which includes extracting the nutrients from their leaves prior to discarding them so that these nutrients can assist in leaf formation in the subsequent growing season.

By way of the shoe example, below is a summary of Aristotle's four causes.

CAUSE	DEFINITION	SHOE EXAMPLE
Material Cause	The "stuff" needed to bring about the existence of X.	Leather is a kind of material needed to bring about a pair of human shoes.
Formal Cause	The structure or shape of X needed to help X perform its function.	The shape or structure of the human foot is needed for human shoes to perform their function.
Efficient Cause	The entity that provides the shape or structure to the material cause of X.	The shoemaker provides the shape or structure to the leather to produce shoes.
Final Cause	The goal, function, or purpose of X.	The goal, function, or purpose of shoes is to assist in walking.

Table 1.1 *Aristotle's Four Causes*

With the four causes in place, Aristotle's basic philosophy of biology is complete.[27] Table 1.2 provides a summary of the key elements of Aristotle's philosophy of biology.

KEY ELEMENTS OF ARISTOTLE'S PHILOSOPHY OF BIOLOGY	BRIEF EXPLANATION
Fact Gathering	Given a set of questions about entity/feature X under scrutiny, empirical investigation of X is required to obtain relevant facts.
Logical Organization	Use of logical structure (syllogism) in order to understand obtained facts and questions about entity/feature X under scrutiny.
Causal Analysis	Use of four-fold causal account to understand the nature of entity/feature X under scrutiny.

Table 1.2 *Summary of Aristotle's Philosophy of Biology*

---

27 For more on Aristotle's four causes as they relate to his biology, see Frank (1988), Shields (2007), and Mathen (1989, 2009).

VI. BIOLOGICAL FUNCTION/TELEOLOGY

Aristotle's four causes may seem like nothing more than a quaint historical oddity in the light of "our understanding of causation." Such a knee-jerk response, however, may reveal a presumptuous view that there exists an agreed upon account of causation in contemporary discussions. Nothing could be further from the truth.[28] It should come as no surprise, then, that Aristotle's account of final causation has both intrigued and disturbed scholars over the years. This interest and concern over final causation proves especially pressing in Aristotle's attempt to apply it to biological entities. From this perspective, it is crucial to understand Aristotle's concept of final causation in order to understand his philosophical views about biological systems; that is, his philosophy of biology.

What does Aristotle mean by a claim like "X is, or comes to be, for the sake of Y"? In the case of shoes, it might seem reasonable to postulate a final cause (and of course an efficient cause) because the shoemaker intends that his product have a specific function or set of functions. So, with those functions in mind, the shoemaker goes about producing a shoe made of a certain set of materials, a particular shape, and for a particular set of purposes. Notice that, although we could be wrong about what those purposes are, it remains true that some set of purposes are relevant to the finished product (e.g., fashionable, expensive, and/or intentionally ugly). Still, even if it makes sense to incorporate goal-directed language with respect to *artifacts*, does it make sense to talk of "purposes," "functions," and "final causes" with respect to biological entities? Since nature does not have a corresponding "shoemaker," it is not clear that it makes much sense to claim that there is purpose in nature. Yet, Aristotle thinks that it is indispensable to employ teleological concepts (i.e., final causes) in order to understand biological systems.

To be sure, Aristotle unambiguously defends the claim that purpose—the idea that processes tend (always or for the most part) toward a particular end state—is a part of nature. He says, "Action for an end is present in things which come to be and are by nature" (*Physics* II.8, 199ᵃ8; trans. Hardie and Gaye, 1984). The following argument captures his motivation:

> **P1.** Either purpose is part of the causal make-up of natural phenomena or luck, chance, and/or coincidences are solely part of the causal make-up of natural phenomena.
>
> **P2.** If luck, chance, and/or coincidences are solely part of the causal make-up of natural phenomena, then it must be the case that natural phenomena do

---

28 For the subtleties surrounding the contemporary debates on causation, see Salmon (1994) and Dowe (2000). An extensive bibliography on many of the contemporary philosophical issues surrounding causation can be found at the following two websites: http://plato.stanford.edu/entries/causation-process/ and http://plato.stanford.edu/entries/causation-metaphysics/.

not exhibit complexity, constancy, beauty and goodness in their behavior and organization.

P3. It is clear from empirical observation that natural phenomena exhibit complexity, constancy, beauty and goodness in their behavior and organization.[29]

P4. It follows then that it is not the case that luck, chance, and/or coincidences are solely part of the causal make-up of natural phenomena.

C1. Thus, purpose is part of the causal make-up of natural phenomena.[30]

Recall that earlier, as part of his philosophy of science, we saw Aristotle's resistance to thinking that there could be a science of chance or luck.[31] This resistance is captured in **P3**. Within the context of causation, Aristotle stresses that material causes are not likely to be adequate to make sense of the kind of organization and beauty present in the natural world. In what can be viewed as his defense of **P3** above, Aristotle tells us:

> For surely it is not likely either that fire or earth or any such element should be the reason why things manifest goodness and beauty both in their being and in their coming to be, or that those thinkers should have supposed it was; nor again could it be right to ascribe so great a matter to spontaneity and chance. When one man said, then, that reason was present—as in animals, so throughout nature—as the cause of the world and of all its order, he seemed like a sober man in contrast with the random talk of his predecessors. (*Metaphysics*, I.3, 984^{b}11–18; trans. Ross, 1984)

The above argument reinforces Aristotle's opposition to thinking that chance events could bring about the kind of precise organization he observed in the natural world; again, a defense of the truth of **P3** above. Moreover, Aristotle adds that there is a beauty and goodness in the existence of natural phenomena and the coming-into-existence of such phenomena which cannot be made sense of by random material processes. This reveals that Aristotle's defense of teleology includes both natural and

---

29 For ease of reading, I have avoided rendering **P3** with a double negation. Technically, however, the falsity of the consequent in **P2** does require the negation of the negation in **P2**. Instructors can assist in explicating this point if needed.
30 In part, this is my argument reconstruction of Aristotle's account at *Physics* II.8, 198^{b}85–199^{a}8. Gotthelf (1987, p. 224) offers his own rendition of this argument, but it is silent on the normative aspect (e.g., goodness and beauty) of Aristotle's account.
31 In this chapter, I cannot do justice to Aristotle's discussion concerning incidental final causes as they relate both to luck and spontaneous generation. I can only direct the reader to Depew's (2009) outstanding analysis.

normative elements. The natural part includes the complexity and organization of natural phenomena and the normative part captures the (aesthetic) beauty and goodness of both the existence and coming-to-be of such complexity and organization.[32]

VII. OBJECTIONS TO ARISTOTLE'S TELEOLOGY

Still, given his allegiance to purpose/final causation as an integral part of his philosophical thinking, Aristotle had better be able to defend this adherence. What this means is that Aristotle needs to offer not only a robust explanation of final causation, but he should also be sensitive to objections to his account. In doing so, he will have further vindicated, either directly or indirectly, the truth of **P3** above (on p. 16). As we shall see, Aristotle anticipates a number of objections that not only reveal further aspects of his biological thinking, but also give additional credence to his concept of final causation. There are, then, five objections to Aristotle's version of teleology that will be discussed below. This analysis will include Aristotle's reply to each and the extent to which he is successful in his replies. It will not only reveal that the first four objections are not lethal to Aristotle's acceptance of final causation, but also that the fifth objection puts serious stress on his endorsement of it. Ultimately, Aristotle's defense of final causation will prove difficult for him, but it will not be for the reasons that usually gain center stage.

Objection #1: Final Causation Is Merely a Convenient Way of Speaking

The first criticism is that final causation is not really a cause but a convenient way of speaking. Accordingly, Aristotle must really mean that natural phenomena *appear* to be purposive akin to human purposive actions, although they are actually necessary brute events that occur. If 'chance' is the only other term available to describe the activities of the natural world, then 'purpose' is a better word choice because of the appearance of human-like conscious/deliberative goal-directed activity in non-human natural phenomena. The point is that Aristotle can legitimately only avail himself of this useful linguistic point about purpose—not that there is some unique kind of teleological cause.

REPLY TO OBJECTION #1

Aristotle actually acknowledges this kind of criticism (*Physics*, Book II. 8, 198^{b}16–23). There are two responses he offers to it. One response he gives is as follows:

---

32 The normative dimension of Aristotle's concept of final causation will be set aside in favor of the natural dimension. For more on Aristotle's concept of function and its normative dimension, see Gotthelf (1989) and Depew (2008).

> It is absurd to suppose that purpose is not present because we do not observe the agent deliberating. Art does not deliberate. If the ship-building art were in the wood, it would produce the same results by nature. If, therefore, purpose is present in art, it is present also in nature. The best illustration is a doctor doctoring himself: nature is like that. It is plain then that nature is a cause, a cause that operates for a purpose. (*Physics*, Book II. 8, 199^{b}26–33, trans. Hardie and Gaye, 1984)

Aristotle is offering a two-pronged argument from analogy against this first criticism. In the same way that goal-directedness (that for the sake of which) is present in art-objects, it is present in nature. In the same way that the goal-directedness of art-objects is not conscious, goal-directedness in nature is not conscious. What Aristotle likely means here is that, when you observe a ship sailing, you do not see any conscious deliberation on the part of the ship (even though it is the product of conscious design) as it performs its goal of sailing (that is, moving through water). Nature too, thinks Aristotle, possesses the unobserved goal-directedness like an art-object.

Now, as a transition to the natural world, Aristotle says that if the ship-building skill were in the wood, then a ship would naturally—in a developmentally sequential fashion—come into existence. Here, Aristotle is trying to show that the sort of sequential unfolding that is present in consciously designed artifacts is also present in nature without the need of a designer. This is why, at this point in the analogy, he can claim that the purpose present in art-objects is also present in living systems. Again, the notion of sequential and developmental organization underlies this entire analogy and gives some credence to it. Thus, at this point, Aristotle thinks he has secured the presence of unobserved final causes.

If this is all Aristotle has as part of his defense, then he is in trouble, because the art-object analogy by itself does not reveal the nature of final causes in natural phenomenon. Aristotle, it seems, is cognizant of this and moves on to his second analogy. So, to further buttress and extend the art-object analogy, he claims that nature is a cause in much the same way that a doctor heals himself. This second analogy is designed to show that the unobserved goal-directedness in nature is like the doctor doctoring himself. Thus, goal-directedness in nature is an unobserved feature which has the quality of being able to care for itself in some fashion. What this means is that living systems—as opposed to earth, air, fire, and water—are able to "self-organize" by maintaining/controlling a restricted range of sequential orderly growth. This point is implicit in Aristotle's attempt to reject fire as the unique cause of nourishment and growth in living systems (*De Anima*, Book II. 7, 416^{a}10–19; trans. Hett). His main complaint is related to control. Fire will burn continuously so long as there is something to burn; so long as the something is combustible, fire does not discriminate. In contrast, biological systems express limitation and control with respect to both nourishment and growth. Although Aristotle concedes that fire is a necessary condition for organized and limited sequential development, he distinguishes the soul—as final and formal cause—as that which ensures this requisite self-organizing control; *and he does so without invoking intentions.*

What are we to make of this argument? Maybe, as some scholars have suggested (e.g., Shields, 2007, p. 72), we should not be very thrilled by the first part of this first argument. Primarily, the worry is that Aristotle is not entitled to move from intentional-state examples (e.g., doctors) to non-intentional-state examples (e.g., nature). The reason is that the pattern of organization and goal-directedness that is present in intentional-state examples is the product of minds—the very feature that is not present in nature. So, whatever sort of order may be present in the natural world, it is not at all obvious that intentional-state goal-directedness is part of it. The upshot is that Aristotle's argument from analogy falls flat, reinforcing the claim that final causation is only a convenient way of speaking.

FURTHER REPLY TO OBJECTION #1

There may be a way to save Aristotle from this insightful critique. First, note that the art-object part of Aristotle's argument is designed to show that it is eminently reasonable to ascribe final causes to natural entities because we do so for artifacts. Again, if you observe the movements of a ship or the structure of a pot, it is clear that they are for the sake of some use or benefit—one can see this without necessarily invoking an external agent as the source of the final cause. This is what Aristotle means when he claims that "art does not deliberate." In the same way, when observing the movements of a spider or the parts of some organisms, one can invoke goal-directedness without referencing a designer.

Now, one may be quite willing to grant this much of the argument. What about the doctor analogy? Here, critics might be taking Aristotle's analogy in a direction that one need not go. Specifically, a doctor doctoring himself or a captain navigating his ship is best understood within the domain of maintaining self-organization. On this reading, Aristotle is deemphasizing intentionality in favor of species-specific self-correcting and limiting organization. Having secured the presence of final causes by way of the art-object analogy, Aristotle can now employ the doctor analogy to show that nature is able to maintain itself, much like a doctor is able to maintain himself, in a way that is unique to it. This interpretation can be gleaned from the following passage:

> Each animal is thought to have a proper pleasure, *as it has a proper function*; viz. that which corresponds to its activity. If we survey them species by species, too, this will be evident; horse, dog, and man have different pleasures, as Heraclitus says "asses would prefer sweepings to gold"; for food is pleasanter than gold to asses. So the pleasures of creatures different in kind differ in kind, and it is plausible to suppose that those of a single species do not differ. (*Nicomachean Ethics*, X.6, 1176^{a}4–9; trans. Ross, 1984; my italics)

Different species have different functions and distinct pleasures, Aristotle tells us in the above passage. With respect to the doctor analogy argument under consideration, doctors (as human beings) consciously go about maintaining themselves because they are the kind of beings that employ rational/intentional means. Nature (here Aristotle means the particular living organisms of the natural world), also, maintains itself (as a doctor does) in non-intentional/non-rational (or less rational) ways that benefit its species and are unique to its species.[33] Empirical observation bears this out; marsupials care for themselves and their young rather differently than seed-dropping annuals and perennials. Thus, if emphasis is given to *unique ways of self-maintenance*, Aristotle's doctor analogy seems more plausible than one might have thought.[34] Indeed, the nutrition/growth argument can be viewed as being additional support for understanding the doctor analogy in terms of limited and non-intentional sequential growth and self-organization. What follows from this is that Aristotle can claim that his reliance on final causation is more than mere fanciful speech. Rather, he can insist that final causation is part of our natural world and only differs in kind (and by degree) with respect to the species under consideration. Looking at the doctor analogy from the self-maintenance and self-organization perspectives solidifies this claim. So, at the very least, Aristotle can be viewed to have blocked the first objection against his allegiance to final causation.[35] The further corollary is that the truth of **P3** (on p. 16) is still worthy of being endorsed.

Objection #2: Final Causation Relies upon Panpsychism

The second objection to final causation is that it is incoherent because Aristotle is relying on a kind of *panpsychism*—the view that all things have a mind or mind-like qualities. The concern here is that Aristotle's account of final causation is absurd, if it is the case that he ascribes intentional states to all of the non-human entities of the natural world. B.F. Skinner charges Aristotle with just this criticism: "Aristotle argued that a falling body accelerated because it grew more jubilant as it found itself nearer its home" (1971, p. 6).[36] For example, so long as basic material elements are in place (e.g., water, soil, nutrients, etc.), an acorn eventually becomes an oak tree. Yet, if Aristotle is claiming that an acorn *desires, wants, needs, is jubilant*, etc., to become an oak tree in the same way that I intentionally desire to get a beer out of the refrigerator, then his account appears to be relying on an absurd view; for there is little or no evidence in support of the claim that the non-human natural world (for Aristotle) is

---

33 We would do well to leave open the possibility that non-human animals are in possession of intelligence to a degree. See Lennox (1999) for a persuasive defense that this is the case.
34 In terms of the doctor analogy, David Sedley hints at this interpretation when he argues that "if the world is indeed such a self-regulating system, it can justly be compared to a doctor treating himself" (Sedley, 1991, p. 192).
35 This reading offers *prima facie* evidence that Aristotle is not out of line in defending his claim that final causation is a part of nature by way of an artifact analogy. Of course, one would expect more from Aristotle in terms of a rigorous defense of final causation. Indeed, as we shall see, he does not let us down.
36 This quotation is taken from Shields (2007, p. 79).

embedded with intentional states.[37] The truth of such a criticism would undoubtedly render final causation/teleology moribund.

REPLY TO OBJECTION #2

In the quotation below, Aristotle's account of final causation reveals why the panpsychism criticism is misplaced. He uses an artifact analogy to make his point. First, he argues that human artifacts have a purpose. This is true, he thinks, because it is evident in the way that early and final stages of artifact-production reveal how the early stages are for the sake of the overall function of the artifact as it is brought to completion. Aristotle completes his argument from analogy by explaining the actions of non-human animal entities:

> This is most obvious in the case of animals other than man: they make things neither by art nor after inquiry or deliberation. That is why people wonder whether it is by intelligence or by some other faculty that these creatures work,—spiders, ants, and the like. By gradual advance in this direction we come to see clearly that in plants too that is produced which is conducive to the end—leaves, e.g. grow to provide shade for fruit. If then it is both by nature and for an end that the swallow makes its nest and the spider its web, and plants grow leaves for the sake of the fruit and send their roots down (not up) for the sake of nourishment, it is plain that this kind of cause is operative in things which come to be and are by nature. And since nature is two-fold, the matter and the form, of which the latter is the end, and since all the rest is for the sake of the end, the form must be the cause in the sense of that for the sake of which. (*Physics*, II.8, 199ª20–32; trans. Hardie and Gaye, 1984)

It is clear that Aristotle is not wedded to the view that the natural world (non-human entities) is embedded with intentional states. He makes clear that people need to understand that final causation is present in non-human organisms without the need of reason or some other deliberative faculty. The key to understanding all of this is, again, sequential unfolding (see Depew, 2008, p. 386). The nature of final causation is tethered to each distinct kind of species. So, there is intentional final causation that is restricted (for the most part) to humans and non-intentional final causation that will come in various forms depending upon the non-human entity under consideration. What this reveals is that Aristotle is arguing for the presence of non-intentional final causation in non-human entities by way of the similar early-stage-to-late-stage-purposefulness (i.e., sequential development) of other activities of natural systems (e.g., plants)—not only human art activities. Thus, the charge of panpsychism is entirely off the mark, leaving Aristotle's defense of final causation intact.

---

37 Even if non-human animals have intentional states, this criticism would still be relevant to non-animal biological entities (e.g., plants).

Objection #3: Final Causation Is Anthropocentric

This is a kind of human chauvinism—the idea that the ultimate purpose of all things in nature is for the sake of human needs. According to this objection, beyond any sort of biological goal-directed end, Aristotle endorses the view that there is an additional end of nature, namely the final goal of existing for the sake of human kind. Yet, if this is Aristotle's considered view, then his theory of final causation would be less than persuasive because it reveals a kind of unwarranted bias in favor of humans.

Here it appears that Aristotle is in a bit of trouble. He makes the following claim in his *Politics*:

> For some animals bring forth, together with their offspring, so much food as will last until they are able to supply themselves; of this the vermiparous or oviparous animals are an instance; and the viviparous animals have up to a certain time a supply of food for their young in themselves, which is called milk. In like manner we may infer that, after the birth of animals, plants exist for their sake, and that the other animals exist [*for the benefit of man*], the tame for use and food, the wild, if not all, at least the greater part of them, for food, and for the provision of clothing and various instruments. Now if nature makes nothing incomplete, and nothing in vain, *the inference must be that she has made all animals for the sake of men*. (*Politics*, I.8, 1256^{b}10–22; trans. Jowett, 1984; my italics and bracketed change)[38]

Aristotle begins his argument by noting that some non-human animals provide enough nourishment for their young until they are able to care for themselves; plants exist for the sake of animals; and animals (both wild and domesticated) for the benefit of humans. Aristotle's reasoning here is related to biological development. In the same way that organisms provide for the development of their young, it stands to reason that these very same organisms aid in the development of other organisms. Plants not only ensure their own survival (season after season), but they also assist in the nutritional lives of many other animals. Indeed, this also seems apparent in predator-prey relationships. And all these animals, along with maintaining their species line, ensure the benefit of the human species by way of food, shelter, clothing, etc.

This line of reasoning could be viewed as a precursor to our modern food chain/energy exchange models. To this extent, Aristotle's "Father of Biology" appellation is on the mark. The problem, however, is final causation. Aristotle is not just observing how the natural world unfolds. Rather, he seems to be claiming that this hierarchical food chain includes teleological necessity at each stage. So, not only do plants have

---

38 'Vermiparous' refers to those organisms (e.g., flies, locust, gnats) that produce or breed worm-like young; 'oviparous' refers to those organisms (e.g., birds, platypuses, reptiles) that lay eggs, and 'viviparous' refers to those organisms that give birth to live young (e.g., elephants, humans, horses).

the goal of survival and reproduction, but they also have the added goal ("for the sake of") and benefit of existing for other organisms. Moreover, he concludes rather quickly that if we accept that nature produces everything for a point, then it follows that man is that purpose of all of nature's creations. Nothing could appear more clearly as a case of anthropocentrism than his last comments in the above quotation. Notice that Aristotle has to insist on a dual-final cause picture—each species has its own final cause and also a final cause in support of other non-related species—in order to secure that all of natural creation is for the sake of humans. As Sedley puts it, "[T]hings are so arranged that the entire contents of the natural world, including not only plants and animals but perhaps even seasons and weather, *exist and function primarily for the benefit of man*.... This is the view I wish to attribute to Aristotle" (Sedley, 1991, p. 180; my italics). If this anthropocentrism is really the core of Aristotle's teleology, then there is little that can be done to save final causation.

REPLY TO SEDLEY AND OBJECTION #3

Sedley's claim above masks two senses of final cause. One sense is related to the bringing to fruition of a goal as a result of an entity's soul structure. For example, an acorn develops into an oak tree because this goal is part of the internal structure of the nutritive soul of the acorn. Also, mothers that nurse their young have the internal soul structure to do this such that the benefit they provide is also part of this internal structure. We can think of this sense of teleology as *intrinsic teleology*. In contrast, a second sense of final causation is related only to benefit. On this view, 'X is for the sake of Y' translates into 'X is a benefit to Y.' We can understand this weak sense of teleology as *incidental teleology* because it is not necessarily the case that X has the internal soul structure to benefit Y, even if X benefits Y in a regular way. For example, beaver dams benefit the feeding practices of other animals, but it is hardly part of the internal structure of beaver souls to benefit the feeding practices of other animals. What Sedley is claiming, then, is that Aristotle defends the view that the final cause of all animals is an *intrinsic teleological* benefit to humans.

With this distinction between intrinsic and incidental teleology in place, although it is true that some kinds of non-human animals and plants have the incidental goals of benefiting humans (and vice-versa), they do not have the intrinsic goals of benefiting human beings (and vice-versa). In order to claim that Aristotle's sense of final causation is anthropocentric, it must be shown that he attributes intrinsic ends to non-human living beings that are beneficial to humans. Yet, as it has been argued (see reply to Objection #2), Aristotle does not attribute such intrinsic ends to non-human living beings. If this reading is correct, then it is reasonable to resist the claim, as put forth by Sedley, that Aristotle's teleology is anthropocentric.

Still, Sedley does not hang his hat entirely on the above *Politics* passage. Rather, he points out that Aristotle makes a distinction between (1) the Human Earthly World (Aristotle uses the term *anthropeia*, which refers to the area of philosophy related to human affairs) and (2) the Godly World (Aristotle uses the term *theoria*, which refers to divine activity).[39] Additionally, he reminds the reader that Aristotle thinks that the Earthly World is subservient to the Godly world. With this distinction in place, Sedley offers an argument from analogy that draws upon this subservient relationship. The following passage from Aristotle guides Sedley's argument from analogy:

> All things are ordered together somehow, but not all alike,—both fishes and fowls and plants; and the world is not such that one thing has nothing to do with another, but they are connected. For all are ordered together to one end. (*Metaphysics*, XII.10, 1075ª15–18; trans. Ross, 1984)

The one end of which Aristotle speaks is God. So, at this point of the argument, Aristotle is claiming that all things are related to the Godly World. Sedley, in order to illustrate Aristotle's anthropocentrism, makes the following move: "But the good orderly arrangement which he associates with 'nature as a whole' can, in attenuated form, be traced all the way down to the sublunary [Earthly] world ..." (Sedley, 1991, p. 193). From this claim, Sedley's argument from analogy is basically that Aristotle believes that there is a weaker Earthly World version of the orderly arrangement that is present in the Godly World (Sedley, 1991, p. 196). Similar to the way God is the end state that all of its subordinates attempt to emulate, Aristotle also thinks (on Sedley's interpretation) that man is the end state that benefits from all of the Earthly entities beneath him. Thus, according to Sedley, Aristotle's account is clearly anthropocentric with respect to the Earthly World (Sedley, 1991, p. 195). It seems, then, that we must accept that P3 (on p. 16) is false because the constancy, complexity, and beauty of the natural world (for Aristotle) is intrinsically and teleologically linked to the benefit and needs of humans.

FURTHER REPLY TO SEDLEY

As a quick reminder of Aristotle's view, all parts of the natural (biological) non-human world have, as part of their natural constitution, the desire to emulate the eternal nature of God by way of reproduction—since they can't live forever literally, they can approximate divine immortality by way of sexual or asexual reproduction. Humans also share this nature, but we also express another aspect of God and that is thought. Although we are not pure thought like God, Aristotle thinks that we can approximate this kind of divine thinking by way of a life of contemplation.[40] Notice that there is nothing about benefiting God in this discussion, but according to Sedley, all other Earthly World creatures are present to benefit humans.

This point about benefit to humans is crucial to Sedley's use of the Godly world analogy. From the fact that all the inhabitants of the universe have an intrinsic

---

39 For further discussion of Aristotle's philosophy of human affairs, see Miller (1997, p. 12) and for the biological account of humans as political animals, see Depew (1995). For Aristotle's discussion regarding divine activity, see Shields (2007, pp. 220–29) and Norman (1969).
40 For Aristotle's discussion on the contemplative life, see Book X of his *Nicomachean Ethics*.

function to emulate an aspect of God, it does not follow that any notion of benefit to God is relevant to this emulation. But for Sedley to insist that non-human Earthly World entities exist to benefit man, he must secure the concept of "acquiring benefit" from the Godly World side of the analogy. Unfortunately, for Sedley, this benefit linchpin is not present. The reason is that Aristotle's God (The Unmoved Mover) is in no need of any kind of benefit from any kind of entity in the universe. The reason for this is that it is an eternal being without matter. It is able to bring about motion without itself moving, thus the title, "Unmoved Mover." Since it has no matter, Aristotle does not think that it can undergo any kind of change. In this sense, the Unmoved Mover's nature is fixed and cannot be influenced by any outside forces. As, Aristotle puts it, the Unmoved Mover, "thinks that which is most divine and precious, and it does not change; for change would be change for the worse" (*Metaphysics*, XII.9, 1074^{b}25–27; trans. Ross) "… it is clear that it is impassive and unalterable" (*Metaphysics*, XII.7, 1073^{a}11; trans. Ross). Moreover, with respect to goodness, Aristotle stresses that "God's essential actuality is life most good and eternal. We say therefore that God is a living being, eternal, [and] most good" (*Metaphysics*, XII.7, 1073^{b}27–29; trans. Ross, 1984; bracketed addition mine). Given these features of the Unmoved Mover (especially that it is unchanging and superlatively good), it is reasonable to see why Aristotle thinks that this being "can in no way be otherwise than it is" (*Metaphysics*, XII.7, 1072^{b}8). Since being benefited would be to undergo a change by way of acquiring more benefit (something that is not possible for the Unmoved Mover), it is clear that Sedley cannot secure any sort of value-laden benefit from the Godly World.

Given the above account, Sedley is not entitled to use the Godly World analogy because there is no need-of-benefit present in it. The result is that the very hierarchical structure that Sedley thinks allows him to conclude that humans are primarily benefited by all other Earth Worldly beings, in the same way that God is primarily benefited by all of the entities of the universe, is not a view held by Aristotle. Thus, Sedley's argument from analogy remains questionable.

Where does this leave us with respect to this anthropocentricism objection? Clearly, Sedley is correct that Aristotle argues that the entities of the universe emulate God in whatever limited way they can. It may be the case that modern naturalists are unimpressed by such a claim, but they certainly cannot accuse Aristotle of being anthropocentric. Thus, although it may appear that the charge of anthropocentrism against Aristotle reveals a serious flaw in his notion of final cause, the objection does not have as much merit as one might think. Therefore, it is not obvious, at any rate, that this objection is a knock-down argument against Aristotle's conception of final causation.[41] This being the case, it further follows that there is no need yet to rush to the rejection of final causation.

---

41 For more on the subtleties surrounding this anthropocentrism debate, see Johnson (2008).

Objection #4:
Final Causation Relies upon Backward Causation

The fourth objection to Aristotle's account of final causation is that it is unacceptable because he relies on a metaphysically implausible principle of backward causation. The idea behind this criticism is that Aristotle's concept of final causation is archaic to the extent that he appears to believe that end states that have not yet come into existence are able to bring about their own existence. Buller endorses this objection when he says, "Aristotelian goal-directed causality ... appeared to put the cart before the horse—explaining a cause before its effects—and thus to require 'backward causation'" (Buller, 1999, p. 5).[42] For example, to claim that the end-state of an oak tree is able to bring about itself (when it has not yet grown) is to rely on a kind of metaphysics about causation that is implausible. If this objection is on the mark, then Aristotle's theory of final causation would not be tenable from a contemporary biology and physics perspective. Indeed, the reasonableness of the reality of final causation would be nearly impossible to accept.

REPLY TO OBJECTION #4

Aristotle is well aware of such a concern and addresses it directly in his *Generation of Animals* by distinguishing two senses of 'prior,' **Prior[1]** and **Prior[2]**, as part of his account of organism development (*Generation of Animals*, II.6, 742ª19–22; trans. Platt, 1984). To understand this distinction, think of the transition from an acorn-to a sapling-to an oak tree. They are as follows: **Prior[1]** → This is the metaphysical sense of prior. This sense of prior is best understood as **the completeness sense**. Whatever is the complete state that X will become, this state is prior in being in the sense of a completed being. For example, the essence or the completeness of an acorn is an oak tree. Given that something complete is more valuable than that which is incomplete, that which is complete is metaphysically prior to that which is incomplete or less complete. This is why Aristotle thinks that the end or final cause is metaphysically or essentially prior to that which generates it.

Prior[2] → This is **the developmental sense** of prior. That which exists for the sake of an end would be the acorn since it is prior to the oak tree in development—that is, the acorn is what necessarily must exist in order to generate the fully developed oak tree. This means that the acorn begins the process of development **first** and then (eventually) the oak tree develops **after** the initial "unfolding" of the acorn. It is in the sense of biological development/unfolding that the acorn (that which exists for or generates the end) is prior to the oak tree (the developmental end state).

---

42   This quotation is taken from Johnson (2008, p. 167, fn. 12). Johnson also mentions on the same page the biologist, C.S. Pittendrigh, who also levies the backward causation objection against Aristotle as follows: "the recognition and description of end-directedness does not carry a commitment to Aristotelian teleology as an efficient causal principle" (Pittendrigh, 1958, p. 394).

The above distinction between the completeness sense of priority and the developmental sense of priority clearly reveals that Aristotle should not be saddled with the criticism that he endorses backward causation. The reason for this is that Aristotle does not claim that the completeness sense of priority is causally efficacious with respect to bringing about end states. Rather, he makes clear that the developmental sense of priority is the causally efficacious sense of priority. Thus, since final causes are housed within the completeness sense of priority, they do not engage in any sort of causation—let alone backward causation.

Thus, in order for the parts of organisms to play their functional role in the developing organism, it is necessary that something that is in a complete state, namely the soul, be present to ensure both the movement and achievement of the relevant organism ends.[43] The soul, as the form of the animal, has metaphysical priority in that it is the complete state of the organism—it is that for the sake of which everything else in the organism does what it does. But Aristotle makes clear that the biological parts of the organism have efficient and material causal priority:

> Therefore all the organic parts whose nature is to bring others into being must always themselves exist before them, for they are for the sake of something else, as the beginning for the sake of the end; all those parts which are for the sake of something else but are not of the nature of beginnings must come into being later. (*Generation of Animals*, II.6, 742^{b}3–7; trans. Platt)

Again, Aristotle is eager to point out that the parts of animals have causal priority (as material and efficient causes) in bringing about the end(s) for the sake of which they exist; indeed, these various parts of organisms (some are part of the beginning of development, while others come later) are all instruments of the soul, which is both a final and formal cause. For the sake of our discussion, although Aristotle does not hesitate to given ontological priority to final/formal causes, he also does not shy away from giving causal priority both to material and efficient causes. Nowhere does Aristotle claim that final causes function as efficient or material causes. Thus, we can reasonably infer that the criticism of backward causation is a misreading of Aristotle's notion of final causation.[44] Not surprisingly, then, the reasonable acceptability of final causation remains a tenable possibility.

---

43 Importantly, the soul ensures movement, but neither does itself move nor does it developmentally cause movement. Of course, if the soul engages in either developmental causation or movement, then the charge of backward causation would have to be taken as a serious objection to final causation.

44 Not every aspect of Aristotle's discussion about causal and metaphysical priority is discussed here. Regardless, there is enough of his analysis on display to show that the backward causation criticism is dubious.

Objection #5:
Final Causation Can Be Reduced

The last objection that will be addressed is the claim that final causation is reducible to a more basic phenomenon. For now, as an objection to Aristotle's teleology, the reductionism objection holds that final causation can be reduced to non-final causation phenomena, such as those entities that belong to chemistry or physics (or earth, air, fire, and water for Aristotle). According to this criticism, then, final causes can be reduced to material and efficient causes. More generally, then, *reductionism* is the view that higher-order properties or entities can be reduced to their constituent parts. For example, it is true that when hydrogen and oxygen molecules combine to form water, liquidity is a feature of this interaction. This feature of liquidity, however, is reducible to chemical bonding—water as liquidity can be reduced to the chemical bonding that brings it to fruition. Similarly, one could claim that final causes can be appropriately reduced to efficient and material causes (i.e., earth, air, fire, and water; let us ignore "law-talk," since this is absent in Aristotle's lexicon).[45] The implication is that final causes do not have the ontological prowess that Aristotle seems to think they have, yielding a direct rejection of the truth of **P3** (on p. 16) on the grounds that final causes can be eliminated in favor of material causes. In a sense, the suggestion here is that Aristotle would be forced to accept that matter can, by chance, self-organize—a view that he clearly would not deem reasonable.

REPLY TO OBJECTION #5

It should be no surprise that Aristotle is aware of this type of criticism.[46] He tells us straightaway that "for surely it is not likely either that fire or earth or any such element should be the reason why things manifest goodness and beauty both in their being and in their coming to be, or that those thinkers should have supposed it was ..." (*Metaphysics*, I.3, 984^{b}11–13; trans. Ross, 1984). Final causation is here understood in terms of the end-state, which captures that which is good or beautiful for the organism, being achieved. Clearly, he rejects that final causation can be reduced to or substituted by material elements. By way of his discussion of the formation of parts of animals, Aristotle explains himself with the aid of an artifact analogy:

---

45   As a generalization, some contemporary reductionist views involve, for example, the reduction of one theory (TB) to another theory (TA) on the grounds that the laws proposed by (TB) are actually equivalent and/or derived from (TA). For more on Aristotle's defense against the reductionism of his day, see Gotthelf (1987).

46   This claim may be an overstatement. There is much debate amongst Aristotle scholars about whether reductionism is really a target at which Aristotle was aiming. At any rate, it is safe to say that Aristotle was not preoccupied with engaging reductionism-type criticisms like many philosophers are today. See Meyer (1992) and Depew (2008) and their references for the philosophical nuances of this debate.

And just as we should not say that an axe or other instrument or organ was made by the fire alone, so neither shall we say that foot or hand were made by heat alone. The same applies also to flesh, for this too has a function. While, then, we may allow that hardness and softness, stickiness and brittleness, and whatever other qualities are found in the parts that have life and soul, may be caused by mere heat and cold, yet, *when we come to the principle in virtue of which flesh is flesh and bone is bone, that is no longer so; what makes them is the movement set up by the male parent, who is in actuality what that out of which the offspring is made is in potentiality.* This is what we find in the products of art; heat and cold may make the iron soft and hard. But what makes a sword is the movement of the tools employed, this movement contains that principle of the art. For the art is the starting point and form of the product; only it exists in something else, whereas the movement of nature exists in the product itself, issuing from another nature which has the form in actuality. (*Generation of Animals*, II.1, 734^{b}28–735^{a}4; trans. Platt, 1984; my italics)

Basically, Aristotle argues that there comes a point where basic chemistry cannot assist in making sense of the coming to be of an organism. He uses an artifact analogy to make his case against reductionism. In the same way that an art-object's (e.g., an axe's or sword's) completed state cannot be entirely made sense of by its material constitution, because an artist is needed to act as the efficient and formal cause of the art-object, the complete state of biological parts (e.g., flesh or bone) require a completed external efficient and formal cause to bring them into existence. His reason for thinking this is that matter cannot—with the help of more matter—bring about the relevant end state because it lacks the recipe (i.e., formal cause) to do so. The metal and wood that comprise an axe cannot form themselves, *by themselves*, in a way that could produce a functioning axe. What this means—and it is important to mark it again—is that art-objects cannot self-organize in terms of manifesting either individual (ontogenetic) or lineage-level (phylogenetic) developmental-sequential unfolding. The organization of axe-parts cannot produce an axe without an external source aiding in the individual development or orderly construction of an axe. In contrast, thinks Aristotle, organisms and species have orderly developmental unfolding patterns that come about by way of an internal recipe or formula (i.e., formal cause).

Similarly, the proportion of earth, air, water, and heat that is relevant to bone production cannot organize itself, *by itself*, in order to produce functioning bones. What is needed, thinks Aristotle, is a completed version of an organism that has and is able to pass along the formula (*logos*) of part-production by way of reproduction; this is precisely what the final and formal causes provide in terms of metaphysical priority (**Prior[1]** noted earlier).[47] This formula, in the guise of a unified formal and final cause,

---

47 To the contemporary reader, this line of reasoning should seem odd. In order to fend off reductionism, Aristotle has created a kind of infinite regress of completed organisms in a correspondingly always-furnished-world. This may work out great for him, but likely a source of frustration

which is present in the male seed/sperm, ensures that the requisite part-formation information is present so that matter can go about actualizing its potential to be a certain end state. Thus, according to Aristotle, it is not possible to reduce entirely the parts of animals to smaller material parts.

What becomes clear in the above discussion is that only complete biological entities have the final/formal cause present in them (by way of seed/sperm) to make sure that parts are able to come about as they should in order to produce another complete organism.[48] This requires the presence of a pre-existing final and formal causes—causes that cannot be reduced to or made identical with material and efficient causes due to their temporal and ontological priority. Thus, even if Aristotle did not directly challenge a reductionism-type objection to his version of teleology, he does, nevertheless, have the resources to fend it off.

So, what is the nature of Aristotle's notion of final cause? Drawing on this same passage and emphasizing Aristotle's conception of male semen, Gotthelf explains (in terms of different kinds of motion) that final causation

> is, essentially, a potential for form, a potential that exists *distinct from* and *not reducible to* any sum of qualitative and locomotive potentials. For, if it were reducible, so would the motion be, and then heat and cold could, in the proper sequence of actions, produce 'the logos [formula] by which one thing is flesh and another bone.' (Gotthelf, 1987, p. 217; bracketed addition mine)

So, according to Gotthelf, final causation is a distinct and irreducible potential for form. The reason for this, thinks Gotthelf, is that if it were not a distinct and irreducible potential for form, then the very kind(s) of motion that governs the movements of material elements (earth, air, fire, water) would be sufficient to account for both larger parts of organisms (e.g., bone and flesh) and overall organism shape and design. What this means is that there must be a unique kind of motion present in semen that accounts for the unique kind of irreducible and distinct potential of final causes. Indeed, Aristotle tells us what this unique "stuff" is:

> All have in their semen that which causes it to be productive; I mean what is called vital heat [*pneuma*]. This is not fire nor any such force, but it is the breath included in the semen and the foam-like, and the natural principle in the breath, being analogous to the element of the stars. Hence, whereas fire

---

for the modern reader. In terms of contrasting Darwin from Aristotle, Depew (2008, pp. 389–90) persuasively shows that Aristotle could not embrace Darwin's "descent from a common ancestor" primarily because of the infinite regress of organisms and the ever-present furnished world. On this issue, strict Aristotelians and contemporary Darwinians will have to part company; indeed, short of dogmatic posturing, it is hard to see how this aspect of Aristotle's natural teleology can be taken seriously. For related issues, see Henry (2006).

48 See Code (1997) for further discussion. Also, see Ariew (2007) for additional arguments Aristotle gives in defense of teleology.

generates no animal and we do not find any living thing forming in either solids or liquids under the influence of fire, the heat of the sun and that of animals does generate them. Not only is this true of the heat [*pneuma*] that works through the semen, but whatever other residue of the animal nature there may be, this also has still a vital principle in it. (*Generation of Animals*, II. 3, 736b34–737a6; trans. Platt; bracketed additions mine)

Aristotle thinks that there is a special kind of heat, called *pneuma*, which is akin to the heat of the stars and sun. Much like the sun and stars are able to produce life by way of their life-producing heat, he thinks that the semen in male biological systems contains the same kind of life-force heat in the form of *pneuma*. This *pneuma* carries the irreducible and distinct kind of life-force associated with final causation. Thus, according to Aristotle, final causation is the irreducible potential for form that exists by way of the special life-producing *pneuma*. It is this *pneuma* that gets passed along by the complete organism (by way of the semen) and accounts for why a growing embryo usually develops into a complete entity like the one that brought it into existence. Clearly, the life force as final cause/soul is present in the *pneuma*-surrounding-semen as well as the formula (*logos*) for the building of an organism.[49] As Aristotle somewhat subtly puts it, "[I]t is clear both that semen possesses soul, and that it is soul, potentially" (*Generation of Animals*, II.1, 735a8–9). Matter simply does not possess the final cause as an organizing/formula principle, thinks Aristotle, and this accounts, in part, for the irreducible nature of final causes.

At this point, the reader may be wondering about the discovery of DNA and the corresponding implications to Aristotle's overall analysis, especially the reductionism discussion. This is a reasonable concern because one could argue that Aristotle's anti-reductionism defense relies primarily upon the irreducible self-organization formula that comprises formal causation. Yet, it seems that formal causation must be viewed as a rather antiquated notion given that the DNA molecule accounts for the "formula" without the need for formal causation. As will be summarized in the conclusion below, it is not at all clear that Aristotle's account of formal causation and defense against reductionism can withstand the reductionist implications of the findings related to the DNA molecule and molecular biology.

It is time to take stock of what has been accomplished in this section. Not only have most of the pressing objections to Aristotle's notion of final causation been explained, but each objection has also been critically evaluated. Briefly, Table 1.3 below provides a summary of the objections to Aristotle's brand of teleology and the corresponding tentative resolutions that have been reached in this section.

---

49 Aristotle actually claims that semen is a compound of *pneuma* and water (*Generation of Animals*, II. 2, 736a1–2).

OBJECTIONS TO ARISTOTLE'S TELEOLOGY	TENTATIVE RESOLUTIONS
#1: Final causation is a convenient way of speaking.	This objection fails in terms of Aristotle's focus on distinct ways of self-maintenance and self-organization.
#2: Final causation smacks of panpsychism.	This misguided objection ignores Aristotle's discussion of the non-intentional goal-directed lives of non-human organisms.
#3: Final causation is anthropocentric.	Sedley's objection fails because he ignores the distinction between "being a benefit for man" and "being for the sake of man" and that his reliance on the god-world analogy ignores the fact that Aristotle's god is in no need of benefit.
#4: Final causation relies on backward causation.	Once appropriate attention is given to Aristotle's different senses of priority (completeness and developmental), it is clear that this objection is off the mark.
#5: Final causation can be reduced.	With the help of *pneuma*, Aristotle attempts an answer to the reduction objection, but it is not clear that it can withstand it; especially in the light of DNA. This objection is *prima facie* on the mark.

Table 1.3 *Objections and Resolutions to Aristotle's Final Causation (Teleology)*

Although much more could be said about the nature and necessity of final causation, we have enough here to conclude that Aristotle does have the resources to combat the reductionism objection.[50] Whether or not we should be persuaded in the existence of the irreducible life-force (i.e., final causation) brought about by the special celestial-type heat of *pneuma* and the infinite regress of completed organisms in an always completed world is a legitimate concern. Put another way, it seems reasonable to think that the "information" or "formula" present in genetic material can be substituted for final causation, rendering the need for final causation otiose. Additionally, if it is accepted that life came about from incomplete living resources in a world very different from the one we currently inhabit, then Aristotle's natural teleology loses more of its charm; alas, the truth of **P3** (on p. 16) becomes very difficult to accept. Be that as it may, it is worth pointing out that these may be the only

---

50 For more on this front, see Gotthelf (1987) and Charles (1988).

objections that genuinely put pressure on Aristotle's concept of final causation.[51] We will have a chance to explore some of this by way of the contemporary debates surrounding genetic reductionism and biological function.

VIII. CONCLUSION

In this section, five major objections to Aristotle's concept of final causation were examined. It has been argued that only the last objection, the objection from reductionism, could do any serious damage to Aristotle's account. The linguistic, panpsychic, backward causation, and anthropocentric objections reflect, to a degree, a lack of careful attention to Aristotle's arguments and explanations. With this in mind, it should be no surprise, as we shall see, that contemporary philosophers and biologists continue to draw on aspects of Aristotle's biology—including a willingness to entertain seriously the need to include purpose within contemporary evolutionary explanations.

This chapter has offered a glimpse into Aristotle's philosophy of biology. In order to pull this off, attention was given to some of the influential ideas and individuals that helped to form Aristotle's own thinking on things biological. Moreover, Aristotle's general philosophy of science was put on display in order to show its conceptual influence on his biological works. Additionally, the details about his logical and causal thinking were explicated in order to see clearly their presence in his biological thinking. Finally, given that final causation permeates much of Aristotle's philosophy, some of the major objections to it (along with Aristotle's responses) were examined. This analysis reveals that Aristotle has much in his arsenal to deflect most of these objections. Although we may not be willing to follow Aristotle down all of his philosophical paths, there is little doubt that he deserves to be considered, as Darwin himself indirectly acknowledged, the first biologist and first philosopher of biology in Western philosophy. Of course, such praise would be to give Aristotle his due.

---

51 Another conceptual concern that emerges as a result of evolutionary thinking is a rejection of Aristotle's belief that entities/species have natural essences (i.e., essentialist/typological thinking) and non-natural states and that the non-natural states are the result of impinging environmental variation on natural essences. For example, rocks have a natural state of falling to the center of the earth, but have non-natural states of being elsewhere due to external interferences; likewise with biological systems: they too have a natural state and the variation we see in the world is a result of various external interferences. From this perspective, biological diversity is the product of intruding environmental variation/interferences on natural essences. It is only when population thinking emerged in the post-Darwinian world that the idea that species do not have essences started to gain credibility. This is yet another reminder of the lingering influence that Aristotle's thinking had on the history and philosophy of biology. Chapter 5, on the species debate, will explore more of this discussion. Also, see Sober's (1980) classic discussion of population thinking and what it means for the plausibility of essentialist/typological thinking.

CHAPTER REVIEW: DISCUSSION AND QUESTIONS

1. Given what has been said in section I and what Darwin himself says, why do you think that Darwin gives Aristotle so much praise? [pp. 2–3]
2. How did the medicine and chemistry of Aristotle's day influence his own reflections on biology? (Note: Your answer should include making sense of Aristotle's mixed causal-empiricist methodology.) [pp. 3–5]
3. Why does Aristotle include soul as part of his account of science and physics? [pp. 5–6]
4. Explain Aristotle's use of his honey-water example. What is the example designed to reveal? Use bed/bed-maker to help make sense of the honey-water example. Can you use your own example to help this discussion along? Is Aristotle's use of soul relevant here? [pp. 6–7]
5. What does Aristotle's discussion about humans and apes tell you about him as a scientist/biologist? [pp. 8–9]
6. Using your own example (like the squirrel example), can you make sense of the logical analysis part of Aristotle's scientific method? If you are able, try to offer your own formal argument reconstruction as well! [pp. 10–12]
7. Use your own example to make sense of Aristotle's four causes. Why do you think we should/should not take this causal framework seriously (focus on the first three causes)? [pp. 12–14]
8. At around pp. 15–16, I offer an argument reconstruction as to why Aristotle takes final causation seriously. Can you explain this in your own words? (Make sure your answer includes the relationship between chance events and biological entities.)
9. Why should Aristotle provide a solid defense of his reliance on final causation? [p. 17]
10. Give a general explanation of each of the five objections to Aristotle's concept of final causation. [pp. 17–32]
11. With respect to the first objection, are you persuaded by the explanation of Aristotle's doctor analogy and the further defense of it? [pp. 18–20]
12. What does Aristotle say to reveal the implausibility of the panpsychism objection? [p. 21]
13. Are you persuaded by the replies to Sedley's version of the anthropocentrism objection? Are there any additional replies that Sedley could offer? [pp. 23–25]
14. After explaining Aristotle's different senses of causal priority, can you see why the backward causation objection fails? [pp. 26–27]
15. How does Aristotle reply to the reductionism example? Use the bones/flesh example to make your answer clear. What work is *pneuma* doing in this discussion? [pp. 29–31]
16. Even granting Aristotle's salvage efforts, why does the objection from reductionism seem to cause problems for Aristotle's notion of final causation? Your answer should include both *pneuma* in the light of DNA and Aristotle's conception of how things come to exist in the world as opposed to our contemporary understanding of how "things" come into the world. [pp. 31–32]

REFERENCES

Anagnostopoulos, Georgios. "Aristotle's Life." In *A Companion to Aristotle*, ed. Georgios Anagnostopoulos, 3–13. West Sussex, UK: Wiley-Blackwell, 2009.
Ariew, André. "Teleology." In *The Cambridge Companion to the Philosophy of Biology*, ed. David L. Hull and Michael Ruse, 160–81. Cambridge: Cambridge University Press, 2007.
Barnes, Jonathan. *Aristotle: A Very Short Introduction*. Oxford: Oxford University Press, 2000.
———, ed. *The Cambridge Companion to Aristotle*. Cambridge: Cambridge University Press, 1995.
———, trans. "Aristotle: Posterior Analytics." In *The Complete Works of Aristotle*, vol. 1, ed. Jonathan Barnes. Princeton: Princeton University Press, 1984.
Berryman, Sylvia. "Teleology without Tears: Aristotle and the Role of Mechanistic Conceptions of Organisms," *Canadian Journal of Philosophy* 37, 3 (2007): 351–70.
Buller, David, J. *Function, Selection, and Design*. Albany: SUNY Press, 1999.
Charles, David. "Aristotle on Hypothetical Necessity and Irreducibility." *Pacific Philosophical Quarterly* 69, 1 (1988): 1–53.
Code, Alan. "The Priority of Final Causes over Efficient Causes in Aristotle's PA." In *Aristotelische Biologie: Intentionen, Methoden, Ergebnisse*, ed. Wolfgang Kullmann and Sabine Föllinger, 127–43. Stuttgart: Franz Steiner Verlag, 1997.
Connell, Sophia. "Toward an Integrated Approach to Aristotle as a Biological Philosopher." *Review of Metaphysics* 55, 2 (2001): 297–322.
Cosans, Christopher C. "Aristotle's Anatomical Philosophy of Nature." *Biology and Philosophy* 13, 3 (1998): 311–39.
Darwin, Charles. "Letter to William Ogle" (Feb. 22, 1882). In *The Life and Letters of Charles Darwin*, vol. II, ed. Francis Darwin. http://www.fullbooks.com/The-Life-and-Letters-of-Charles-Darwinx29408.html.
Depew, David, J. "Incidentally Final Causation and Spontaneous Generation in Aristotle's *Physics* II and Other Texts." In *Was ist Leben?: Aristoteles Anschauungen zur Entstehung und Funktionsweise von Leben*, ed. Sabine Föllinger, 285–97. Stuttgart: Franz Steiner Verlag, 2009.
———. "Consequence Etiology and Biological Teleology in Aristotle and Darwin." *Studies in the History and Philosophy of the Biological and Biomedical Sciences* 39, 4 (2008): 379–90.
———. "Humans and Other Political Animals in Aristotle's *Historia Animalium*." *Phronesis* 40, 2 (1995): 156–76.
Dowe, Phil. *Physical Causation*. New York: Cambridge University Press, 2000.
Edelstein, Ludwig. *Ancient Medicine*. Baltimore: Johns Hopkins University Press, 1967.
Falcon, Andrea. *Aristotle & the Science of Nature*. Cambridge: Cambridge University Press, 2005.
Frank, Lewis. "Teleology and Material/Efficient Causes in Aristotle," *Pacific Philosophical Quarterly* 69, 1 (1988): 54–98.
Gotthelf, Allan. "Darwin on Aristotle." *Journal of the History of Biology* 32, 1 (1999): 3–30.
———. "Understanding Aristotle's Teleology." In *Final Causality in Nature and Human Affairs*, ed. Richard F. Hassing, 71–82. Washington, DC: Catholic University Press, 1997.

———. "The Place of the Good in Aristotle's Natural Teleology." In *Nature, Knowledge, and Virtue: Essays in Memory of Joan Kung* [*Apeiron* special issue, vol. 22, no. 4], ed. Richard Kraut and Terry Penner, 113–39. Edmonton, AB: Academic Printing and Publishing, 1989.

———. "Aristotle's Conception of Final Causality." In *Philosophical Issues in Aristotle's Biology*, ed. Allan Gotthelf and James G. Lennox, 204–41. Cambridge: Cambridge University Press, 1987.

Gottlieb, Paula, and Elliot Sober. "Aristotle on 'Nature Does Nothing in Vain.'" *HOPOS: The Journal of the International Society for the History of Philosophy of Science* 7 (Fall 2017): 246–71.

Grant, Edward. *A History of Natural Philosophy*. Cambridge: Cambridge University Press, 2007.

Grene, Marjorie. *A Portrait of Aristotle*. Chicago: University of Chicago Press, 1963.

Grene, Marjorie, and David Depew. *The Philosophy of Biology: An Episodic History*. Cambridge: Cambridge University Press, 2004.

Hankinson, R.J. "Philosophy of Science." In *The Cambridge Companion to Aristotle*, ed. Jonathan Barnes, 109–39. Cambridge: Cambridge University Press, 1995.

Hardie, R.P., and R.K. Gaye, trans. "Aristotle: Physics." In *The Complete Works of Aristotle*, vol. 1, ed. Jonathan Barnes. Princeton: Princeton University Press, 1984.

Henry, Devin. "Organismal Natures." *Apeiron* 41, 3 (2009): 47–74.

———. "Aristotle on the Mechanism of Inheritance." *Journal of the History of Biology* 39, 3 (2006): 425–55.

Johnson, Monte Ransome. *Aristotle on Teleology*. Oxford: Oxford University Press, 2008.

Jowett, Benjamin, trans. "Aristotle: Politics." In *The Complete Works of Aristotle*, vol. 2, ed. Jonathan Barnes. Princeton: Princeton University Press, 1984.

Lang, Helen S. *The Order of Nature in Aristotle's Physics*. Cambridge: Cambridge University Press, 1998.

Lennox, James. "Aristotle's Biology and Aristotle's Philosophy." In *A Companion to Ancient Philosophy*, ed. Mary Louise Gill and Pierre Pellegrin, 292–315. Oxford: Blackwell, 2006.

———. *Aristotle's Philosophy of Biology*. Cambridge: Cambridge University Press, 2001a.

———. *Aristotle: On the Parts of Animals*. Oxford: Clarendon Press, 2001b.

———. "Nature Does Nothing in Vain …" In *Aristotle's Philosophy of Biology*, ed. James Lennox, 205–23. Cambridge: Cambridge University Press, 2001c.

———. "Aristotle on the Biological Roots of Virtue: The Natural History of Natural Virtue." In *Biology and the Foundation of Ethics*, ed. Jane Maienschein and Michael Ruse, 1–31. Cambridge: Cambridge University Press, 1999.

Leunissen, Mariska. *Explanation and Teleology in Aristotle's Science of Nature*. Cambridge: Cambridge University Press, 2010.

Lloyd, G.E.R. *Aristotelian Explorations*. Cambridge: Cambridge University Press, 1999.

———. *Magic, Reason, and Experience*. Cambridge: Cambridge University Press, 1979.

Mathen, Mohan. "Teleology in Living Things." In *A Companion to Aristotle*. ed. Georgios Anagnostopoulos, 335–47. West Sussex, UK: Wiley-Blackwell, 2009.

———. "The Four Causes in Aristotle's Embryology." In *Nature, Knowledge, and Virtue: Essays in Memory of Joan Kung* [*Apeiron* special issue, vol. 22, no. 4], ed. Richard Kraut and Terry Penner, 159–79. Edmonton, AB: Academic Printing and Publishing, 1989.

Meyer, Susan Sauvé. "Aristotle, Teleology, and Reduction." *The Philosophical Review* 101, 4 (1992): 791–825.

Miller, Fred. D., Jr. *Nature, Justice and Rights in Aristotle's Politics*. Oxford: Oxford University Press, 2007.

———. "Aristotle's Philosophy of Soul." *The Review of Metaphysics* 53, 2 (1999): 309–37.

Norman, Richard. "Aristotle's Philosopher-God." *Phronesis* 14, 1 (1969): 63–74.

Ogle, William. *Aristotle on the Parts of Animals*. New York: Garland, 1882/1987.

Peck, A.L., trans. *Aristotle: History of Animals, Books I–III*. Cambridge, MA: Harvard University Press, 1965.

———, trans. *Aristotle: Generation of Animals*. Cambridge, MA: Harvard University Press, 1942.

Pittendrigh, C.S. "Adaptation, Natural Selection, and Behavior." In *Behavior and Evolution*, ed. Anne Roe and George G. Simpson, 390–419. New Haven: Yale University Press, 1958.

Platt, Arthur. "Aristotle: Generation of Animals." In *The Complete Works of Aristotle*, vol. 1, ed. Jonathan Barnes. Princeton: Princeton University Press, 1984.

Preus, Anthony. *Science and Philosophy in Aristotle's Biological Works*. New York: Georg Olms Verlag Hildesheim, 1975.

Reeve, C.D.C. *Practices of Reason*. New York: Oxford University Press, 1995.

Ross, W.D., trans. "Aristotle: Metaphysics." In *The Complete Works of Aristotle*, vol. 2, ed. Jonathan Barnes. Princeton: Princeton University Press, 1984.

———, trans. "Aristotle: Nicomachean Ethics." In *The Complete Works of Aristotle*, vol. 2, ed. Jonathan Barnes. Princeton: Princeton University Press, 1984.

Salmon, Wesley. *Causality and Explanation*. New York: Oxford University Press, 1998.

Salter, David. *Holy and Noble Beasts: Encounters with Animals in Medieval Literature*. Oxford: D.S. Brewer, 2001.

Sedley, David. "Is Aristotle's Teleology Anthropocentric?" *Phronesis* 36, 2 (1991): 179–96.

Shields, Christopher. *Aristotle*. London/New York: Routledge, 2007.

Sober, Elliott. "Evolution, Population Thinking, and Essentialism." *Philosophy of Science* 47, 3 (1980): 350–83.

Solmsen, Friedrich. *Aristotle's System of the Physical World*. Ithaca: Cornell University Press, 1960.

Van Der Eijk, Philip J. *Medicine and Philosophy in Classical Antiquity*. Cambridge: Cambridge University Press, 2005.

Chapter 2

CHARLES DARWIN
The Grandeur of a Philosophical Naturalist

As buds give rise by growth to fresh buds, and these, if vigorous, branch out and overtop on all sides many a feebler branch, so by generation I believe it has been with the great Tree of Life, *which fills with its dead and broken branches the crust of the earth, and covers the surface with its ever branching and* beautiful ramifications."

—Charles Darwin, Origin of Species *(pp. 171–72, my emphasis)*

I. INTRODUCTION: THE NEWTON OF A BLADE OF GRASS?

One important principle that gained scholarly legitimacy in the post-Newtonian world was *mechanism*, the idea that the production of material things—including all biological phenomena—could reasonably be understood to come about entirely from mechanical laws.[1] For example, consider Newton's Law of Cooling. According to this law, the rate of change of the temperature of any object is proportional to the difference between the object's temperature and the temperature of its ambient environment.[2] What this means is that, for example, the rate at which my bowl of tomato soup cools is determined by the current temperature of my bowl of soup and the bowl of soup's immediate environment. So long as one has the relevant temperatures, the mechanics of heat exchange is all that is needed to make sense of the rate at which my bowl of soup will cool. Using Aristotelian language, mechanism is the view that only efficient causes (see Ch. 1) are needed to make sense of the material world. So, in the same way that Newton was able to make sense of the physical world by way of his various mechanical laws, it could be argued that

---

1 This is not to suggest that mechanism was not alive and well before Newton. For more on this history, see Des Chene (2002), Berryman (2009) and Gaukroger (2007 and 2011).
2 For more on Newton's actual equations and experiments related to his heat-transference equation, see Cheng and Fuji (1998) and Winterton (1999).

the biological world has its corresponding mechanical laws that make sense of all that pertains to biological phenomena.

In the wake of the Newtonian Revolution and the corresponding mechanistic world-view that had swept through much of the Western world, the eighteenth/nineteenth-century philosopher Immanuel Kant sharply proclaimed: "It is absurd to hope that another Newton will arise in the future who shall make comprehensible by us the production of a *blade of grass* according to natural laws which no design has ordered" (Kant, 2010, p. 168; my italics). Much like Aristotle, Kant thought that biological systems exhibited irreducibly complex goal-directed functions, which cannot be explained and understood entirely by mechanical processes—even if celestial phenomena can.[3] Although Kant does not explicitly invoke any sort of divine influence with respect to the natural world, he clearly does not think that mechanical laws by themselves can fully make sense of both the self-organization and self-maintenance that is replete in the natural world—even in a blade of grass!

Ironically, Charles Darwin would eventually be proclaimed just such a Newton of the biological world after the publication of his ground-breaking (and hastily written) *The Origin of Species* (hereafter *Origin*). The immense and profuse impact of this work across academic and non-academic arenas is difficult to overestimate.[4] In a nutshell, the implication of Darwin's arguments is that man is best understood as one amongst the animals—a biological species that is part of nature, not some special being that merely interacts with nature.[5] It is not hyperbole to suggest that such a view about humans was no less shocking than the view that humans live on a planet that is not the center of the universe. Indeed, what Copernicus and Newton accomplished with respect to the nature and dynamics of the celestial realm, Darwin correspondingly accomplished in terms of the nature and dynamics of humans in the earthly realm. Whether or not Darwin would have appreciated it, "the Newton of a blade of grass" designation is most fitting—as we will see.

Beyond profiling this "grassy Newton," this chapter will not only offer a glimpse into the major influences on Darwin's thinking, but will also provide an admittedly adumbrated reconstruction of the main arguments in defense of *natural selection* in

---

[3] I am not suggesting that Kant's concept of biological teleology is identical to that of Aristotle's. I am, however, claiming that both of them could not accept that material stuff could be understood to self-organize to create the kind of biological complexity that we see in the world. For more on these issues see Rosenberg (2006), Grene and Depew (2004, ch. 4), Ginsborg (2004), Ruse (2003, ch. 2), McLaughlin (1990), Lennox (1993), and Cornell (1986).

[4] Recently, Waters (2003), Lewens (2007), and Reznick (2010) have offered excellent discussions on both the philosophy and implications of Darwin's ideas. Note that Alfred Wallace, an avid naturalist, had produced a similar theory to that of Darwin's and Darwin wanted to make sure his own rendering came to light first. For a glimpse into Wallace's work, see Quammen (2008), Rosen (2007), and the Wallace on-line collection: http://wallace-online.org/. For more scholarly work on Wallace and additional citations, see McKinney (1966), Jones (2002), and Bulmer (2005).

[5] This claim is implicit in the *Origin* and explicit in Darwin's *Descent of Man* (2004), but it was clear that Darwin's peers understood the implications of the *Origin's* principles with respect to the human species. For more on these issues, see Bizzo (1992).

the *Origin*. To this end, Section II of this chapter begins with a look at Darwin's early life and his bio-philosophical development. Section III offers an overall sketch of the general argumentative structure of the *Origin*. Section IV provides a summary reconstruction of the philosophy of science Darwin came to endorse. Section V provides the framework of how Darwin's defense of the probability of natural selection will be handled. Section VI takes a brief detour in order to situate Darwin's *Origin* within the broader religious milieu in which it was received. Section VII, which has three sub-sections, offers a rendition of Darwin's overall defense of natural selection in the *Origin*. Section VIII provides an evaluation of the elements of Darwin's defense of natural selection and concludes that he bit off a bit more than he should have chewed, but that such an enthusiastic chomp hardly diminishes his overall defense. Finally, Section X concludes with an assessment of Darwin's efforts and the incalculable gratitude that biology and the philosophy of biology owe to the range and depth of his labor.

II. LIFE AND TIMES: DARWIN'S TRANSFORMATION INTO A NATURALIST

Much like Aristotle, Darwin was a tireless worker and came from a family of medical practitioners—Darwin's father, Robert Darwin, was a doctor as was his grandfather, Erasmus Darwin, who developed his own theory of the origins of life that gave significant weight to competition and sexual selection.[6] To a certain extent, then, biology was a part of Darwin's upbringing. As he grew older and his external influences became narrower, his interests in those things biological grew steadily as he came to "see" the wonders of the biological world. His elation, upon observing some of the lush South American tropics and the various organisms inhabiting the tropical vegetation during his *Beagle* voyage, leaves little doubt regarding his interests, eagerness to engage nature, and the "new vision" that was thrust upon him ("Darwin" in Keynes, 1988, p. 37).

Ultimately, these interests and corresponding delights sustained him during his later years in life.[7] Indeed, even though the latter part of his life was wracked with illness, Darwin tells us the psychological effect of his joyful toil:

> My chief enjoyment and sole employment throughout life has been scientific work; and the excitement from such work makes me for the time forget, or drives quite away, my daily discomfort. I have therefore nothing to record during the rest of my life, except the publication of my several books. (2003, p. 69)

---

6 See Smith and Arnott (2005).
7 Darwin tells in his autobiography: "My habits are methodical, and this has been of not a little use for my particular line of work. Lastly, I have had ample leisure from not having to earn my own bread. Even ill-health, though it has annihilated several years of my life, has saved me from the distraction of society and amusement" (2003, p. 74).

This tireless effort, to be sure, is evident in his eight-year infatuation with producing a proper taxonomy—scientific classification of organisms—of a particular sub-species of barnacles. The precision of his empirical work is quite palpable, even from his early work on this organism:

> In October, 1846, I began to work on "Cirripedia." When on the coast of Chile, I found a most curious form, which burrowed into the shells of Concholepas and which differed so much from all other Cirripedes that I had to form a new sub-order for its sole reception. Lately an allied burrowing genus has been found on the shores of Portugal. To understand the structure of my new Cirripede I had to examine and dissect many of the common forms; and this gradually led me to take up the whole group. (1887, vol. I, p. 80)

Given the above accounts and the great many things that have resulted from Darwin's ideas, one would have expected Darwin to have been an outstanding student right from the start. This, however, is not the case. Born in 1809 in Shrewsbury, a county town of Shropshire, England, Darwin's early schooling was not very eventful. He neither excelled in school nor did poorly, and was considered by his own teachers to be a rather ordinary student. To be sure, Darwin took himself to be more an outdoor sportsman than academic (Lewens, 2007, pp. 9–18). This lack of interest in formal education was evident in both his prematurely terminated stint at medical school at the University of Edinburgh (1825–27) and his self-proclaimed "sadly wasted" time at Cambridge University (1828–31). Yet, in the midst of all this "academic mediocrity" and a somewhat reluctant (and ironic) leaning toward becoming a clergyman, Darwin's intellectual curiosities got the best of him. He was rather taken by the botany lectures and field trips of the Cambridge botanist and Anglican priest John Henslow. It was Henslow who eventually convinced Darwin to be a full-time naturalist because he detected both the intense interest and empirical care with which Darwin engaged nature. Additionally, his interest was piqued by the lectures of and field trips with the Cambridge geologist, Adam Sedgwick, and he found considerable joy in his own beetle collecting hobby.

It is reasonable, then, to think of Darwin as more than a mere fancier of the natural world (Bowler, 1990).[8] In addition, he became increasingly interested in various provocative theories about the natural world, which he would eventually fit—like a jig-saw puzzle—into his *Origin of Species*.

Much like Aristotle, Darwin employed a mixed causal-empiricist methodology (see Section II of Ch. 1). What this means is that he undertook an approach to understanding biological systems from a particular causal framework that both reinforced and was itself reinforced by his actual field studies (de Beer, 1995). Specifically,

---

8 'Pre-Beagle Darwin' refers to Darwin before his five-year British Navy mapping expedition (to South America and the now famous Galápagos Islands) on the ship of war known as the H.M.S *Beagle*.

beyond the biogeographic data that he gathered during his famed *Beagle* voyage (1831–36), there was a set of ideas being espoused by "naturalist" thinkers that he could not ignore.[9] For example, he was (1) impressed by the ideas of Jean Baptiste Lamarck, a French biologist who argued for the limitless variation of organisms; (2) charmed by William Paley's design arguments for the existence of a divine creator in his *Natural Theology*; (3) fascinated by the possibility of "deep geologic time" as defended by Charles Lyell in his *Principle's of Geology*; and (4) ultimately convinced by Thomas Malthus' sobering account of the dangers of human population growth in the wake of limited resources in his *Essay on the Principle of Population*. In part, these four major intellectual encounters in Darwin's life helped shape his *Origin* and the "beautiful ramifications" that were to follow.[10]

III. A SKETCH OF DARWIN'S ARGUMENT IN THE *ORIGIN*

Darwin tells us that his *Origin* is comprised of "one long argument" (1958, p. 435). In order to appreciate the depth and scope of his discussion in this text, it will help first to have a general sketch of what he hopes to accomplish and what is the overall structure of his argument. This will make the relevance of Darwin's philosophy of science (discussed in the next section) and the quality of his overall defense apparent to the reader.

What, then, is this extended argument and what are the implications of it for understanding biology? With a bit of panache, Darwin tips us off with his answer in the introduction to his *Origin*:

> I can entertain no doubt, after the most deliberate study and dispassionate judgment of which I am capable, that the view which most naturalists entertain, and which I formerly entertained—namely, that each species has been independently created—is erroneous. I am fully convinced that species are not immutable; but that those belonging to what are called the same genera are linear descendants of some other and generally extinct species, in the same manner as the acknowledged varieties of any species are the descendants of that species. Furthermore, I am convinced that Natural Selection has been the main but not the exclusive means of modification. (1958, p. 30)

It is not trivial to decipher what Darwin is trying to convey in terms of argumentation. He clearly endorses the following statements:

---

9   I am here using 'naturalist' in a non-technical sense. It is simply referring to those thinkers who had an influence on Darwin's reflections on the natural world.
10 There is an embarrassment of riches regarding Darwin scholarship; this includes his letter correspondences (http://darwinonline.org.uk/EditorialIntroductions/Freeman_LifeandLettersandAutobiography.html) to both friends and fellow scholars and his notebooks (http://www.english-heritage.org.uk/daysout/properties/home-of-charles-darwin-down-house/darwins-notebooks/).

(1) Species have not been independently created [by a divine being].

(2) Species are not unchangeable [i.e., not immutable].

(3) Currently existing species are products of a history of previously existing species.

(4) Natural selection is the primary mechanism by which species are and have been modified.

But how do 1–4 fit together in the form of a coherent argument, since the *Origin* is supposed to be one long argument? As the rest of this analysis will reveal, the 'one long argument' notion is a little misleading. There are numerous arguments presented in the *Origin* that are designed to defend a number of points. Still, the truth of (3) sanctions the truth of (2); that is, if it is true that currently existing species are products of a history of previously existing species, then it follows that currently existing species are the product of changes which are incompatible with the idea that species are immutable. Additionally, the truth of (2) renders particular versions of (1) likely; if it is false that species are immutable, then any special-creation theory that endorses the belief in the creation of immutable species would correspondingly be false. If this line of reasoning is on the mark, then it is reasonable to ask: what justifies the truth of (3)? Darwin's answer is (4); that is, the power of natural selection justifies accepting that currently existing species are products of a long history of previously existing species. Of course, one can further ask: what justifies the truth of (4)? Well, it is the details within numerous arguments in the *Origin* that reveal the truth of natural selection as the primary mechanism by which species are and have been modified.

What all of this reveals is that if Darwin can show that (3) and (4) are true, then the falsity of (1) and (2) must fall by the wayside. **The Main Argument** of the *Origin*, then, would look something like this:

P1. If, primarily, as a result of natural selection, currently existing species are the product of a long history of related extinct species, then it is not the case that species are unchangeable, and it is not the case that species have been independently created by a divine being.

P2. Primarily, as a result of natural selection, currently existing species are both changeable and the product of a long history of related extinct species.

C1. Thus, it is not the case that species are unchangeable and it is not the case that species have been independently created by a divine being.

The *Origin* can best be viewed as an attempt to show that species are changeable (that is, that **P2** is true). What this means is that Darwin has taken upon himself the task of showing that species are historically contingent and related entities as primarily a result of the power of natural selection. The implication for biology, thinks Darwin, is that it is wrong to view species as unalterable and created "as-is" by a divine designer. As we will see, there are numerous arguments that Darwin offers in the *Origin* to bolster his defense of why species change (the truth of **P2**) and ultimately reveal the soundness of **The Main Argument** noted above. To foreshadow a little, Darwin will argue that species change in the light of (1) a very old Earth, (2) the rather close connection between environmental change and organism change, (3) the reasonableness of the analogy between artificial selection (i.e., domestic breeding) and natural selection, and (4) the advantages that accompany a world in which there is almost constant battles for survival within and across species. It is the combination of (1)–(4) that motivates Darwin to think that natural selection is the major factor in the production of the currently existing species that capture the outer branches of the bushy tree of life. To appreciate his defense, a general idea of the philosophical underpinnings that guide this remarkable text is needed. To begin, then, the next section will directly examine Darwin's philosophy of science.

IV. THE ORIGIN OF DARWIN'S PHILOSOPHY OF SCIENCE

There are two influential thinkers I have left off the earlier list: John Herschel and William Whewell. Both of these men espoused a philosophy of science—that is, an account of what are the hallmarks of a good scientific theory—that greatly impressed Darwin. Remember, Darwin was eager to seek out and incorporate interesting ideas into his own work. To this end, some of the ideas about the nature of explanation by these two philosophers of science proved irresistible conceptual tools for his *Origin of Species* project. I noted above that Darwin is best understood as a mixed causal-empiricist like Aristotle. It stands to reason, then, that causation and empirical observation played central roles in Darwin's philosophy of science. This, indeed, is the case even though Darwin did not endorse Aristotle's four-cause account. Still, a brief explanation of both 'cause' and 'empirical' needs to be put forth in order to appreciate Darwin's peculiar thinking about biology.

First, Darwin's reflections about causation were influenced by Herschel, a prominent astronomer and philosopher of science in Darwin's day.[11] Herschel argued that it is the mark of a true cause—a *vera causa*—that it not only makes sense of a set of phenomena under investigation, but that it gains credibility from unexpected areas of inquiry. In terms of what constitutes a good scientific theory, Herschel makes two compelling points. First, he stresses that an inductive argument is powerful when its conclusion is supported by evidence, but also when such evidence comes from

---

11   I do not think that 'prominent' accurately captures Herschel's achievements. His work in mathematics and the construction of telescopes alone suggest an almost da Vinci-type person. At any

unrelated—even hostile—areas of inquiry. Second, Herschel tells us that there is little else (in terms of evidence) that could be more weighty than to have unrelated evidence from a previously unexpected and/or hostile source (Herschel, 1831, p. 170).

Additionally, Whewell's account of induction also had a lasting effect on Darwin's thinking about evidence and scientific methodology. Whewell offers the following of what he calls the "consilience of inductions":

> [T]he evidence in favor of our induction is of a much higher and forcible character when it enables us to explain and determine cases of a *kind different* from those which were contemplated in the formation of our hypothesis. The instances in which this has occurred, indeed, impress us with a conviction that the truth of our hypothesis is certain. (Whewell, 1847/1967, p. 65; italics in original; my underline)

In this passage, Whewell is pointing out that a very impressive inductive argument will not only support a specific hypothesis, but will also "jump" over to yield insights into other areas of inquiry.[12] The truth with respect to such a fruitful hypothesis will be considered, thinks Whewell, certain. Whewell summarizes his criteria as follows:

> 1. Our hypothesis must be capable of distinctly and appropriately connecting the phenomena.
>
> 2. Generalizations resulting from such hypotheses, if correctly obtained and eminently simple, are theories.
>
> 3. Theories which explain phenomena, detached from those which were used in the generalization, are highly probable, and advance to certainty as the number of unexplained phenomena is diminished. (Whewell, quoted from Laudan, 1981, p. 171)

Although possibly an obvious point, Whewell requires in (1) not only that a cogent hypothesis precisely reveals the underlying relationships amongst the phenomena under investigation—recall Newton's Law of Cooling—but also that it must be determined

---

rate, there is little doubt that Darwin held Herschel in high esteem. Darwin offers the following respectful praise about Herschel: "He never talked much, but every word which he uttered was worth listening to" (2003, p. 28). It is also reasonable to understand why he sent Herschel a copy of his *Origin of Species*. He tells us precisely why: "I should excessively like to hear whether I produce any effect on such a mind" (Charles Darwin to C. Lyell, November 23, 1859 found in Carey [2004]). For more on the Darwin-Herschel connection, see Cannon (1961), (Ruse, 1975a), Schweber (1989), Lewens (2007), and Warner (2009).

12 For a recent attempt at unifying different areas of inquiry, see Wilson (1998) for an updated rendition of consilience.

(2) only after careful observation and study of the phenomena under investigation. The production of this kind of precision characterizes Newton's Law of Cooling as described earlier.[13]

Additionally, Whewell points out that a theory gains credibility if it is "eminently simple." Here, Whewell is requiring the criterion of *parsimony* with respect to theory allegiance. This is the view that, when confronted with competing hypotheses, one should endorse the hypothesis that explains everything its competing hypotheses explain, but does so by invoking the fewest number of background assumptions, rules, or entities (this criterion is also known as Ockham's Razor).[14] Put another way, one should embrace the simpler theory so long as it explains everything the more complicated theory explains.

For example, consider the case of the origin of crop circles (Newall, 2005). Let's assume that theory 1 claims that crop circles are the product of humans and extraterrestrials working together, while theory 2 argues that crop circles are the result of a kind of hoax perpetrated by humans. Theory 1 is more complicated than theory 2 because it includes an additional entity (namely extraterrestrials) that theory 1 does not include. Since the evidence shows that humans can produce such elaborate and precise drawings, there is no need to invoke an additional set of entities like extraterrestrials. So, the criterion of parsimony demands that theory 2 should be endorsed over theory 1 since it invokes a fewer number of entities and still explains everything that theory 1 explains. Theory 2 is simpler.[15]

Lastly, Whewell reminds us that a theory moves in the direction of certainty when it can explain previously unexplained phenomena. This is the criterion of *fecundity*; that is, a theory is considered fruitful if, and only if, it not only explains what it sets out to explain, but helps make sense of unresolved or poorly understood phenomena and/or opens up new areas of study. For example, Newton's theory of universal gravitation is fecund because it not only resolved problems associated with understanding falling bodies, but it also proved to help make sense of the influence of planets on other planets, motion of tides, and the orbital rotation of the moon around the Earth.

One remaining concern is how does a scientist go about offering the kind of evidence that would satisfy the demands of both Herschel and Whewell? Herschel, drawing on the success of Newtonian mechanics, suggests that arguments from analogy assist in verifying true causes. In the example at hand, in the same way that we

---

13 Any undergraduate physics major will tell you that things are not so simple! As it turns out, Newton's Law of Cooling holds up only within certain parameters, but this is a technicality that we need not let bother us at this juncture of our discussion.

14 For more on why a defense of parsimony is needed, see Sober (2016) and Mizrahi (2016).

15 Much more could be said about the parsimony issue and the background assumptions that motivated Whewell and others to endorse it. Additionally, this chapter could have been woven around the parsimony issue, resulting in a different way of approaching the historical and philosophical aspects of Darwin's project in the *Origin*. This route was not chosen in favor of this more historically guided way of introducing students to Darwin's work. At any rate, adventurous minds should consult Sober (2008 and 1988) and his citations for more on the history and philosophy of parsimony as it relates to biology.

are able to infer a kind of force that keeps a stone-filled pouch moving in its circular pattern, as a result of being whirled around by a tautly held string, a corresponding type of force must be present to account for why a bigger stone, namely the moon, continuously maintains its circular motion around the Earth. It is the result of empirical observation and inferences of certain phenomena, argues Herschel, that justifiably allows one to make similar compelling observations and inferences about other less obviously related phenomena. Thus, the fifth criteria of the Herschel-Whewell philosophy of science is that well-crafted arguments from analogy are required to give ontological justification for one's hypothesized true causes (Herschel, 1831, p. 149). Table 2.1 provides the core elements of the Herschel-Whewell philosophy of science that Darwin thoroughly endorsed.

CRITERIA	EXPLANATION
True Cause	Evidence in support of the true cause(s) for a given hypothesis under consideration is required.
Surprise and Hostility	Evidence from unexpected and/or hostile areas of inquiry should yield further support of hypothesis under consideration.
Fecundity	Existing evidence should not only validate hypothesis under consideration, but should also shed light on some other, different, and/or poorly understood area of inquiry.
Parsimony	Other things being equal, hypothesis X should be accepted over all other competing hypotheses so long as X explains everything its competitors explain and does so with fewer assumptions, rules, and/or entities.
Arguments from Analogy	Arguments from analogy are needed to reveal the existence of hypothesized true cause(s).

Table 2.1 *The Herschel-Whewell Criteria for a Good Philosophy of Science*

It is not too much of a leap, presumably, to claim that the language of "strength," "certainty," and "compels assent" in Herschel and Whewell's account of good science influenced what Darwin meant when he claimed that *natural selection is probable to the highest degree* (see next section) by the time he revised the sixth edition of the *Origin*.[16] As we will see, it is the two-pronged Herschel-Whewell philosophy of science that Darwin perceptively and subtly incorporates into his *Origin*.

---

16 Note that a major assumption of this section is that there is little substantial difference between Herschel and Whewell's philosophies of science. No doubt, this is an oversimplification and a bit misleading. For more on this interesting historical confluence of ideas, see Depew and Grene (ch. 6, 2004), Hull (2000), and Ruse (1975a).

V. NATURAL SELECTION: PROBABLY TRUE OR TRUE TO THE HIGHEST DEGREE PROBABLE?

One of the main purposes of the *Origin*, as pointed out in **The Main Argument** (see Section III for a reminder of this argument), is to reveal the eminent plausibility of *natural selection* as a real force of nature. After gathering a bit of steam in the first two chapters, Darwin offers, in chapter three, his understanding of natural selection within the context of variation and the precariousness of life and its hardships. The thrust of this discussion is that 'natural selection' refers to the preservation of useful variations—however miniscule they may be—that (1) keep individual organisms alive in their complex environments, (2) allow such organisms to pass along their useful variations to their offspring, and (3) aid offspring in their struggle to survive (1958, p. 75). A closer look into the *Origin* will help make sense of these claims and the corresponding, wide-ranging biological and philosophical implications that were to follow.

In this section and the next, I will turn to the main arguments in the *Origin*. The goal here is to reveal the care with which Darwin took to defend his ideas within the framework of the Whewell-Herschel criteria discussed in the previous section and his defense of the mutability of species (that is, **P2**) of **The Main Argument**. I will also weave into the discussion the role of some of the other influences on Darwin's thinking. This approach will not only make clear Darwin's own arguments, but the background justification for some of the moves he makes.

Fortunately for us, Darwin offers a clear summary conclusion of his *Origin*. This is of great assistance in reconstructing his overall argument. Yet, during his lifetime, Darwin produced five additional editions of the *Origin*, making modifications in the light of the various criticisms he encountered and new biological data that came his way. Interestingly, in the first edition, Darwin offers the following conclusion: "**Natural selection ... it seems to me to be in itself probable**" (C. Darwin, 1985, p. 443). In the sixth edition, Darwin concludes much more forcefully: "**The theory of natural selection ... seems to be in the highest degree probable**" (C. Darwin, 1958, p. 443). Notice that, in the first edition, Darwin cautiously concludes that natural selection is probably true. Let us call this The Weak Conclusion. In the sixth edition, he appears to throw caution to the wind when he announces that natural selection is probable to the highest degree. Let us call this The Strong Conclusion. Because the second conclusion is stronger than the first, evidence for it had better be stronger. Notice that if The Strong Conclusion is a reasonable inference given the data and arguments that Darwin provides, it follows that The Weak Conclusion would also be justified. Let us determine which conclusion wins the day—if either does—in the light of the overall justification that Darwin furnishes.

To begin, Darwin is offering an **inductive argument**. In effect, he is claiming that if his premises are true, then his conclusion is probably true or to the highest degree probably true. More specifically, Darwin is offering what approximates the sort of inductive reasoning known as *inference to best explanation*. According to this sort of inferential reasoning, a particular hypothesis, **H**, is superior to a set of competing alternative hypotheses, **Hx**, if and only if, **H** makes better sense of the relevant data/evidence than **Hx** *despite* the possibility that **H** and/or **Hx** could be false in the light of correct data.[17] For example, the hypothesis, **H**, a massive meteor/comet impact 66 million years ago caused the extinction of most dinosaur species (and most other species), is thought to be the inference to the best explanation in the light of the existing data (see Pope, D'Hondt, and Marshall 1998). Alternative hypotheses, **Hx**, like disease agents, human intervention, mere temperature fluctuation, large scale volcanic activities, or immense plate tectonic events do not make better sense of the existing data than **H**. Although it is possible that one of the **Hx** hypotheses is correct and **H** is incorrect (or all of the hypotheses are incorrect), there is little evidentiary corroboration to infer the likelihood that any one of these **Hx** hypotheses is superior to **H**. Thus, **H** is the inference to the best explanation.[18]

It is just this sort of inferential reasoning that Darwin employs in defense of his theory of natural selection in the *Origin*. As Lipton (2000, p. 184) tells us, "Darwin inferred the hypothesis of natural selection because, although it was not entailed by his biological evidence, natural selection would provide the best explanation of that evidence." His argument for the strong conclusion, which will be filled in as we move along in the discussion, has the following structure:

The Strong Argument for Natural Selection

P1. If the various accounts about the facts regarding the natural world are true, then the theory of natural selection seems to be in the highest degree probable.

P2. The various accounts about the facts regarding the natural world are true.

C1. The theory of natural selection seems to be in the highest degree probable.

This argument may seem straightforward at first glance, but it is not. Immediately, it is not at all clear that **P1** is true. Even if one granted Darwin the truth about all the facts regarding the natural world, it could still be reasonable to reject the *to-the-highest-degree-probable* claim regarding natural selection. Why is this the case?

---

17   It should be noted that not all scholars house inference to best explanation (also called *abduction*) within the domain of induction. For a good discussion of inference to best explanation, see Lipton (1991) and for the abduction discussion, see Douven (2011).

18   The idea that competing hypotheses and their corresponding auxiliary assumptions can be tested with respect to the relevant evidence and against competing hypotheses and their corresponding auxiliary assumptions is another way to think about how one hypothesis can be determined to be predictively superior to others. For an insightful analysis of the concept of testability of hypothesis, see Sober (1999). (Note that this is a demanding article from Sober given the amount and variety of philosophical terrain covered. It will likely require instructor assistance for undergraduates—but well worth the effort!)

Primarily, if one offers a very strong conclusion, then one will have the arduous task of defending that conclusion with corresponding weighty support. Failure to provide compelling evidence and argumentation would result in one's argument lacking justification. This is the logical context in which Darwin finds himself by his "highest degree probable" claim. In terms of overall structure, the important upshot of the truth of **P1** of **The Strong Argument for Natural Selection** is that the truth of **P2** (that currently existing species are malleable and the product of a long shared history of descent owing to natural selection) of **The Main Argument** of the *Origin* will have been made manifest.

If Darwin can employ the biological facts of the world to harness a bundle of arguments in defense of his strong conclusion, then we would be obligated to accept his defense of natural selection being true to the highest degree probable; that is, this is the inference to the best explanation. So, evidence and persuasive argumentation could reasonably reveal the power of Darwin's overall analysis. Let us, then, grant the truth of **P1** of **The Strong Argument for Natural Selection**, but be on alert that Darwin has a formidable task.

Even with the truth of **P1** conceded, it is **P2** that requires careful scrutiny. Here I will have to evaluate some of the background assumptions that motivate Darwin to make certain claims about the natural world. Interestingly, as we will see, Darwin does not shy away from his own ignorance or the general scientific ignorance about a particular issue. Still, even in the light of various degrees of knowledge gaps, he boldly pushes for the reasonableness of his considered views. For instance, in terms of making sense of variation in organism behavior and physical form, Darwin offers the following summary judgment:

> Our ignorance of the laws of variation is profound. Not in one case out of a hundred can we pretend to assign any reason why this or that part has varied. But whenever we have the means of instituting a comparison, the same laws appear to have acted in producing the lesser differences between varieties of the same species, and the greater differences between species of the same genus.... Whatever *the cause* of each slight difference between the offspring and their parents—and *a cause* for each must exist—we have reason to believe that it is the steady accumulation of the beneficial differences which has given rise to all the important modifications of structure in relation to the habits of each species. (1958, pp. 158–60)

Forthrightly, Darwin admits to the ignorance concerning the causal processes (associated with reproduction) which yield organism variation. Yet, he stresses that observational comparisons compel one to accept that certain "biological laws" (including natural selection) must be at play with respect to both (1) the closeness of observable differences of related species (e.g., varieties of dogs or pigeons) and (2) the considerable distance between observable features of species of the same genus (e.g., lion and panther or lion and domestic cat). He pushes on to suggest that the slow (geological sense of slow) piling-up of useful slight differences of individual organisms—useful differences that are then passed on from generation to generation (the details of which were not well understood during Darwin's lifetime)—makes sense of all the important present-day observable structures and behaviors—it is the best explanation. Well, is Darwin's confidence justified in the wake of his candid concession of ignorance? Restated, can Darwin really maintain his "highest degree probable" claim given his appeal to ignorance associated with certain biological facts? Let us find out.

VI. A NECESSARY DEVIATION: THE DIVINE CHALLENGE TO NATURAL DESIGN

Before turning to Darwin's arguments in defense of his vision of biology, there is one major lingering concern that requires attention because it is relevant to Darwin's overall defense. Recall that **The Main Argument** of the *Origin* defends the view that species change and that species are not the products of the whims of a divine being, a view shocking to, and widely resisted by, religious Christians (e.g., the theologian, Charles Hodge, in Schaff and Prime, 1874, pp. 318–20).

Darwin was acutely aware that the potential cacophony of defiance, like that of Hodge, would drown out any kind of reasonable offering he could give. In response, he employs a strategy of openness and humility.

In the last chapter of the *Origin*, Darwin cautiously employs this strategy by acknowledging that there are many objections to his account. For instance, the notion that the various features of biological systems could be fully understood without introducing some sort of divine intervention seemed preposterous to many of his contemporaries. In fact, **P2 of The Strong Argument for Natural Selection** above (i.e., Darwin's claim that the various accounts about the facts concerning the natural world are true) would have been deemed false by many of Darwin's contemporaries for the very reason that it ignores the "real" fact that complexity, organization, and regularity require a designer—God. William Paley, an Anglican priest who died about five years before Darwin's birth and with whose work Darwin would later be keenly familiar, after describing the workings and parts of the eye, makes this divine designer point explicit:

> Were there no example in the world of contrivance except that of the eye, it would be alone sufficient to support the conclusion which we draw from it, as to the necessity of an intelligent Creator. (Paley, 1802, p. 59)

Paley dismisses the possibility that chance events of the natural world could produce the kind of complexity and structure one finds, for example, in the eye: "What does chance ever do for us? In the human body, for instance, chance, i.e. the operation of causes without design may produce a wen, a wart, a mole, a pimple, but never an eye" (Paley, 1802, p. 49). Like Kant and Aristotle before him, Paley is pointing out that random material processes could not possibly account for the complexity and organization of certain biological phenomena.

There is little doubt that Darwin was aware of Paley-type arguments from design. Darwin acknowledges this kind of design challenge and provides his own rendition of it as he proceeds to defend the truth of **P2** of **The Strong Argument for Natural Selection**:

> Nothing at first can appear more difficult to believe than that the more complex organs and instincts should have been perfected, not by means superior to, though analogous with, human reason, but by the accumulation of innumerable slight variations, each good for the individual possessor. (1958, p. 435)

Darwin is clearly sensitive, as the above quotation reveals, to the fact that it is difficult for people to believe that the complex physical make-up of organisms is *not* the product of a divinely reasoning being, but the mere aggregation of a great many small and beneficial biological variations. It is this religious challenge that helps make sense of Darwin's overall project in the *Origin* and should be registered as an integral leitmotif of this text. As most of us know, this challenge is still with us today in the guise of "intelligent design" creationism. We will have a chance to look at the details of this modern clash between nature as a designer and a superior or supreme-being as a designer in Chapter 8. For now, let us turn to Darwin's defense of natural selection as an alternative to a divine artisan.

VII. DARWIN'S "ONE LONG ARGUMENT"[19]

How could Darwin offer a cogent reply to the religious challenge and make manifest the Whewell-Herschel inspired philosophy of science behind such a reply? Restated, what is Darwin's defense of **P2** of **The Strong Argument for Natural Selection** (i.e., Darwin's belief that his factual claims about the biological world are true) and does this defense ultimately vindicate the truth of **P1** (the claim that if the facts about the biological world are true, then natural selection is true to the highest degree possible)? To begin to answer these questions, let us take a look at the defense of the reality of natural selection as presented by Darwin in the following somewhat lengthy passage:

> If under changing conditions of life organic beings present individual differences in almost every part of their structure, and this cannot be disputed; if there be, owing to their geometrical rate of increase, a severe struggle for life at some age, season, or year, and this certainly cannot be disputed; then, considering the infinite complexity of the relations of all organic beings to

---

19   The book title with this very same name is Ernst Mayr's wonderful *One Long Argument: Charles Darwin and the Genesis of Modern Evolutionary Thought* (1993).

each other and to their conditions of life, causing an infinite diversity in structure, constitution, and habits, to be advantageous to them, it would be a most extraordinary fact if no variations had ever occurred useful to each being's welfare, in the same manner as so many variations have occurred useful to man. But if variations useful to any organic being ever do occur, assuredly individuals thus characterized will have the best chance of being preserved in the struggle for life; and from the strong principle of inheritance they will tend to produce offspring similarly characterized. This principle of preservation, or the survival of the fittest, I have called, Natural Selection. It leads to the improvement of each creature in relation to its organic and inorganic conditions of life. (1958, pp. 129–30)

Notice that Darwin points out that 'natural selection' is synonymous with 'principle of preservation.' This is noteworthy because this principle refers to a suite of facts about the world that helps make sense of why features of organisms are as they are. What this means is that 'natural selection' refers to the very same suite of facts. These are the facts that are relevant to the truth of **P2** of **The Strong Argument for Natural Selection**. Darwin points out these facts by way of a multi-layered argument that is ultimately defended in various stages throughout the *Origin*. The argument looks like this:

The Preservation Argument

P1. If (1) the planet is old and (2) it has undergone various environmental changes, and (3) there is a struggle for life as a result of organisms reproducing at high rates at certain ages, seasons, or times of the year, then the unlimited combinations of interactions amongst and between organisms and their corresponding environments will produce an unlimited supply of physical and behavioral variation, some of which will be beneficial to some organisms, and these beneficial variations will be passed on to the relevant offspring (principle of inheritance).

P2. If it is true that the unlimited combinations of interactions amongst and between organisms and their corresponding environments will produce an unlimited supply of physical and behavioral variation, some of which will be beneficial to some organisms, and these beneficial variations can be passed on to the relevant offspring (principle of inheritance), then these offspring will have the best chance to be preserved in the struggle for life.

P3. It is obviously true that (1) the planet is old and (2) it has undergone various environmental changes, and (3) there is a struggle for life as a result of organisms reproducing at high rates at certain ages, seasons, or times of the year.

P4. It follows that the unlimited combinations of interactions amongst and between organisms and their corresponding environments will produce an unlimited supply of physical and behavioral variation, some of which will be beneficial to some organisms and it is also true that these beneficial variations will be passed on to their offspring.

C1. Thus, such individual organisms will have the best chance to be preserved in the struggle for life.

If Darwin can show that the facts embedded in **P3** are obviously true—and he seems quite confident that they are—then **The Preservation Argument** will be proven sound *and* the truth of **P2** of **The Strong Argument for Natural Selection** will also have been established. The implication is that Darwin will have delivered a cogent defense of **The Strong Argument for Natural Selection**. Ultimately, the soundness of both **The Preservation Argument** and **The Strong Argument for Natural Selection** vindicate the truth of **P2** of **The Main Argument** of the *Origin* and thereby divulging the soundness of **The Main Argument**. What evidence, then, does Darwin provide? In what follows, a summary of the facts and corresponding arguments will be provided in the hopes of revealing to what extent we should be persuaded by Darwin's "highest degree probable" claim about the truth of natural selection.

The Tree-of-Life: An Antiquated Earth Needed

One daring image of Darwin's theorizing in the *Origin* that is rather pronounced is his *tree-of-life conception*. Basically, Darwin thought that all existing organisms could be traced back to a single or few original stock species. From this initial stock, the tree of life began to unfold and continued to unfold into the more-or-less multitude of branches of organisms that currently exist (see Darwin's diagram on p. 56).

There are three striking implications of this tree image that put Darwin at odds with some of his contemporaries. First, Darwin was rejecting the commonly held view that organisms varied only a little from their fixed species form. According to this common view, all token zebras are slight variations of their corresponding zebra type. If Darwin is correct about his tree of life image, then it must be the case that there had to have been tremendous species-level changes to account for the transition from one or a few original species to the remarkable modern-day plethora of species. From this perspective, for example, there must have been transitional species that preceded and paved the way for the contemporary zebra species. Second, Darwin had to contend with the general consensus that organisms are highly (if not perfectly) designed to cope with their local environments. If this is correct, then most organismal changes could only prove deleterious. In contrast to this optimality thesis, Darwin endorsed the view that organisms did change and such changes included behavioral and physical variations, some of which proved to be enhancements in the

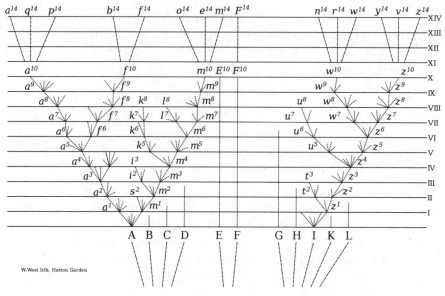

Darwin's Tree-of-Life Sketch

wake of local environmental pressures, while others proved to be diminutions. Third, Darwin also embraced a certain picture of the geological record, claiming that the forces that have shaped the Earth were not the product of a single cataclysmic event (*catastrophism*), but rather the slow accumulation of tiny changes (*uniformitarianism*).

It is this third point that Darwin engages in the geology chapters (chs. 10–13) of the *Origin*. He knew full well that, given the incredible geological variation on the planet and the abundance and diversity of life around him, *a great deal of time* would be needed to allow for the tree-of-life hypothesis, that is, his belief that there was a transition from a single or a few species to the incredible modern variety of species. Indeed, it is the truth of this third point that would begin to give credence to the first two noted above, and ultimately assists in supporting the validity of natural selection. It is here that the work of the geologists Charles Lyell and Andrew Ramsay proved to be invaluable to Darwin.[20] The Ramsay part of this connection is stressed by Reznick, upon reminding the reader that Darwin argued for an old-Earth by way of "the total thickness of fossil strata found in the geological record" (Reznick, 2010, p. 276). So, drawing from the geologic time-scale methodology of Ramsay, Darwin speculated upon the Earth's age. He did so by determining how much time it would likely take for the geological strata he was evaluating to have formed. To buttress Ramsay's guidelines, Darwin included the tenets of Lyell's *uniformitarian geologic principles* to

---

20 As noted earlier, Darwin was an impressive geologist in his own right. For more on Darwin the geologist, I would direct the reader to Judd (1909), Herbert (2005), and chapters 4–7 of Grant and Estes (2009).

approximate the age of the once forest land of Sussex Weald.[21] Darwin assumes the following principles of Lyell's account as part of his overall approximation:

1. *Uniformity of Law*: The laws of nature are constant across time and space. For instance, Newton's first law of motion states that an entity X will maintain a constant velocity unless X is acted upon by another entity Y. This law of nature is thought to hold across time and space. Additionally, we can make inferences about past, present, and future physical events in the light of lawful uniformity.

2. *Uniformity of Methodology*: The appropriate hypotheses for explaining the geological past are those hypotheses that draw upon contemporary data. For example, the explanation of the motion of the tides today is used to explain the motion of the tides in the distant past.

3. *Uniformity of Kind*: Past and present causes are all of the same kind, have the same energy, and produce the same effects. The energy and effects of waves on the shoreline today are the same kinds of waves with the same energy and effects on the shorelines of the past.

4. *Uniformity of Degree*: Geological circumstances have remained steady, ongoing, and gradual over time (Hooykaas, 1963).[22] For example, slow and regular cumulated changes account for phenomena like mountains and valleys.

Given these four principles, Reznick (2010, pp. 277–79) reminds us that the "important take-home message is that Darwin argued, correctly, that the Earth is indeed ancient, and old enough for his proposed theory of evolution to account for the history of life." In this same discussion, Reznick makes a number of points. First, Darwin's calculations about the age of the Earth reveal that his estimate should have been approximately 100 million years. Reznick infers from this that Darwin's argument is nonetheless persuasive despite being off by a factor of three. This means that, as part of his account of making sense of the history of life by way of his "descent-with-modification, tree-of-life metaphor," Darwin was entitled to 200 million years less than he thought, but a 100 million year-old Earth is still an ancient one. At the very least, then, Darwin could claim that the Earth is extremely old; to be sure, from Darwin's perspective, old enough to account for his tree-of-life proposition. The upshot is that Darwin could claim that **(1)** of **P1** of **The Preservation Argument** is true.

---

21 Darwin's endorsement of Lyell's ideas should come as no surprise. He tells us, "The science of geology is enormously indebted to Lyell—more so, as I believe, than to any other man who ever lived" (2003, p. 44).

22 These four points are also explained in the penultimate chapter of Stephen Jay Gould's *Time's Arrow, Time's Cycle* (1987, pp. 99–179).

A Changing Earth and Organism Changes: Forging the Tree-of-Life Hypothesis

As mentioned in the beginning of the chapter, Darwin was moved by Lamarck's observations regarding species origin and organism change. Broadly, Lamarck thought that extant organisms both originated from and are related to simpler organisms. The variety of existing species, argued Lamarck, is the product of corresponding diverse environments. Organisms that acclimated to environmental changes by way of using or not using various features were able to pass along beneficial acquired characteristics to the next generation. In this way, changes in an organism's lifetime were catalysts with respect to modified, yet related, subsequent species.[23]

For Darwin, the continuous generation of species variation and the historical connectedness of life put forth by Lamarck were welcome elements to his reflections on both environmental change and species change.[24] To see this, remember that (2) of **P1** of **The Preservation Argument** states that the Earth is a place that has undergone considerable environmental changes. At first glance, this should seem like an unremarkable claim. Simply observing one's own garden over a few years is enough to reveal dynamic circumstances. My own backyard has proven to be an inviting place of residence for birds, bees, wasps, spiders, squirrels, frogs, rabbits, and skunks. I have observed parts of old willow trees plummet to the ground, grass come and go, bushes replaced by other bushes, etc.

Still, when one considers the Earth as a place of dynamic change from the perspective of Darwin's tree-of-life hypothesis, (2) of **P1** of **The Preservation Argument** is not a trivial claim. Darwin needs to point out to his reader that the Earth—from a geological perspective—is both an old and dynamic place. Why give prominence to such an issue? The answer is two-fold. First, having defended the idea that the planet is very old, Darwin also needed to show that it had undergone considerable changes in order to drive home the point that change—and not stasis—is an integral part of the Earth's history. Second, one could imagine a naysayer granting Darwin and Lyell their old Earth, but insist that age itself is not indicative of substantial geologic and geographic change, rendering the tree-of-life hypothesis suspect. For Darwin, age and substantial geologic change are needed to make full sense of his tree-of-life hypothesis; that is, changes in both species and character traits require environmental changes for selection pressure to be present and thus for natural selection to have variation upon which to work.

Interestingly, even Lyell's theistic tendencies could not hide his geological and biological assessment of the planet (Lyell, 1830, pp. 384–85). In brief, Lyell does not shy away from pointing out the connectedness of life on Earth and how the conditions

---

23 This is likely a caricature of Lamarck's influence and insights into organism and species change (see Galera, 2016).
24 This is not to say that Darwin endorsed all of Lamarck's theory of evolution. For a good comparison and contrast of Darwin's and Lamarck's theories of evolution, see Mayr (1982).

of the various sorts of geological entities have played a role in organism changes. It is just this combination stressed by Lyell that Darwin emphasized in numerous places. For example, by close examination of the South American coasts, Darwin stresses that slow changes and existing sediment formations either preserved or prevented the preservation of fossils. Not only does this account make sense of why certain areas have or have not preserved ancient life forms, but it is also a reminder to the reader of the constant gradual changes that have taken place over vast amounts of time and continue to affect local environments (1958, pp. 301–02). With this Lyellian-type geological knowledge at hand, we can begin to understand what Darwin means when he tells us about the effect of these environmental changes with respect to species modification:

> Species ... probably change much more slowly, and within the same country only a few change at the same time. This slowness follows from all of the inhabitants of the same country being already so well adapted to each other, that new places in the polity of nature do not occur until after long intervals, due to the occurrence of physical changes of some kind, or through immigration of new forms. Moreover variations or individual differences of the right nature, by which some of the inhabitants might be better fitted to their new places under the altered circumstances would not always occur at once. (1958, p. 299)

Changes are slow both in geology and correspondingly in species. Other factors in species change are chance events like the infusion of new subspecies or environmental changes. The point here is that Darwin was sensitive to emphasize the gradual changing of environments and to what extent these changes are relevant to species change in conjunction with other natural factors. This combination of organismal and environmental change is what Darwin needed to reveal the connectedness of life in his tree-of-life metaphor. As Grant and Estes (2009, p. 61) stress, Darwin's careful attention (during his *Beagle* voyage) to geologic and geographic changes, as they pertain to the production and modification of species, reveals a careful empiricist and penetrating theoretician. Specifically, it is the geologic and geographic relationship to the appearance and distribution of organisms on which Darwin was fixated and how this makes sense of common descent via natural selection that capture Darwin's naturalist instincts. Darwin succinctly summarizes all this:

> The innumerable past and present inhabitants of the world are connected together by the most singular and complex affinities, and can be classed in groups under groups, in the same manner as varieties can be classed under species and sub-varieties under varieties, but with much higher grades of difference. These complex affinities and the rules for classification, receive a rational explanation on *the theory of descent, combined with the principle of natural selection, which entails divergence of character and the extinction of intermediate forms.* (1876, p. 14; my italics)

In the same way that Darwin detected similarities in the variety of species he observed on the different Galápagos Islands, Darwin infers more generally that all of life shares such similarities and differences, but only to a much greater degree. His insightful empirical work in South America, along with the excellent geological and biogeographical instincts he cultivated prior to the *Beagle* trip, assisted in an array of evidence in support of his tree-of-life hypothesis that would be placed on display in the *Origin*.

At this point, we can submit that Darwin—with considerable assistance from Lamarck and Lyell—would have been quite confident to claim that he has defended the truth of **(2)** of **P1** of **The Preservation Argument**; namely, that the Earth has undergone various environmental changes and that there is a correlation between these environmental changes and species changes.

Recall that an integral aspect of Darwin's philosophy of science is *fecundity*; his ideas should prove fruitful to other areas of inquiry. In terms of making sense of the incomplete fossil record, as we saw above, Darwin points out why coastal and oceanic fossilization is a complicated matter connected to the degree of sediment accumulation. Not only did this help make sense of his tree-of-life hypothesis, but it also shed some light on gaps in the fossil record in certain environments. So, Darwin not only bolstered the overall understanding of life on Earth from a biological point of view, but he also assisted in making sense of some geological perplexities. This is, no doubt, fecund.

Yet, something was missing and Darwin knew it.

Domestic Breeding: If Man Can Do It, Then Why Can't Nature?

With Lyell's framework in hand, a good bit of field biology and biogeography under his belt, and the variety of eye-opening data accumulated from the *Beagle* voyage, Darwin began to piece together his overall theory. Still, he knew that he needed a way to convince people of the reality of natural selection in a manner that was true to his philosophy of science. Don't forget that one of the features of his philosophy of science is the need to use *arguments by analogy* in order to show the reasonableness of his metaphysical claims. For even if an old and changing Earth is granted to Darwin, variation amongst species (e.g., different species of snakes or birds) in the light of different changing environmental conditions still might not convince those who are looking for a specific set of mechanisms that reveal such profound species-level changes (e.g., transition from a tapir-like animal to modern-day horse). Alas, this is not an unreasonable demand of so startling a claim as the tree-of-life hypothesis!

It is his reflections on domestic breeding practices that move Darwin to offer the following empirical argument in defense of the existence and power of natural selection. The following three passages illustrate Darwin's precision and passion regarding this domestication argument. He starts off with human breeding of domesticated animals and plants:

Passage 1

The great power of this principle of selection is not hypothetical. It is certain that several of our eminent breeders have, even within a single lifetime, modified to a large extent their breeds of cattle and sheep.... Breeders habitually speak of an animal's organization as something plastic, which they can mold almost as they please.... When a race of plants is once pretty well established, the seed-raisers do not pick out the best plants, but merely go over their seed-beds, and pull up the "rogues," as they call the plants that deviate from proper standard. With animals this kind of selection is, in fact, likewise followed; for hardly any one is so careless as to breed from his worst animals. (1958, pp. 50–51)

Notice that Darwin begins by dismissing the view that the power of natural selection is only an idea with little empirical support. So, if Darwin can show (with the empirical evidence from the work of breeders) that artificial selection is a powerful mode of species change, then he can argue for a corresponding kind of change as a result of natural selection. In effect, he will have killed two birds with one stone; that is, not only will he have established the existence of natural selection, but also that natural selection is likely responsible for the tree-of-life to the highest degree possible because of how potent a force it is.

Now, having made clear that the features of organisms are malleable, can be modified by proper breeding, and that favored features can be retained in the same way that some plants are selected over others, Darwin moves on to make the analogy between man selecting and nature selecting:

Passage 2

As man can produce, and certainly has produced, a great result by his *methodical and unconscious means of selection*, what may not natural selection effect? Man can act only on external and visible characters: Nature, if I may be allowed to personify the natural preservation or survival of the fittest, cares nothing for appearances, except in so far as they are useful to any being. She can act on every internal organ, on every shade of constitutional difference, on the whole machinery of life. Man selects only for his own good: Nature only for that of the being which she tends. Every selected character is fully exercised by her, as is implied by the fact of their selection.... Under nature, the slightest differences of structure or constitution may well turn the nicely balanced scale in the struggle for life, and so be preserved. How fleeting are the wishes and efforts of man! how short his time! And consequently how poor will be his results, compared with those accumulated by Nature during whole geological periods! Can we wonder, then, that Nature's productions should be far "truer" in character than man's productions; that they should be infinitely better adapted to the most complex conditions of life, and should plainly bear the stamp of far higher workmanship? (1958, p. 91; my italics)

Having acknowledged that humans have accomplished some impressive feats by way of both conscious and unconscious breeding manipulation of animals and plants, Darwin declares how much more powerful nature is by comparison.[25] The result is that human sculpting is far inferior to nature's sculpting because the former must contend with both short human life spans and various sorts of comparatively suboptimal biases (e.g., economic value of a particular trait). He concludes his argument from analogy as follows:

Passage 3

I can see no limit to this power, in slowly and beautifully adapting each form to the most complex relations of life. The theory of natural selection, even if we look no farther than this, seems to be in *the highest degree probable*. (1958, p. 443; my italics)

In terms of his philosophy of science, Darwin's use of artificial selection satisfies both the argument from analogy requirement and the surprise/hostility criterion. The argument from analogy criterion is obviously satisfied, given his move to focus on the superior (though similar) power of natural selection within the context of human manipulation of plants and animals. Yet, the use of human manipulation of animals and plants in his argumentative strategy is a surprise (somewhat akin to Aristotle's view of nature as a doctor-doctoring-himself). So long as one grants geologic time, the idea of nature-as-sculptor as a somewhat mirror-image of humans-as-sculptors clearly catches one off-guard in its persuasive straightforwardness.[26] Human-manipulated organisms are commonly encountered. Today, we even find ourselves in the midst of "designer babies" and gene-therapy (e.g., normal allele placed in defective cells). Darwin has drawn on our shared experiences to make it difficult to gainsay this part of his analogy—humans can and have manipulated animals and plants in a great variety of ways in a rather short amount of time. He then simply asks us to imagine what nature could produce over vast amounts of time. He concludes that nature could surely do

---

25 An example of unconscious selection would be the increased size of seeds and corresponding vigorous plants as a result of deeper planting. By comparison, the wildtype of the same sort of seeds are smaller in size and produce less vigorous plants. Also, variety of color of plants and seeds is greater as a result of artificial selection of specific colors of seeds and plants. Thus, conscious artificial selection of certain features results in the unconscious selection of other potentially beneficial and harmful features.

26 Largent (2009) reasonably suggests that the Whewellian demand for parsimony, in the sense of simplicity and harmoniousness of explanation, is also satisfied by the argument from artificial selection.

much more than the rather banal biddings or fancies associated with human pursuits. What could be more forthright and trouble-free than this line of reasoning?

Below is a formal reconstruction of Darwin's "**Argument from Artificial Selection**":

The Argument from Artificial Selection (Argument from Analogy)

P1. If some of the variations in organisms are products of humans modifying animals and plants for their own benefit and pleasure, then there is no obvious reason why the principles of artificial selection, which have acted so efficiently under domestication, should not have acted under nature.

P2. Some of these variations in organisms (i.e., domesticated organisms) are the product of humans modifying animals (and plants) for the benefit and pleasure of humans. ("This process of selection has been the great agency in the production of the most distinct and useful breeds.") [**Artificial Selection**]

P3. "There is no obvious reason why the principles which have acted so efficiently under domestication should not have acted under nature." [**Argument from Analogy**]

P4. If there is no obvious reason why the principles which have acted so efficiently under domestication should not have acted under nature, then natural selection ... seems to be in the highest degree probable.

C1. Natural selection ... seems to be in the highest degree probable.

P1 and **P4** seem false from the get go. Although Darwin has established the truth of the antecedents, the consequents do not follow in any obvious way. To be sure, humans have intentionally and unintentionally manipulated organisms to secure various sorts of traits for their various whims and projects. Yet, it is just this intentional feature that is missing in the case of nature (recall that this same sort of criticism burdened Aristotle's doctor analogy). So, even though **P2** is true, **P3** does not follow because the intentional and unintentional activities of human artificial selection appear to be vastly different from the non-intentional events pertaining to the natural world.

An immediate response to the above objection could be that it misses the point of the analogy. Specifically, the analogy is trying to give weight to the effects of manipulation on plants and animals in the long and short run. From this perspective, regardless of how impressive and unique intentional influences on plants and animals have been in the short-run, the cumulative effects of hundreds of millions of years on plants and animals could easily produce modifications of even greater magnitude and variety. Thus, giving additional weight to human intentionality to tell against the artificial selection analogy is unwarranted.

Still, the variation and efficiency present in artificial selection are products of human reflection and planning. These intentional states are not present in nature, telling against the magnitude, variation, and precision that Darwin is claiming is correspondingly present in nature. What this suggests, at least at first glance, is that either natural selection is not a force of nature or it is a force whose powers are more circumscribed than Darwin thought. The implication is that **The Argument from Artificial Selection** above is unsound, leaving Darwin's "highest-degree-probable" conclusion in serious jeopardy.

Malthus' Pressure Cooker: The Final Nail in the Coffin?

Interestingly, Darwin was aware of this sort of criticism. Even with geologic time, a changing planet, and the power of artificial selection integrally part of his defense of natural selection, he still did not have much of an engine to drive the kind of change that is exhibited in his tree-of-life image. Put another way, there does not seem to be enough evidence to support the idea that the present diversity of life originates from natural selection working on a few basic body plans eons ago.

Astutely, Darwin recognizes that it is not at all obvious how to make the analogy from artificial selection really work. Indeed, he openly tells us that it was a mystery to him, even with the insights related to Lyell's ideas and his own reflections on the data and organism variation he observed (2003, p. 53). Darwin goes on to explain how he thought he had figured out the problem:

> In October 1838, that is, fifteen months after I had begun my systematic enquiry, I happened to read for amusement Malthus on *Population*, and being well prepared to appreciate the struggle for existence which everywhere goes on from long-continued observation of the habits of animals and plants, it at once struck me that under these circumstances favourable variations would tend to be preserved, and unfavourable ones to be destroyed. The result of this would be the formation of new species. Here, then, I had at long last got a theory by which to work.... (2003, p. 53)

What Darwin found so very useful in Malthus' work is that his ideas regarding human populations were also relevant to non-human organisms. Broadly, Malthus argued that human populations had a tendency to grow geometrically (e.g., 3, 6, 12, 24, etc.), while food and other natural resources (e.g., land and water) only increased arithmetically (e.g., 3, 6, 9, 12, etc.) at best. Clearly, populations would increase beyond the amount of available resources. The result would be a struggle to secure the limited resources as populations really started to explode. As Darwin notes above, he observed the very same struggle playing out between individuals of the same species. The result is that selection would favor those individuals in a population that have the subtle trait variations required for securing the limited resources. For Darwin, this combination of selection and retention of some variations over others not only was the engine for speciation events, but was also the "pressure cooker" he needed to reveal the truth regarding the presence and power of natural selection (1958, pp. 76–79). Regarding the drama of nature's wars, Darwin specifically tells us this:

> Battle within battle must be continually recurring with varying success; and yet in the long-run the forces are so nicely balanced, that the face of nature remains for long periods of time uniform, though assuredly the merest trifle would give the victory to one organic being over another.... *All that we can do, is to keep steadily in mind that each organic being is striving to increase in a geometrical ratio*; that each at some period of its life, during some season of the year, during each generation or at intervals, has to struggle for life and to suffer great destruction. When we reflect on this struggle, we may console ourselves with the full belief, that the war of nature is not incessant, that no fear is felt, that death is generally prompt, and that the vigorous, the healthy, and the happy survive and multiply. (1958, p. 83 and p. 87; my italics)

With the argument from artificial selection and Malthus' pressure cooker woven into his account, Darwin's general defense of the truth of natural selection (i.e., by way of **The Argument from Artificial Selection**) is complete. Again, let us be mindful of Darwin's philosophy of science. If he can show the boons of drawing from unexpected areas of inquiry and can employ analogical reasoning where needed, then his overall argument would be strengthened. Well, as we have seen, Darwin was able to draw from two unexpected areas of inquiry—the breeding of domesticated animals and Malthus' human population/resources analysis—to reveal further the seriousness with which natural selection should be taken. Additionally, Darwin employs the analogy of war (terms like 'battles,' 'destruction,' 'victory,' and 'war of nature') to illustrate further the frequent and brutal daily conditions in which organisms struggle to survive and it is within this "war of nature" that selection does its work. Thus, not only does Darwin think that **The Argument from Artificial Selection** is proven sound, but he thinks his one-long-argument vindicates both **The Preservation Argument** and **The Strong Argument for Natural Selection**. With the soundness of these arguments secured, Darwin can claim to have substantiated the soundness of **The Main Argument** of the *Origin*.

VIII. TAKING STOCK: ASSESSING DARWIN'S ARGUMENT

Let us remind ourselves of what has been accomplished thus far. Table 2.2 reveals Darwin's defense of natural selection.

PRINCIPLES	EXPLANATION
Geologic Time	With the help of Ramsay and Lyell, Darwin defends the view that the Earth is a few hundred million years old.
A Dynamic Earth	Flexing his own geologic muscles and drawing upon Lamarck's insights, Darwin displays for the reader that, not only is the Earth a steadily changing place, but also that organism and species change occurs and has occurred accordingly and to a degree.
Domestic Breeding	If humans can modify features of organisms for their own benefit over a relatively short span of time, we should expect that nature can do considerably more, like produce species-level changes, over geologic time.
Malthusian Pressure Cooker	So long as populations exceed resources, there will be competition for those resources. Only those individuals who possess the requisite traits to cope with resource scarcity will likely survive and reproduce.

Table 2.2 *Four Principles Supporting Darwin's Defense of Natural Selection*

How impressed should we be with Darwin's argument? Natural selection, he tells us, is true *in the highest degree possible* because it follows from the truth of **Principles 1–4** above. If Darwin is correct about the truth of 1–4, then the implication appears to be that the premises of **The Strong Argument for Natural Selection** (see p. 50 for a reminder of the argument) are true, rendering it sound. But a close look at each of the four principles in **Table 2** will reveal that Darwin should have stuck with a "very probable" conclusion rather than his "highest degree probable" conclusion.

Assessment of Principle 1

First, recall that Darwin was aware that, in order for his tree-of-life hypothesis to be taken seriously by his peers, he would need to establish that the Earth is *very* old. His **Principle 1** was able to do this, but only to a degree. Darwin's own thinking, backed by Lyell's insights, suggested his estimate of 300 million years, contrary to the highest estimate in Darwin's day of 40 million.[27] Thus, *his own* evidence suggests that he was not going out on a precarious limb when he articulated **Principle 1** in favor of natural selection.

No doubt, as Reznick also points out, modern radiometric measurements reveal an aged Earth that would have pleasantly surprised Darwin and given him more evidence for **The Strong Argument for Natural Selection** (Reznick, 2010, pp. 278–79). Yet, it

---

27 I am here ignoring Darwin's resistance to William Thomson's (Lord Kelvin's) rate of cooling method, which suggested that the Earth could only have sustained life for about 40 million

is this modern measurement system—and not necessarily Darwin's appeal to an old Earth—that allows Reznick to conclude that an aged Earth could accommodate life's unfolding. The implication, then, is that even if the 300 million year mark is granted to Darwin, it still may not be a sufficient amount of time to capture the actual diversity of the tree of life. Reznick does not give enough weight to this point in his praise of Darwin's reflections. For example, evidence appears to suggest that microorganisms existed around *3.4 billion years ago* in rock structures known as stromatolites (Allwood et al., 2006)! Additionally, the Cambrian explosion—the sudden appearance and rapid surge (in geologic terms) of diverse multicellular body plans (forty-one distinct phyla) in the fossil record—occurred approximately 600 million years ago (Meyer, Nelson, and Chien, 2001). This would require an Earth that is approximately twice as old as the one that Darwin hypothesized. Thus, even if Darwin is allowed to express triumph of **The Strong Argument for Natural Selection** from his own vantage point, it is more reasonably justified in the light of contemporary radiometric measurements that put the Earth at about 4.6 billion years old. Charitably, then, the upshot is that Darwin's 300 million year-old Earth clearly provides support for the endorsement of natural selection being probably true (i.e., **The Weak Conclusion**), but it is not obviously true that he is entitled to an endorsement of natural selection being probable to the highest degree (i.e., **The Strong Conclusion**) in the light of this supporting evidence. As we now know, however, Darwin's cautious aplomb regarding the truth of natural selection relies on more than merely an old-Earth hypothesis.

Assessment of Principle 2

Second, Darwin's point about the connectedness amongst organisms and between organism and environment is difficult to ignore. From the variety of beetles to the variety of domestic and wild cats, and all the other varieties of animals and plants, the degree of similarities and differences amongst living entities suggests a shared history. As Darwin somewhat rhetorically asks: "What can be more curious than that the hand of a man, formed for grasping, that of a mole for digging, the leg of a horse, the paddle of the porpoise, and the wing of the bat, should all be constructed on the same pattern, and should include similar bones, in the same relative positions?" (Darwin, 1958, p. 413). His answer should be no surprise: "The explanation to a large extent is *simple* on the theory of the selection of successive slight modifications,—each modification being profitable in some way to the modified form, but often affecting by correlation other parts of the organisation" (1958, p. 414; my italics). This reliance on selection and common descent (and parsimony) becomes strikingly clear when Darwin offers his assessment of his meticulous embryological/morphological measurement studies of the features of varieties of pigeons and other animals. He thinks he can explain this variety of facts related to the embryo of various organisms.

---

years. See Hattiangadi (1971) for an interesting philosophical take on the Darwin-Kelvin dispute and for more on dating the Earth's age, see Dalrymple (1991 and 2004).

Specifically, he offers an account of how embryonic development is both similar in many cases and distinct in other cases to its adult stage (1958, p. 422). Rather than trying to offer some single account already endorsed by his peers,[28] he relies upon his mixed causal-empiricist outlook to produce a common-descent-and-natural-selection inference to the best explanation:

> In two or more groups of animals, however much they may differ from each other in structure and habits in their adult condition, if they pass through closely similar embryonic stages, we may feel assured that they all are descended from one parent-form, and are therefore closely related. Thus, community in embryonic structure reveals community of descent; but dissimilarity in embryonic development does not prove discommunity of descent, for in one of the two groups the developmental stages may have been suppressed, or may have been so greatly modified through adaptation to new habits of life, as to be no longer recognisable. (1958, p. 426)

Darwin accounts for embryonic variation in two ways. First, any two organisms that have similar fetal structures and pass through similar developmental stages are very likely to be related to a common ancestor. Second, in cases where there is not much similarity of developmental structure and stages between two putatively related organisms, it is likely that natural selection has modified the structures and stages in the wake of various environmental pressures. Thus, Darwin explains that both embryonic similarity and dissimilarity can be explained by way of both common descent and natural selection.

Again, not only do we see Darwin offering additional evidence of why natural selection and common descent should be taken seriously, but also how this evidence reveals his philosophy of science; that is, Darwin's account is fruitful in the sense that it helps make sense of an area of inquiry (namely embryology) that was not well understood prior to his account. This is why Nyhart stresses the point that Darwin's reflections on embryological issues "provided a crucial hinge-point between solving an old problem—the nature of the relationship between embryological development and classificatory affinities—and resituating that problem itself within a framework that decentered its importance" (Nyhart, 2009, p. 201). Put another way, what seemed to be a serious problem in embryology was resolved by Darwin by way of natural selection.[29]

---

28 I direct the reader to Nyhart (2009) for an excellent account of Darwin's cautious handling of embryonic development in the light of the efforts of his predecessors and peers.

29 In general, a good scientific theory is, in principle, one for which evidence can be secured either for or against the theory (i.e., testable). Yet, one possible worry regarding Darwin's ideas related to embryology is that they give the impression of not being evidence-worthy. Common descent appears to be untouched regardless of whether or not there is similarity or dissimilarity between embryological development because natural selection is guiding such variation. In Darwin's defense, he does not claim that dissimilarity in embryological development tells in favor of either common descent or natural selection. Rather, he says that dissimilarity *may be understood* by way of common descent and natural selection. What this suggests is that careful empirical

Yet, even granting Darwin the truth of **Principle 1** (see p. 66) and acknowledging the insightfulness of his work on embryology and morphology, there are serious objections associated with this tree-of-life hypothesis as it relates to organism-organism connectedness and organism-environment connectedness that may tell against **Principle 2**. (see p. 66) Chiefly, there is the problem of an incomplete fossil record and the absence of intermediate/transitional species. Specifically, if life really has unfolded as Darwin suggests, then there should be a fossil trail of transitional species—both living and in the fossil record—that vindicates the elaborate, continuous, and gradual branching of life that Darwin claims must have taken place. The concern, however, is that the fossil record and existing above-ground environments reveal no such elaborate branching pattern of a shared biological history. The implication is that, contrary to Darwin's observed geologic and biogeographic data, the tree-of-life hypothesis is unsubstantiated. The further repercussion is that, even if one accepted natural selection as a force of nature, one need not accept that it is as powerful as Darwin contends. Again, this would tell against the soundness of **The Strong Argument for Natural Selection**.

Darwin was quite sensitive to this objection and went so far as to claim that it is one of the most serious objections to his account (1958, p. 161 and p. 294). His reply to the absence of transitional species and the fragmented fossil record comes in three parts: (1) slow pace of speciation events, (2) extinction by competition, and (3) poor fossilization rates. The first two are related to natural selection as Darwin explains:

> I believe that species come to be tolerably well-defined objects, and do not at any one period present an inextricable chaos of varying and intermediate links ... because new varieties are very slowly formed, for variation is a slow process, and natural selection can do nothing until favourable individual differences or variations occur, and until a place in the natural polity of the country can be better filled by some modification or some one or more of its inhabitants. (1958, p. 165)

Strategically, in the spirit of the "surprise" criterion of his adopted philosophy of science, Darwin responds by sort of flipping the account. He claims that species, to a large extent, do not admit of much variation in short periods of time. His reasoning, which draws upon his reliance on Lyellian gradualism and deep geologic time, is that the emergence of variation is slow to occur in a geologic sense. What this means is that any given snapshot of a population X in an environment Y will most likely show that X is firmly well adapted. It is only when a beneficial variation "pops up" *and* is present in a suitable environment is the force of natural selection made apparent. For example,

---

work should be put forth before dismissing the possibility that embryological dissimilarity tells against common descent. From this perspective, Darwin is still leaving open the possibility that a comparison of any specific sets of embryos (i.e., testable evidence) could tell against or in favor of the closeness of their evolutionary relatedness. Still, it is not too harsh to submit that Darwin could have proceeded with a bit more caution regarding this part of his embryology discussion.

the set of 13–15 ft. tall giraffes could be wiped out in an environment where food is available only on trees more than 15 ft. tall. Under such environmental constraints, the giraffe population will survive (all other things being equal) only if it is able to find alternative food sources, or a more than 15 ft. tall sub-set of giraffes happens to emerge from this population. So, according to Darwin, it is the combination of a useful variation(s) in the corresponding appropriate environment—a combination that geologically comes to pass at a snail's pace, if at all—that produces viable intermediate species. Without this combination, selection will likely eliminate individuals with certain variations. It is this point about elimination that Darwin stresses when he concludes this section by offering three versions of "extinction by competition." The summary of the last of the three makes his point clear:

> Lastly ... if my theory be true, numberless intermediate varieties, linking closely together all the species of the same group, must assuredly have existed; but the very process of natural selection constantly tends, as has been so often remarked, to exterminate the parent-forms and the intermediate link. Consequently evidence of their former existence could be found only amongst fossil remains.... (1958, p. 165)

This part of Darwin's reply is to impress upon the reader to be careful about the sort of skeptical inferences one might draw from taking a snapshot of any set of existing organisms in its existing environment. He points out that just about any existing environment will be represented by the "winners"; that is, the reason that other variations of existing species in the same environment are not seen is because selection has weeded out those who are less fit (1958, p. 295). This is why one's snapshot view of population X in environment Y will likely not reveal the kind of variety one might expect. For example, if you go to Africa and observe zebras in the wild, you will likely not see intermediate zebras; you will likely see the successful set of zebras that remain from all the variant zebras. This does not mean that no species will exhibit transitional variants; it means that such cases will likely not be the norm. Again, the reason why this is not the norm is because selection will eliminate comparatively sub-optimal variants—especially in the case of those species which are in highly competitive, predatory, and/or resource depleted environments. Thus, Darwin can claim that, rather than telling against his account, the absence of viable intermediate varieties *surprisingly* vindicates his position.

But then there should be a fossil trail at the very least, right? Even if very few living transitional forms are present, surely traces of their existence should be extant in some petrified fashion. The final comment about fossils in the above quotation takes us directly to the last of Darwin's rebuttals regarding intermediate/transitional species:

> Only a small portion of the surface of the earth has been geologically explored, and no part with sufficient care, as the important discoveries made every year in Europe prove. No organism wholly soft can be preserved. Shells and bones

decay and disappear when left on the bottom of the sea, where sediment is not accumulating.... But the imperfection in the geological record largely results from another and more important cause than any of the foregoing; namely, from the several formations being separated from each other by wide intervals of time. (1958, p. 299–300)

Darwin offers three separate reasons for the intermittent fossil record. First, he points out that a great deal of this planet has not been explored or has only been poorly explored. What this means is that some transitional forms may very well be preserved and waiting to be found, but simply have not been discovered. Imagine what could be waiting to be discovered deep under Manhattan or the various inhospitable environments around the world! Much more careful work needs to be done, hints Darwin, before it can be concluded that the fossil record is as bleak as it might appear.

Still, Darwin's second point acknowledges the rather unevenness of the fossil record, but he points out that the reason for this is that many organisms decay before fossilization can occur because the right sort of sedimentary accumulation around these decomposing carcasses did not occur. Alternatively, he indicates, some organisms are made of material that is rather unlikely to fossilize (e.g., soft-shelled animals). What these two points suggest is that Darwin recognized and offered evidence to explain an interrupted fossil record.

The third point, which Darwin takes to be his strongest, is that large spans of time between fossilization events reveal why the fossil record is as disrupted as it is. For example, camels and horses shared a common ancestor about 85 million years ago. In this vast amount of time, locating more preserved intermediate ancestors in a geologic time gap, which may have produced many tumultuous geologic events, would either be a stroke of good luck or would mean that appropriate conditions for fossilization of many intermediaries simply did not materialize. Thus, combining the notions of a changing planet and geologic time, Darwin thinks he has adequately addressed this objection.[30]

Darwin is on quite solid ground with respect to this overall reply in defense of **Principle 2**. He correctly points out that much of the Earth has not been studied (this is still the case!). Also, much as in his day, fossils are being found with great frequency today.[31] The reasonable inference, then, is that there may very well be more fossil evidence in favor of transitional forms. Then again, it could just as well be the case that there is not much evidence in favor of transitional forms. This is likely why Darwin does not hang his hat on this first reply alone. He continues to offer additional arguments in support of why both existing organisms and the fossil record should be intermittent. If one takes seriously an antiquated Earth that is in gradual flux (both

---

30   Darwin offers numerous replies to other objections in ch. 10 of the *Origin*. Not all of these can be discussed here. I would urge readers to engage these discussions at their leisure. Also, see Reznick's (2010) geology chapters, chapters 16–19, for an illuminating and systematic handling of Darwin's diverse geology arguments and current findings in geology in the light of Darwin's ideas.

31   For a wonderful tour of, and education about, fossils, see Parker (2007) and Schwartz (1999).

geologically and zoologically) combined with selection pressure, then the absence of transitional forms with respect to many existing species and many fossils is not at all a cause for alarm. Rather, what we observe is the use of the surprise/hostility criterion on Darwin's part; that is, he takes a hostile objection and actually endorses it as additional and surprisingly good evidence in support of natural selection. **Principle 2 has been well defended by just this strategy.** We should be quite persuaded by this line of reasoning on Darwin's part.

Assessment of Principle 3

We can now turn to Darwin's domestic breeding analogy in defense of natural selection. How convincing is it? Part of the answer to this question requires being sensitive to the fact that species-level transformation—that is, the tree-of-life—is the standard set by Darwin. What this means is that Darwin's domestic breeding analogy had better be able to account for his tree-of-life vision. Yet, the obvious problem is that when humans go about tinkering with animals (especially in Darwin's day), the scope of trait modification is almost entirely restricted to within-species features. For example, humans manipulate features of dogs, cats, pigeons, cattle, etc., but they do not engage in species-level changes.[32] One does not see, for instance, the production of a new kind of cat-like organism or a new kind of cow-like or horse-like animal.[33] Rather, one sees taller cats, faster horses, cats with different colored eyes and fur, cows that can produce more milk than under normal circumstances, pigeons with different combinations of colored feathers, etc. What this reveals is that human manipulation of organisms is rather restricted in its scope in comparison to the variety of species on the planet. As Waters recognizes, there is a kind of springing vault in reasoning that Darwin is making, rendering the artificial selection analogy rather speculative (Waters, 2003, p. 137). The relatively small changes that humans are able to produce by way of artificial selection do not translate easily into the relatively large species-level changes associated with the tree-of-life hypothesis. This analogy actually reveals the much more modest view that, even if natural selection is impressive enough to modify features of species, it is not at all obvious that it can bring about the tree-of-life as Darwin tenders.

Let us not, however, be too quick to endorse the above conclusion. True enough, it seems that Darwin is trying to reason across an unbridgeable chasm with respect to artificial and natural selection. The underlying problem is related to variation. His domestication analogy appears to suffer from not being able to show how both the degree and kind of variation could be possible to produce the tree-of-life-common-descent

---

32 Of course, this is not to say that the creation of unique species is not possible, depending upon how one understands 'species.' Note that I have not defined what 'species' means at this juncture of the text. This topic will be addressed separately in ch. 5.

33 Recent work in synthetic biology makes the possibility of new species creation more of a reality than science fiction. See Ananth (2014) and Kaebnick and Murray (2013) for some of the philosophical and biological implications of this burgeoning sub-discipline within biology.

portrayal. Yet, by drawing on his research efforts in geography, comparative anatomy, taxonomy, and embryology, Darwin thinks he has the answer.

For example, Darwin offers a careful account in the *Origin* of how new species emerge (1954, p. 108). In summary, he argues that geographic isolation plays an important role with respect to speciation. His argument comes in three parts. First, Darwin points out that secluded areas like the islands he studied during his *Beagle* voyage allow for uniform changes of individuals of the same species. What this does is create circumstances such that cross-breeding is unlikely. The inference to be made here is that, for example, individuals belonging to species W, which inhabit environment X, will have peculiar enough features in relation to their isolated life-circumstances such that a different species Y from environment Z will either not have access to species W in environment X due to geographical entry barriers (e.g., river, lake, large fallen tree, land upheaval, etc.) or Y will simply not have the requisite sexual attraction to W due to W's peculiar features (and vice-versa).[34] The result is that, in the absence of sexual interest or the presence of geographical barriers hindering cross-breading, distinct species will likely emerge.[35] Second, Darwin points out that modified, old-inhabitants will most likely occupy new niches in an isolated environment because geographical barriers to entry increase the availability of new niches. Again, species will have a chance to develop without being disturbed by outside interlopers. Third, because isolation curbs invaders from occupying niches and cross-breading, existing species variants will have time to develop into distinct species. Recall, since beneficial features arrive rarely and by chance, it could take considerable time for the appropriate trait variations to be made manifest. Once they are present, then they can be selected and retained. In this way, new species can slowly emerge as distinct reproductively isolated species.

As important as isolation is with respect to speciation, Darwin thought there were even more powerful means by which new species are produced. Darwin wastes no time in conveying the overall importance of *large areas of space* with respect to new-species production and modification. This account comes in three parts. Part 1:

> Although isolation is of great importance in the production of new species, on the whole I am inclined to believe that largeness of area is still more important, especially for the production of species which shall prove capable of enduring for a long period, and of spreading widely. Throughout a great and open area, not only will there be a better chance of favourable variations,

---

34 It seems that sexual selection is a sort of distinct cause of evolutionary change for Darwin and provides yet another avenue to account for variation. This is also a reminder that Darwin did not think of natural selection as the only causal factor to account for the tree-of-life. See Huxley (1938) and Ruse (1975b, pp. 238–40) for an analysis of Darwin's account of sexual selection and Rougharden (2004) for a more contemporary discussion about sexual selection.

35 The creation of new species by way of geographical barriers is known as allopatric speciation. Although this cannot be pursued in detail here, Darwin's notebooks suggest that he endorsed this mode of speciation. For more on this, see Kottler (1978).

arising from the large number of individuals of the same species there supported, but the conditions of life are much more complex from the large number of already existing species; and if some of these many species become modified and improved, others will have to be improved in corresponding degree, or they will be exterminated. Each new form, also, as soon as it has been improved, will be able to spread over the open and continuous area, and will thus come into competition with many other forms. (1954, pp. 108–09)

Part 2:

Moreover, great areas, though continuous, will often, owing to former oscillations of level, have existed in a broken condition; so that the good effects of isolation will generally, to a certain extent, have concurred. (1954, p. 109)

Part 3:

Finally, I conclude that, although small isolated areas have been in some respects highly favourable for the production of new species, yet that the course of modification will generally have been more rapid on large areas; and what is more important, that the new forms produced on large areas, which already have been victorious over many competitors, will be those that will spread most widely, and will give rise to the greatest number of new varieties and species. They will thus play a more important part in the changing history of the organic world. (1954, p. 109)

In part 1, Darwin explains that, even more so than isolated areas, wide open areas allow for more speciation to occur because the greater number of existing species makes for a more complex set of circumstances in the sense that one must keep up—or be fortunate enough to be able to keep up—with one's neighbors or be destroyed by the competitive advantages possessed by one's neighbors. The results, thinks Darwin, are various kinds of tit-for-tat arms races amongst these open-area dwellers, resulting in more varieties and species and extended migratory excursions. These open-area movements will further result in interactions with new competitors and, again, create more varieties and, ultimately, new species.

In part 2, Darwin reinstates geographic isolation as a needed component to establishing new species within an open-area backdrop. So, within any open area setting, there will be changes to local areas that simulate the kind of isolation one might find on an island. It is under these circumstances that advantageous features can be molded and retained, as Darwin noted earlier. What actually follows from this second part is that isolation plays a crucial role—as a kind of cementing of useful traits—in the open-areas account of species modification and production. So, according to Darwin, speciation events occur in open-area locales wherein there are many interacting species and varieties along with pockets of isolated areas.

Lastly, in part 3, Darwin reiterates the benefits of open-area species. Notably, he surmises that the greatest variety of species will come to thrive and burgeon over the greatest amount of distances—and do so in a relatively fast amount of geologic time—because they will find themselves under the most intense competition. It is under these dynamics, thinks Darwin, that speciation events occur to the greatest extent and, thus, make sense of the organic diversity that constitutes his tree-of life hypothesis.

These explanations by Darwin are potent magnifications of the truth of common descent and natural selection when geologic time is brought into the artificial selection analogy. This translates into the view that the various sorts of bio-geographical phenomena Darwin observed on both islands and various mainlands make the artificial selection analogy eminently reasonable—especially in the light of geologic time. Isolation by itself and in conjunction with a competitive open-area mainland will likely produce variation, but the scope and magnitude of this variation within these dynamics get amplified considerably when hundreds of millions of years are part of the equation. Thus, the force of Waters' reminder of the serious limitations of Darwin's artificial selection analogy is weakened considerably; that is, Darwin's reasoning by way of his artificial selection analogy should be taken very seriously.

Assessment of Principle 4

Recall that Darwin found in Malthus' work what he thought are the needed facts and model to establish the idea of natural selection. Malthus provided Darwin with the framework in which natural selection, and its overall effect on life, would be made manifest in nature. So long as food and water supplies lag behind population growth, there will be a battle either between or within species for those relatively scarce life-sustaining resources. And, so long as the battles rage, individuals fortunate enough to have those features that give them an advantage will tend to survive, reproduce, and pass along the advantages to their offspring. Interestingly, Darwin connects the Malthusian Pressure Cooker to his tree-of-life hypothesis as follows:

> A corollary of the highest importance may be deduced from the foregoing remarks, namely, that the structure of every organic being is related in the most essential yet often hidden manner, to that of all the other organic beings, with which it comes into competition for food or residence, or from which it has to escape, or on which it preys. (1958, p. 86)

This idea of the connectedness of life via natural selection sorting out eons of resource-securing fracases and mêlées, as Darwin describes, gains credibility once geologic time is included. In fact, once eons of population dynamics in conjunction with resource scarcity dynamics are given weight into the equation of life, natural selection is, proclaimed Darwin, "the doctrine of Malthus, applied with manifold force to the whole animal and vegetable kingdoms" (1958, p. 76).

It is difficult to locate any serious error in reasoning on Darwin's part regarding **Principle 4**. Notably, however, it could be argued that Darwin overstated the competitive framework in his rendering of the Malthusian Pressure Cooker. In contrast, for example, Kropotkin (1970) argued that cooperation, even more so than competition, is the appropriate mechanism to account for species survival. Yet, his reliance on such claims is the result of his geological and geographical studies of Siberia and resistance to the burgeoning social Darwinism that was affecting social policy around the world (Todes, 1989). This brutal environment did not reveal to Kropotkin the bloody battle he found in Darwin's *Origin*. Still, even if it is true that there exist pockets of the natural world that are not as brutal as Darwin proclaims[36] (including recent human populations), one would be hard-pressed not to see these bellicose-like dynamics rather replete and fairly uniform in the natural world.[37] Thus, the conclusion to be drawn is that Darwin's principle four is on the mark.

IX. TWO LINGERING WORRIES

There is little doubt that Darwin's defense of natural selection is impressive. As we have seen, Darwin rigorously supports his hypotheses by both incorporating a philosophy of science structure (comprising the pillars of true cause, surprise/hostility, fecundity, and parsimony) and unleashing an imposing array of biological data that coalesce to produce a formidable foundation in support of the existence and potency of natural selection. Furthermore, even if each one of his four principles does not independently give strong credence to the existence and power of natural selection, when they are taken together, his theory of natural selection seems to command steadfast assent—that is, **The Main Argument** of the *Origin* is the inference to the best explanation.

Still, there are two concerns that should give us pause. First, it is one thing to be able to defend how natural selection molds variation, but it is entirely a different set of issues surrounding (1) (a) the origin and (b) continuation of variation and (2) the

---

36 In fairness to Darwin, he recognized that in extreme climates (e.g., Antarctica), the harsh competition between animals is greatly mitigated, but that in warm and damp environments frequent clashes would remain constant (1958, pp. 86–87). This concession reveals that Darwin could accommodate the data found by Kropotkin by allowing for mutual aid scenarios in desolate harsh climates.

37 This is not to suggest that Malthus' analysis is free of criticism and that Darwin endorsed all aspects of his account. Flew's (1957) analysis of the argumentative structure and cogency of Malthus' arguments in his essay is also helpful. See also Vorzimmer (1969) and Depew and Weber (1996) on the connection between Malthus' account and natural selection. For more on how accurate Malthus' account is with respect to human populations, see Meadow's (2000) positive spin and Fernández-Villaverde's (2001) critical economic analysis.

specific causal factors associated with the retention of successful variations. Darwin did not have specific answers to these concerns (Okasha, 2001, p. 71).

For our purposes, what damage results in this ignorance? Well, it is not clear that the inability to answer the origin part of (1) (a) is of great harm. Other than Gregor Mendel's work, very little was known about the causal factors surrounding variation. Darwin, like Aristotle, knew that it was somehow related to the underlying biology of inheritance, but that was about it. Yet, even if Darwin could not offer the ultimate cause of why there is variation at all, the evidence he gives leaves little doubt regarding what natural selection can, on occasion, do with beneficial variation. As noted earlier, Darwin drew, in part, on the insights of Lamarck in forming his own account regarding beneficial variation. So, **The Strong Argument for Natural Selection** does not encounter great resistance here.

Second, in terms of the continuation of variation of (1) (b), Darwin's ignorance proves to be a more serious obstacle with respect to his defense of **The Strong Argument for Natural Selection**. As Darwin concedes, "Our ignorance of the laws of variation is profound. Not in one case out of a hundred can we pretend to assign any reason why this or that part has varied" (1958, p. 158). This is a rather unsettling concession. The issue is that Darwin needs to offer an account of how variations continue to come about—even when a good set of variations are actualized. The reason for this is that a survival-sustaining set of features will lose its value if every individual in a population has the same set, and only, that set. It is new variation, even in the presence of beneficial features, that keeps species alive; for as the environment shifts, the new variations could very well be the ones that are needed to sustain a particular species. Without new variation, species would likely die out and new species could not form. Ultimately, the power of natural selection (and **The Strong Argument for Natural Selection** for that matter) is at stake and Darwin has no causal evidence to account for the presence of continuous variation.

A second concern is with retention of beneficial variations. Even if a slightly more hooked beak gives a particular bird a reproductive advantage, how does this advantage remain present in the next generation? To a degree, Darwin answered this concern by way of a combination of *blending inheritance* and *Lamarckian acquired characteristics*.[38] Both of these views, however, have been, for the most, discredited or modified by contemporary biology; and even these accounts, as Darwin understood them, do not provide a specific mechanism for trait retention. Even if it is true that

---

38 'Blending inheritance' refers to the presence of a trait that is the product of a combining of parental features. For example, a sixteen-foot giraffe would be the result of an eighteen-foot father and a fourteen-foot mother. 'Lamarckianism' refers to the theory of Jean-Baptiste Lamarck, who argued that an organism could pass along a feature that it acquired in its lifetime. For instance, if a weight-lifting Olympic champion develops great strength in his lifetime, then this acquired strength could be passed on to his children. Again, both of these ideas have been discredited or re-explained in the light of modern genetics (Landman, 1993). The extent to which Darwin took these ideas seriously is debatable. For example, see Vorzimmer (1963) regarding Darwin on blending inheritance and Egerton (1976) and Saunders (1985) on Darwin and the history of the Lamarckian idea of acquired characteristics.

variations exist and some variations are better than other variations, these facts alone do not explain how beneficial variations are retained from one generation to the next. Darwin needs this mechanism to account for the gradual accumulation of beneficial variations over time to account for the tree-of-life via the power of natural selection.[39] He simply does not have such a mechanism at his disposal. To this extent, then, there is a missing piece of the puzzle that Darwin was not able to locate.

X. CONCLUSION AND DARWIN'S INFLUENCE ON THE PHILOSOPHY OF BIOLOGY

This chapter has offered a sketch of how to make sense of Darwin's defense of natural selection in the *Origin*. Darwin thought that he could defend the view that natural selection is probably true to the highest degree in the final edition of the *Origin*, but also cautiously claimed that it is probably true in the original edition. This chapter has revealed that he is not entitled to the strong claim, that is, **The Strong Argument for Natural Selection**. More reasonably, Darwin should have concluded that the existence and force of natural selection is *very likely true*. This is superior to claiming merely that it is probably true and more reasonable than claiming that it is probably true to the highest degree. There is little doubt that he is entitled to this middle-ground conclusion in the light of the variety of data he detonates in the *Origin* and the philosophy of science that he persuasively and subtly discharges. Thus, his account of the power of natural selection to make sense of the great diversity of life is very likely the inference to the best explanation.

Yet, let us make sure not to think of this middle-ground concession as some sort of failure. Rather, the suggestion here is that a more modest, but still powerful conclusion should have been defended. Not only have much of Darwin's ruminations been confirmed by contemporary biology, but they have also helped to pave the way for numerous lines of research as well as philosophical debates. Some of these areas of study and ideas for serious scholarly exchange include the following:

The Modern Synthesis
Reductionism
Teleology/Function
Species
Unit of Selection
Systematics
Evolution and Development
Evolution and Knowledge
Evolution and Ethics

---

39 Darwin did have a peculiar account of sexual reproduction, but it too has been, for the most part, critically rejected. See Winther (2000), Ibraimov (2009), and Deichmann (2010) for further discussions.

Evolution and Psychology
Evolution and Race
Evolution and Religion
Evolution and Culture
Evolution and Sex

If this set of topics counts as a failure (and this is a mere subset of the many issues being re-evaluated in the light of Darwinism), then most scholars can only hope to fail so stunningly well! Indeed, as much as we owe Aristotle the debt of reminding us of how seriously we should take biology, we owe this Newton-of-a-blade-of-grass an even greater debt for allowing us to see just how beautiful the ramifications are of bringing biology to life.

CHAPTER REVIEW: DISCUSSION AND QUESTIONS

1. How should "The Newton of a Blade of Grass" title for Darwin be understood? [pp. 39–40]
2. Why did Kant object to the possibility of such a title for any person? [p. 40]
3. How should we understand Darwin-the-naturalist as a mixed causal empiricist? In what ways are Darwin and Aristotle different regarding this? [pp. 42–43 and note 3]
4. In your own words, explain The Main Argument of the *Origin*. [p. 44]
5. How did Herschel and Whewell separately contribute to Darwin's overall philosophy of science? [pp. 45–48]
6. Explain each criterion for a good philosophy of science in Table 2.1. [p. 48]
7. Explain the inductive argument style called "inference to best explanation" [pp. 49–50]. Can you make sense of its role in Darwin's reasoning after answering question 8 below?
8. What is The Strong Argument for Natural Selection? What does this argument demand of Darwin? [pp. 50–51]
9. What is the divine challenge to Darwin's conception of natural design? [pp. 52–53]
10. How was Darwin sensitive to the divine challenge? [p. 53]
11. Explain—in your own words—what is The Preservation Argument. What does Darwin need to do to prove that it is sound? [pp. 54–55]
12. With the tree-of-life image in mind, why did Darwin need to provide evidence in defense of a very old Earth? In general, what ideas did Darwin take away from Ramsay and Lyell regarding the age of the Earth? [pp. 55–57]
13. What sort of connection was Darwin trying to make with respect to a changing Earth and changing organisms? How is the fecundity requirement satisfied in this discussion? [pp. 58–60]
14. What is The Argument from Artificial Selection? What potential objection jeopardizes this argument? In term of his philosophy of science, what work is this analogy doing? [pp. 63–64]

15. In terms of replying to the potential objection in 14, how does Darwin's reliance on Malthus' views about population help him? How helpful is his philosophy of science here? [pp. 64–65]
16. Even if we grant Darwin his calculation of the age of the Earth, why might this result still tell against his account? How lethal of a blow is his not-so-old-Earth? [pp. 66–72]
17. Why does Waters think that Darwin's artificial selection analogy is speculative? In terms of Darwin's discussion about the production of new species, why are you or are you not persuaded by the reply to Waters in this section? [pp. 72–75]
18. What is the one possible objection to Darwin's reliance on Malthus? Are you or are you not persuaded by the brief reply to this objection? [pp. 75–76]
19. What two lingering worries does Okasha address? Why are they so problematic for Darwin? [pp. 76–78]
20. Given the lingering worries, what alternative middle-ground conclusion is suggested that Darwin should have defended? Why do you or don't you find this alternative suggestion reasonable? [p. 78]
21. If the alternative conclusion should be accepted by Darwin, why does this not damage the overall quality and influence of Darwin's efforts? [pp. 78–79]

REFERENCES

Allwood, Abigail C., Malcolm R. Walter, Balz S. Kamber, Craig P. Marshall, and Ian W. Burch. "Stromatolite Reef from the Early Archaean Era of Australia." *Nature* 441 (June 8, 2006): 714–18.

Ananth, Mahesh. "Review of Gregory E. Kaebnick and Thomas H. Murray, eds., *Synthetic Biology and Morality: Artificial Life and the Bounds of Nature.*" *Journal of Value Inquiry* (2014). DOI: 10.1007/s10790-014-9432-2.

Berryman, Sylvia. *The Mechanical Hypothesis in Ancient Greek Natural Philosophy*. Cambridge: Cambridge University Press, 2009.

Bizzo, Nelio Marco Vincenzo. "Darwin on Man in the 'Origin of Species': Further Factors Considered." *Journal of the History of Biology* 25, 1 (Spring 1992): 137–47.

Bowler, Peter J. *Darwin: The Man and His Influence*. Cambridge: Cambridge University Press, 1990.

Browne, Janet. *Charles Darwin: The Power of Place*. New York: Knopf, 2002.

———. *Charles Darwin: Voyaging*. New York: Knopf, 1995.

Bulmer, Michael. "The Theory of Natural Selection of Alfred Russel Wallace FRS." *Notes and Records of the Royal Society* 59, 2 (2005): 125–36.

Carey, Toni V. "John Herschel." *Philosophy Now* 48 (2004): 32–35.

Cornell, John F. "Newton of the Grassblade? Darwin and the Problem of Organic Teleology." *Isis* 77, 3 (1986): 404–21.

Dalrymple, G. Brent. *Ancient Earth, Ancient Skies: The Age of Earth and Its Cosmic Surroundings*. Stanford: Stanford University Press, 2004.

———. *The Age of the Earth*. Stanford: Stanford University Press, 1991.

Darwin, Charles. *Evolution: Selected Letters of Charles Darwin 1860–1870*. Ed. Samantha Evans Frederick and Alison Pearn. Cambridge: Cambridge University Press, 2008.

———. *The Descent of Man*. New York: Penguin Classics, 2004.

———. *Autobiography of Charles Darwin*. New Delhi: Rupa & Co., 2003.

———. *The Origin of Species by Means of Natural Selection or the Preservation of Favoured Races in the Struggle for Life*. 1st ed. London: Penguin Classics, 1985.

———. *The Origin of Species by Means of Natural Selection or the Preservation of Favoured Races in the Struggle for Life*. 6th ed. New York: Mentor/Penguin, 1958.

———. *The Variation of Animals and Plants under Domestication*. Vol. 1. 2nd ed. New York: D. Appleton and Co., 1876.

Darwin, Francis, ed. *The Life and Letters of Charles Darwin, Including an Autobiographical Chapter*. 3 vols. London: John Murray, 1887.

Deichmann, Ute. "Gemmules and Elements: On Darwin's and Mendel's Concepts and Methods in Heredity." *Journal for General Philosophy of Science* 41, 1 (2010): 31–58.

Depew, David J., and Bruce Weber. *Darwinism Evolving: Systems Dynamics and the Genealogy of Natural Selection*. Cambridge, MA: MIT Press, 1996.

Des Chene, Dennis. *Spirits and Clocks: Machine and Organism in Descartes*. Ithaca: Cornell University Press, 2002.

Douven, Igor. "Abduction." In *The Stanford Encyclopedia of Philosophy*, ed. Edward N. Zalta. Spring 2011 Edition.

Egerton, Frank, N. "Darwin's Early Reading of Lamarck." *Isis* 67, 3 (1976): 452–56.

Fernández-Villaverde, Jesús. "Was Malthus Right? Economic Growth and Population Dynamics." 2001. http://economics.sas.upenn.edu/~jesusfv/pennversion.pdf.

Flew, Anthony. "The Structure of Malthus' Population Theory." *Australasian Journal of Philosophy* 35, 1 (1957): 1–20.

Galera, Andrés. "The Impact of Lamarck's Theory of Evolution Before Darwin's Theory." *Journal of the History of Biology* (2016): 1–18.

Gaukroger, Stephen. *The Collapse of Mechanism and the Rise of Sensibility: Science and the Shaping of Modernity, 1680–1760*. New York: Oxford University Press, 2011.

———. *The Emergence of a Scientific Culture: Science and the Shaping of Modernity 1210–1685*. New York: Oxford University Press, 2007.

Ginsborg, Hannah. "Two Kinds of Mechanical Inexplicability in Kant and Aristotle." *Journal of the History of Philosophy* 42, 1 (2004): 33–65.

Gould, Stephen Jay. *Time's Arrow, Time's Cycle: Myth and Metaphor in the Discovery of Geologic Time*. Cambridge, MA: Harvard University Press, 1987.

Grant, K. Thalia, and Gregory B. Estes. *Darwin in Galapagos: Footsteps to a New World*. Princeton: Princeton University Press, 2009.

Grene, Marjorie, and David Depew. *The Philosophy of Biology: An Episodic History*. Cambridge: Cambridge University Press, 2004.

Hattiangadi, J.N. "Alternatives and Incommensurables: The Case of Darwin and Kelvin." *Philosophy of Science* 38, 4 (1971): 502–07.

Herbert, Sandra. *Charles Darwin, Geologist*. Ithaca: Cornell University Press, 2005.

Herschel, John F.W. *A Preliminary Discourse on the Study of Natural Philosophy*. London: Printed for Longman, Brown, Green, Longmans, and Taylor, 1831.

Hodge, Charles. "Discussion on Darwinism and the Doctrine of Development." In *History, Essays, Orations, and other Documents of the Sixth General Conference of the Evangelical Alliance, Held in New York, October 2–12, 1873*, ed. Philip Schaff, and S. Iraneus Prime, 317–20. New York: Harper & Brothers, 1874.

Hodge, Jonathan, and Gregory Radick, eds. *The Cambridge Companion to Darwin*. 2nd ed. Cambridge: Cambridge University Press, 2009.

Hookyaas, Reijer. *Natural Law and Divine Miracle: The Principle of Uniformity in Geology, Biology and Theology*. Leiden: Brill, 1963.

Hull, David. "Why Did Darwin Fail?" In *Biology and Epistemology*, ed. Richard Creath and Jane Maienschein, 48–63. Cambridge: Cambridge University Press, 2000.

Huxley, Julian. "Darwin's Theory of Sexual Selection and the Data Subsumed by It, in the Light of Recent Research." *The American Naturalist* 72, 742 (Sept.–Oct. 1938): 416–33.

Ibraimov, A.I. "Darwin's Gemmules and Development." *Anthropologist* 11, 1 (2009): 1–5.

Jones, Greta. "Alfred Russel Wallace, Robert Owen and the Theory of Natural Selection." *British Journal for the History of Science* 35, 1 (2002): 73–96.

Judd, J.W. "Darwin and Geology." In *Darwin and Modern Science*, ed. A.C. Seward, 337–84. Cambridge: Cambridge University Press, 1909.

Kaebnick, Gregory E., and Thomas H. Murray, eds. *Synthetic Biology and Morality: Artificial Life and the Bounds of Nature*. Cambridge, MA: MIT Press, 2013.

Kant, Immanuel. *Critique of Judgement*. Trans. J.H. Bernard. Digireads.com, 2010.

Keynes, R.D., ed. *Charles Darwin's Beagle Diary*. Cambridge: Cambridge University Press, 1988.

Kottler, Malcolm K. "Charles Darwin's Biological Species Concept and the Theory of Geographic Speciation: the Transmutation Notebooks." *Annals of Science* 35, 3 (1978): 275–97.

Kropotkin, Peter. *Mutual Aid: A Factor of Evolution*. Manchester, NH: Extending Horizons Books–Porter Sargent Publishers, 1970.

Landman, Otto E. "Inheritance of Acquired Characteristics Revisited," *Bioscience* 43, 10 (1993): 696–705.

Largent, Mark A. "Artificial and Natural Selection in the *Origin*." In *The Cambridge Companion to the "Origin of Species,"* ed. Michael Ruse and Robert J. Richards, 14–29. Cambridge: Cambridge University Press, 2009.

Laudan, Larry. "William Whewell on the Consilience of Inductions." In *Science and Hypothesis: Historical Essays on Scientific Methodology*, ed. L. Laudan, 163–80. Reidel: Dordrecht, 1981.

Lennox, James. "Darwin Was a Teleologist." *Biology and Philosophy* 8, 4 (1993): 409–21.

Lewens, Tim. *Darwin*. New York/London: Routledge, 2007.

Lipton, Peter. "Inference to Best Explanation." In *A Companion to the Philosophy of Science*, ed. W.H. Newton-Smith, 184–93. Oxford: Blackwell, 2000.

———. *Inference to Best Explanation*. London: Routledge, 1991.

Lyell, Charles. *Principles of Geology*, vol. 3. London: John Murray, 1833.

Mayr, Ernst. *One Long Argument: Charles Darwin and the Genesis of Modern Evolutionary Thought*. Cambridge, MA: Harvard University Press, 1993.

———. *The Growth of Biological Thought: Diversity, Evolution, and Inheritance*. Cambridge, MA: Belknap Press of Harvard University Press, 1982.

McKinney, H. Lewis. "Alfred Russel Wallace and the Discovery of Natural Selection." *Journal of the History of Medicine and the Allied Sciences* 21, 4 (1966): 333–57.

McLaughlin, Peter. *Kant's Critique of Teleology in Biological Explanation*. Lewiston, NY: Edwin Mellen Press, 1990.

Meadows, Donella H. "We Still Haven't Proved Malthus Wrong." April 25, 2000. https://www.alternet.org/story/2767 global_citizen%3A_we_still_haven%27t_proved_malthus_wrong.

Meyer, Stephen C., P.A. Nelson, and Paul Chien. "The Cambrian Explosion: Biology's Big Bang." 1–50. http://www.discovery.org/articleFiles/PDFs/Cambrian.pdf.

Mizrahi, Moti. "Why Simpler Arguments Are Better." *Argumentation* 30, 3 (2016): 247–61.

Newall, Paul. "Ockham's Razor." *The Galilean*. June 2005. http://www.galilean-library.org/site/index.php/page/index.html/_/essays/philosophyofscience/ockhams-razor-r55.

Nyhart, Lynn K. "Embryology and Morphology." In *The Cambridge Companion to the Origin of Species*, ed. Michael Ruse and Robert J. Richards, 194–215. Cambridge: Cambridge University Press, 2009.

Okasha, Samir. "Darwin." In *A Companion to the Philosophy of Science*, ed. W.H. Newton-Smith, 68–75. Oxford: Blackwell Publishers, 2001.

Ospovat, Dov. *The Development of Darwin's Theory*. Cambridge: Cambridge University Press, 1995.

Parker, Steve. *World Encyclopedia of Fossils & Fossil-Collecting*. Leicester: Anness Publishing, 2007.

Pope, Kevin O., Steven L. D'Hondt, and Charles R. Marshall. "Meteorite Impact and the Mass Extinction of Species at the Cretaceous/Tertiary Boundary." *Proceedings of the National Academy of Sciences, USA* 95, 19 (Sept. 1998): 11028–29.

Quammen, David. "The Man Who Wasn't Darwin." *National Geographic* 214, 6 (Dec. 2008): 106.

Reznick, David N. *The Origin Then and Now: An Interpretive Guide to the Origin of Species*. Princeton: Princeton University Press, 2010.

Rosen, Jonathan. "Missing Link: Alfred Russel Wallace, Charles Darwin's Neglected Double." *The New Yorker* 82, 49 (Feb. 12, 2007): 76–81.

Rosenberg, Alex. *Darwinian Reductionism: Or, How to Stop Worrying and Love Molecular Biology*. Chicago: University of Chicago Press, 2006.

Roughgarden, Joan. *Evolution's Rainbow: Diversity, Gender and Sexuality in Nature and People*. Berkeley: University of California Press, 2004.

Ruse, Michael. *Darwin and Design: Does Evolution Have a Purpose?* Cambridge, MA: Harvard University Press, 2003.

———. "Darwin's Debt to Philosophy: An Examination of the Influence of the Philosophical Ideas of John F.W. Herschel and William Whewell on the Development of Charles Darwin's Theory of Evolution." *Studies in the History and Philosophy of Science* 6 (1975a): 159–81.

———. "Charles Darwin's Theory of Evolution: An Analysis," *Journal of the History of Biology* 8, 2 (Autumn 1975b): 219–41.

Saunders, S.R. "The Inheritance of Acquired Characteristics: A Concept That Will Not Go Away." In *What Darwin Began*, ed. L.R. Godfrey, 148–61. Boston: Allyn and Bacon, 1985.

Schwartz, Jeffrey H. *Sudden Origins: Fossils, Genes, and the Emergence of Species*. New York: John Wiley and Sons, 1999.

Smith, C.U.M., and Robert Arnott. *The Genius of Erasmus Darwin*. Aldershot, UK: Ashgate Publishing, 2005.

Sober, Elliott. "Why Is Simpler Better?" *aeon* (2016). https://aeon.co/essays/are-scientific-theories-really-better-when-they-are-simpler.

———. *Did Darwin Write the Origin Backwards? Philosophical Essays on Darwin's Theory*. Amherst: Prometheus, 2011.

———. *Evidence and Evolution: The Logic behind the Science*. Cambridge: Cambridge University Press, 2008.

———. "Testability." *Proceedings and Addresses of the American Philosophical Association* 73, 2 (1999): 47–76.

———. *Reconstructing the Past: Parsimony, Evolution, and Inference*. Cambridge, MA: MIT Press, 1988.

Spengler, Joseph J. "Was Malthus Right?" *Southern Economic Journal* 33, 1 (July 1966): 17–34.

Todes, Daniel P. *Darwin without Malthus: The Struggle for Existence in Russian Evolutionary Thought*. Oxford: Oxford University Press, 1989.

Vorzimmer, Peter. "Darwin, Malthus, and the Theory of Natural Selection." *Journal of the History of Ideas* 30, 4 (Oct.–Dec. 1969): 527–42.

———. "Charles Darwin and Blending Inheritance." *Isis* 53, 3 (1963): 371–90.

Walters, S.M., and E.A. Stow. *Darwin's Mentor: John Stevens Henslow 1796–1861*. Cambridge: Cambridge University Press, 2001.

Waters, C. Kenneth. "The Arguments in the *Origin of Species*." In *The Cambridge Companion to Darwin*, ed. Jonathan Hodge and Gregory Radick, 116–39. Cambridge: Cambridge University Press, 2003.

Whewell, William. *The Philosophy of the Inductive Sciences*. Facsimile of the 2nd ed., vol. 2. London, 1847; New York: Johnson Reprint, 1967.

Wilson, E.O. *Consilience: The Unity of Knowledge*. New York: Knopf, 1998.

Winterton, R.H.S. "Newton's Law of Cooling." *Contemporary Physics* 43, 3 (1999): 205–12.

Winther, Rasmus, G. "Darwin on Variation and Heredity." *Journal of the History of Biology* 33, 3 (2000): 425–55.

Chapter 3

THE UNIT(S) OF SELECTION

A tribe including many members who ... were always ready to give aid to each other and sacrifice themselves for the common good, would be victorious over most other tribes; and this would be natural selection.

—*Darwin*, Descent of Man, and Selection in Relation to Sex *(1871, p. 166)*

I. INTRODUCTION: HISTORICAL REMINDER AND THE KEEPER OF GENES

The gap between Aristotle's and Darwin's conceptual and fieldwork efforts and the corresponding reflections and results produced by modern biology is a space that could not be filled in the previous historical sections (see Lyons, 2011 and Depew and Weber, 1996). Suffice it to say that the discovery of hereditary material, the findings of the structure and nature of the DNA molecule, the subsequent unification and mathematization of genetics and evolutionary theory (known as "the modern synthesis"), the variety of philosophical analyses regarding conceptual issues in biology, and the continuing ground-breaking discoveries by contemporary field biologists and geneticists have contributed tremendously to our understanding of biological systems (e.g., Carroll, 2005 and Kaebnick and Murray, 2013). It is not hyperbole to suggest that both Aristotle and Darwin would be quite pleased to see the remarkable advances that have occurred since their respective philosophical and empirical achievements.

Although most of us may not be familiar with the details of these biology achievements that emerged from this history and those that are before us today, almost all of us know that an integral part of why we look the way we do is the result of the *genes* we have inherited from our parents. We also know that each of our *DNA* is unique in a way that has transformed crime scene investigation practices (Alas, some of us may very well recall the fallout of the (in)famous O.J. Simpson murder trial!). Indeed, many of us can remember the frequent news flashes regarding yet another *hereditary* discovery uncovered by the *genetic sequencing* efforts of those scientists working on the human *genome* project. We might even be familiar with the role of *allele combinations* that account for how medicine understands various sorts of diseases, how botanists create plant hybrids, and how agricultural scientists are modifying the farm animals and crops we eat. Let us, additionally, not forget those science buffs, who joyfully embrace terms like '*nucleic acid,*' '*genetic code,*' '*gene splicing,*' and '*chromosome mutation.*' In short, most of us have been raised around the language of "the power of genetics" (even without the help of a Bio 101 course) to the point that it is part of our own casual lingo. The geneticist, Richard Dawkins, embellishes all of this when he tells us that the fluffy specks spread in the air by a willow tree:

> are, literally, spreading instructions for making themselves. They are there because their ancestors succeeded in doing the same. It is raining instructions out there; it's raining programs; it's raining tree-growing, fluff-spreading, algorithms. That is not a metaphor, it is the plain truth. (Dawkins, 1996, p. 111)

What would motivate Dawkins to press this "DNA truth" in such an over-the-top fashion? To answer this question, we must briefly go back to the publication of Wynne-Edwards' *Animal Dispersion in Relation to Social Behavior* (1962) and George C. Williams' *Adaptation and Natural Selection* (1966). While studying the social and reproductive behavior of birds, Wynne-Edwards argued for the reasonableness of natural selection functioning at the level of the group such that the optimum breeding rate is determined by the overall competition between local populations. Ultimately, Wynne-Edwards thought that there are certain group-level traits (associated with, for example, display sites, roosting sites, and migratory homing) that benefit the group at the expense of the individual. Wynne-Edwards explains that "what is actually passed from parent to offspring is *the mechanism* for responding correctly in the interests of *the group* in a wide range of circumstances" (1962, p. 144; my italics). For example, if it is in the survival interest of the group that a reduction in the reproductive rate should occur due to a shortage of resources, then the group will make the necessary behavioral changes for the benefit of the group. This occurs, argues Wynne-Edwards, because there is a group-level "population-control trait" naturally selected at the level of the group as a result of competition between groups for the scarce resources (Wiens, 1966).

So what's the big deal? The big deal is that Darwinism had to contend with the reality that some individuals sacrifice themselves for the good of the group to which they belong. This sacrificial behavior (in the form of some individuals not reproducing for the sake of the group) was predominantly observed within certain insect populations. The issue is that such behavior rubs against the putative view that natural selection only benefits *the individual* in its quest for survival and reproductive success. In terms of trying to make sense of this sacrificial and social behavior in insects, the evolutionary biologist, W.D. Hamilton (1964) showed that the perceived social behavior present in many insect populations was really the result of an odd genetic quirk (known as *haplodiploidy*) in which sisters share a greater proportion of their genes with each other than with their related brothers.

The result is that sisters sacrifice themselves and/or their reproductive possibilities for their queen and raise the queen's daughters, thus ensuring that their own genetic advantage persists into the next generation. The implication of these findings is that so-called cooperative behavior was strongly tethered to genetic relatedness. The further upshot is that, thanks to Hamilton's remarkable work, considerable weight was given to natural selection favoring and ensuring genetic survivorship and replication over the survival and reproductive success of individuals and groups. Yet, Wynne-Edwards hypothesis that selection, at times, favors groups over other units seems to contradict Hamilton's findings. *This* was the big deal!

It was, in part, allegiance to the mathematical genetics work of Hamilton that led to the strong backlash by biologists to Wynne-Edwards' group selection interpretations: enter the critical response of Williams. Thinking in terms of adaptations, Williams argued that natural selection would favor only those entities that constituted adaptive features. Since adaptations take many generations to remain relatively fixed in a population, neither the individual nor the group survives long enough to be *the unit of selection*. Given the focus on adaptations, Williams determined that Wynne-Edwards-type group or species-level selection arguments were rarely validated out in the world, if at all. Rather, it is only the gene complex, the unit that can ensure its presence in future generations, which can legitimately be the proverbial bulls-eye for natural selection (Williams, 1966).

It is fairly soon after this scholarly exchange regarding the unit favored by natural selection that Dawkins makes his presence felt as the "Keeper of Genes." To understand his impact on the discussion, imagine you were told that your own existence, including your behavior, mind, and physical features (and the existence of just about every living thing on this planet), is nothing more than the activity of genes "figuring out" how to make sure they spread themselves to future generations? Well, taking the "genic selectionism" baton from Williams, Dawkins moves in just this gene's-eye-view direction when he poetically proclaims:

> They are in you and me; they created us, body and mind; and their preservation is the ultimate rationale for our existence. They have come a long way, those replicators. Now they go by the name of genes, and we are their survival machines. (Dawkins, 1976, p. 21)

The above passage seems like a hard pill to swallow. The idea that we are nothing more than machines doing the dirty work of selfish genes seems hard to believe. Yet, such a view is a reminder of the long reach of Darwin's **Strong Argument for Natural Selection** (see ch. 2). Notably, Darwin's concept of natural selection continues to be the core idea with respect to the nature of evolutionary change. Although obliquely hinted by Darwin in some of his works (e.g., Darwin, 1958, p. 88), the many interesting findings by contemporary biologists regarding genetics and the relationship between biological systems and their environments have revealed a rather complex network within which natural selection functions. One major philosophical controversy that has emerged, since Darwin first speculated about it (see opening quotation) and Dawkins later demonstratively conjectured, concerns "the level" at which natural selection functions in order to make sense of biological systems as evolving entities within nature's complex network. It should come as no surprise that Dawkins thinks that the level that natural selection functions is the genetic. Is this true? Wynne-Edwards, in contrast, thinks that there are group-level characteristics that are frequently the units of selection. Is this correct? Darwin, alternatively, favored the individual as the proper unit of selection? Is this on the mark? In the form of a question, this concern about "unit(s) or level(s)" can be expressed like this: *Does natural selection primarily range over groups, individuals, genes, or something else?* We will consider this as the "unit(s) of selection" question.

It should be stressed that the formulation of the unit(s) of selection question above is not intended to suggest that the debate is merely an empirical issue. There are numerous conceptual puzzles that require analysis in order to make sense of this debate. For instance, some scholars in the units of selection debate are focused on providing the conditions under which any unit/level could be a unit/level of selection without taking a stand as to whether those conditions are manifest out in the world. So, in thinking about the question that governs this chapter, the reader should be prepared to engage a mix of empirical and conceptual issues.

It is important to pause and be clear about the use of 'primarily' in the above question. Even if it is true that certain biological facts might reveal selection working at a particular biological "level," it remains to be determined whether the level under consideration constitutes *the primary level* at which selection functions. So, in the same way that a single Swallow does not a spring make, it correspondingly follows that a few instances of renegade "selfish genes," sparse cases of sacrificial individuals, or rare occurrences of altruistic groups do not necessarily reveal *the primary level* at which selection functions. Of course, this suggests that a definition of 'primarily' is needed, and we will see that this is indirectly offered by various scholars in this debate.

II. LAYING OUT THE UNIT(S) OF SELECTION DEBATE

In the previous section, the reader should have noticed the use of terms like 'selfish,' 'sacrificial,' and 'altruistic' to make sense of different sorts of self-regarding and other-regarding behavior. This choice of words is no accident; for, as will be made clear in the next few sections, one way of providing an answer to the unit(s) of selection question has focused on making sense of the *behavior* of living entities that does not appear to benefit the individual expressing the behavior, but is beneficial to others on the receiving end of the behavior—that is, biological *altruism* (Sober, 2000, p. 91).

In contrast, the natural world is well-supplied with individuals securing their respective survival and reproductive success, frequently at the expense of their family and group members. So, according to some scholars, apparent other-regarding behavior is more of a smoke-screen or veneer waiting to be uncovered by the persistent ethologist. As Ghiselin unequivocally states, "Scratch an 'altruist' and watch a 'hypocrite' bleed" (1974, p. 247).

Reasonably, one could insist that it is through the behavior of the self-interested individual and its interaction with its environment that most accurately captures what is the *primary* unit of selection. Hence, why not insist that the individual is the *primary* unit of selection? Yet, cases of behavioral sacrifice for a group are not entirely uncommon in the natural world. For example, soldier/worker bees behave in such a way as to sacrifice their own lives and reproductive success for the benefit of the hive to which they belong. Consequently, should we follow those biological pundits (e.g., Wilson, 1975 and Wade, 1976 and 1980) who have insisted that selection primarily favors the group over the individual? Alternatively, individuals and individual groups come and go, while copies of their hereditary material continue from one generation to the next. From this perspective, genuine sacrifice *qua* altruism is rare or short lived and it is competition at the genetic level that primarily captures the winnowing force of natural selection. Genes, then, are really "out for themselves" and talk of sacrificial altruism is, upon close inspection, a sham or mere façade.[1] Accordingly, following the lead of Williams and Dawkins, shouldn't we view "the gene" as the *primary* unit of selection?

So far, by way of the notion of biological altruism, we have been exposed to the individual, the gene, and the group as legitimate contenders regarding over what natural selection *primarily* "selects." One reason for the persistence of defenders of these various camps is reflective of the question(s) trying to be answered.[2] Lloyd (2001, p. 274) offers an excellent analysis of why diametrically opposed answers have emerged in the literature. She points out that, depending upon the specific question being answered and/or valued, corresponding answers follow. For the sake of this discussion, we can assume that determining the *primary* unit(s) of selection is subsumed under the four questions (see Table 3.1) that Lloyd thinks govern the unit(s) of selection debate.

Before explaining **(Q1)–(Q4)**, it is important to understand the point of Table 3.1. As the chapter unfolds, separate arguments will be presented in defense of the individual, the gene, the group, multi-levels, and the developmental system as the primary unit of selection. Upon examining each theory, a corresponding table, like Table 3.1, will be put on display so that the reader can understand what question(s) is of greatest importance to particular authors and the corresponding theory they defend. For example, as will be evident in the next few pages, Darwin defends the individual as the *primary* unit of selection because he values **(Q3)** in the debate. Thus, for Darwin, **(Q3)** is *more relevant* than either **(Q1)**, **(Q2)**, or **(Q4)**. This explanatory approach will be used throughout the chapter.

---

1 Note that this discussion of altruism is about the costs and benefits of the *behavior* of living systems, *not* the psychology associated with giving and receiving benefits; do not confuse biological altruism with psychological altruism. For more on this point, see Sober (2002) and Ananth (2005).

2 At this point, some inquisitive students might be wondering why natural selection can't simultaneously (to varying degrees) be working at all of these levels. Not to worry; this possibility will be addressed later in the chapter.

QUESTIONS OF IMPORTANCE	EXAMPLE OF POSSIBLE ANSWERS	DEGREE OF RELEVANCE
(Q1): What is the *interactor*?	Individuals or Groups	More Relevant or Less Relevant
(Q2): What is the *replicator*?	Genes, Individuals, or Groups	More Relevant or Less Relevant
(Q3): What is the *beneficiary*?	Genes, Individuals, or Groups	More Relevant or Less Relevant
(Q4): What entity *manifests adaptations*?	Genes, Individuals, or Groups	More Relevant or Less Relevant

Table 3.1 *Unit(s) of Selection Questions*

In Table 3.1, **(Q1)** refers to the entity that directly engages with the external environment and constitutes a difference-maker with respect to survival and reproductive success. Drawing on the work of Hull (1980), Lloyd explains that 'interactor' refers to "that entity which interacts, as a cohesive whole, directly with its environment in such a way that replication is differential—in other words, an entity on which selection acts directly" (Lloyd, 2001, p. 269). Those who value **(Q1)** over **(Q2–Q4)** would insist that the answer to **(Q1)** is also the answer to the unit(s) of selection question posed at the beginning of this chapter. For example, if the individual is considered to be the cohesive-whole-interactor, then the individual would be the *primary* unit of selection. In contrast, those who value **(Q2)** over **(Q1)** and **(Q3–Q4)**, argue that the *primary* unit of selection will be the entity that is able to replicate itself with substantial precision and pass along its intact near-duplicate form to future generations. Usually, defenders of **(Q2)** insist that only genetic material is capable of such precise replication and proliferation and, therefore, deserves the status of being the *primary* unit of selection. Alternatively, champions of **(Q3)** insist that the unit of selection is that entity which benefits most by its activities. So, if the group gains the most as a result of the benefits it accrues from the activities of individuals that comprise it, then the group gains *primary* ascendency as the unit of selection. Then again, defenders of **(Q4)** think that the entity that manifests adaptions is best suited to lay claim as the *primary* unit of selection. As Sober puts it, "When a population evolves by natural selection, what, if anything, is the entity that does the adapting?" (Sober, 1984, p. 204; quoted from Lloyd, 2001, p. 272; see also Sober, 2000, pp. 90–91). For instance, if adaptations occur at the level of the individual, gene, or group, then that level will constitute the *primary* unit of selection.[3]

---

[3] Note that **(Q1–Q4)** could be answered in different ways. The units have been chosen in this way merely to illustrate how these questions could be answered. This will become clearer as the chapter unfolds.

The above paragraph should suggest to the reader that the unit(s) of selection debate is sort of a tug-of-war as to which question(s) **(Q1)–(Q4)** should be most valued and thus reign supreme in this dispute. Yet, as Table 3.1 above shows, even if the most important questions are agreed upon, there is still ample room for debate as to what entity constitutes the *primary* unit of selection. With this tug-of-war in mind, the goal of this chapter is two-fold: (1) to canvass the unit(s) of selection debate by making clear why proponents of any one of these perspectives defend the viewpoint that they do; this will include the newer kids in town, multilevel selection theory and developmental systems theory, as possible alternatives and (2) to provide a few critical replies to each alternative. It should be noted at the outset that there are more nuances and arguments to this debate than can be delineated in this chapter. The hope is to whet your appetite and move you to consult the references at the end of this chapter.

III. SETTING THE STAGE: DARWIN ON INDIVIDUAL AND GROUP SELECTION

The standard view of evolution, as suggested by Darwin in the *Origin*, is that natural selection operates at the level of individuals competing against one another within their respective populations (Okasha, 2009). As Darwin somewhat rhetorically asks, "can we doubt (remembering that many more *individuals* are born than can possibly survive) that *individuals* having any advantage, however slight, over others, would have the best chance of surviving and of procreating their kind?" (Darwin, 1958, p. 89; my italics). From this view, changes in the overall structure of a given population can be understood to be the result of individuals competing against one another for limited resources and reproductive benefits. It is natural selection working through the individual, thinks Darwin, that makes best sense of evolutionary change. In terms of the units of selection debate, we can reasonably infer that Darwin would endorse the claim that natural selection *primarily* ranges over the individual.

Nature, however, is quite indifferent to our attempts to organize its happenstances. Darwin puzzled over this annoying truth in his attempt to reconcile individual selection with his observation that some individuals performed actions that were not only detrimental to their survival and reproductive success, but also beneficial to the group to which they belonged. For example, Darwin openly discusses how it is possible for natural selection not to have pruned away insect sterility given that it does not directly benefit an individual to be sterile (Darwin, 1958, pp. 275–80). To resolve this difficulty, Darwin has to employ "community" benefit/altruism in the light of the continued presence of sterility in certain populations. He concludes that, if natural selection is relevant to such an odd state of affairs, then it must be the case that community X gains an advantage over community Y due to the contributions provided by sterile individuals of community X. Darwin acknowledges, then, that *individual selection* must make room for *community/group selection* on occasion. Thus, individual selection is, for Darwin, compatible with group selection (Darwin, 1958,

p. 276). Observe that, although natural selection can range over groups at the expense of individuals, Darwin could still claim that natural selection still *primarily* ranges over the individual. Table 3.2 below provides a summary of how **(Q1)**–**(Q4)** are answered by those who defend the individual as the primary unit of selection. Some of these answers will be clearer as the chapter unfolds.

QUESTIONS OF IMPORTANCE	UNIT(S) THAT ANSWERS EACH QUESTION	DEGREE OF RELEVANCE
(Q1): What is the *interactor*?	Individual	Less Relevant
(Q2): What is the *replicator*?	Gene and Individual	Less Relevant
(Q3): What is the *beneficiary*?	Individual and the Group (to a lesser extent)	More Relevant
(Q4): What entity *manifests adaptations*?	Individual and the Group (to a lesser extent)	Less Relevant

Table 3.2 *The Individual and the Unit(s) of Selection*

Thus far, then, the entity that gains the benefits at the expense of its corresponding benefactors determines which entity natural selection favors. As mentioned above, from Darwin's perspective, **(Q3)** is the question to which he is responding and this is captured in Table 3.2 above as the more relevant question with respect to the unit(s) of selection debate. This is not to say the other questions were of no value to Darwin. Rather, usually or primarily, Darwin observed that individuals are benefiting themselves over other individuals in the group to which they belong. This is why Darwin thought that natural selection ranges, for the most part (i.e., primarily), over individuals. On occasion, however, individuals receive less than they give (biological altruism). This sometimes results in a greater benefit to the group to which these individuals belong that concomitantly is a difference maker with respect to competing groups. With respect to these circumstances, Darwin would acknowledge that the answer to **(Q3)**—what is the beneficiary?—would be the group. At times, then, nature appears to accommodate sacrificial-type altruistic behavior on the part of individuals. This captures why Darwin also thought it was necessary to incorporate group selection as a legitimate instance of natural selection working at the level of the group (see chapter epigraph). Still, it is important to emphasize that, if Darwinism is focused on what is the primary unit of selection, then the individual is still (primarily) the unit of selection.

Additionally, as we saw in the previous chapter, Darwin took adaptations seriously by way of his principle of natural selection. Although he embraces the importance of heredity and the transmission of beneficial traits to the next generation, he knew nothing about genes and DNA, so he would have not considered **(Q2)** and would have likely answered "the individual" to **(Q1)** and **(Q4)**. As we will see, contemporary Darwinians (e.g., Gould and Brandon) will openly take Darwin's lead and defend "the individual" as *the primary* unit of selection and as the answer to **(Q1)**, **(Q3)**, and **(Q4)**.

IV. THE GENETIC TURN: THE NONSENSE OF INDIVIDUAL AND GROUP SELECTION?

After the triumphant efforts c. 1860 of Gregor Mendel (Cox, 1999 and Deichmann, 2010), genetics gained ascendancy in the 1930s and 1940s. Indeed, what is called "The Modern Synthesis," is the forging of evolutionary biology to an almost exploding field of population genetics (Merlin, 2010). One of the results of this genetics-driven biology is that biologists have not been very sanguine about taking seriously either the individual or the group as the unit of selection. Rather, those who embrace this "gene's-eye-view" insist that genes are primarily the unit of selection (for instance, Hamilton, 1964 and Williams, 1966). In particular, as we already observed at the beginning of this chapter, Dawkins has argued that only genes (as replicators)—and neither individuals nor groups—are the primary units of selection.

To make sense of this talk of genes, an admittedly adumbrated and piecemeal "Bio 101" reminder may be useful. Among other living things, organisms are comprised of **cells**. These cells house a **nucleus**, which is the control or command center of the cell. Inside the nucleus is compact thread-like material known as **chromosomes** (the long spirals in Figure 3.1 below). This thread-like material is made up of both deoxyribonucleic acid (**DNA**) and proteins. Specifically for our purposes, the structure or shape of DNA is known as a "double helix" configuration. 'Double' refers to the two strands of chromosomes that are present in the nucleus and 'helix' refers to the spiral or crisscrossing pattern created by the two twisted strands of chromosomes. You might recall that the design looks like this:

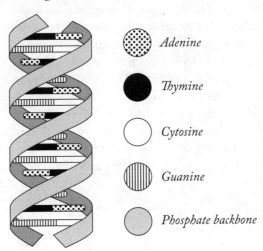

Figure 3.1 *Spiral Design of DNA Molecule*

Each of the twisted chromosomes is supported by a sugar and phosphate backbone. Attached to these backbones are projecting **"bases"** that are labeled "A," "C," "T," and "G." These bases represent "the code" of DNA. The different combinations of this code on each side of each chromosome connect with each other. For example, "A" connects with "T" and "C" connects with "G." These connected bases are known as **base pairs** and (among other things) assist in the production of protein. It is specific protein that eventually functions to produce particular traits. As shown in Figure 3.2 below, a **gene** is a segment of DNA (or a segment of base pairs) on chromosomes that codes for a specific protein and contributes to the formation of a particular trait.

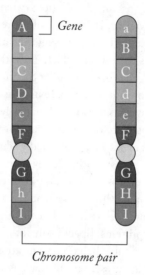

Figure 3.2 *Genes on Chromosomes*

As indicated, each segment of the chromosomes codes for a specific trait. Different versions of the same gene are possible and these different versions are known as **alleles**. For example, near the top of the left chromosome of Figure 3.2, let's assume that the lowercase b segment of DNA codes for blue eyes and near the top of the right chromosome the uppercase B segment of DNA codes for brown eyes. The gene for brown eyes can come about from two allele combinations. One version of this gene includes alleles "BB" and another version of the gene is alleles "Bb." So, the gene for brown eyes has two possible sets of allele combinations: (1) "BB" and (2) "Bb" and (without further information) either one could be present at the position or **locus** of the uppercase B segment near the top of the right chromosome. In contrast, near the top of the left chromosome, there is only one set of alleles that code for the protein that facilitates the production of blue eyes, namely "bb" and it must be the combination of alleles that is present in this position or locus.[4]

---

4 This rather thin explanation of "genes" is intended to remind students of a few points, but in no way captures the impressive details related to underlying chemistry, protein synthesis, DNA

With this (admittedly) terse account of DNA in mind, we can return to the unit(s) of selection debate. Specifically, defenders of the gene's-eye-view (like Dawkins) argue that natural selection and competition is really occurring at the level of alleles at the same locus. Using the eye color example, the presence of "BB," "Bb," and "bb" is the result of competition between other alleles at these particular loci. In contrast, alleles at the other loci are not competing with the eye color alleles. It is competition occurring at this level that motivates gene selectionists, like Dawkins, to argue that the gene is the primary unit of selection.

Additionally, Dawkins makes a distinction between **replicators** and **interactors**. The idea is that individuals interact with the environment, but it is genetic material that does the actual replication wherein competition with other genetic material is at the forefront. Thus, it is the combination of allelic replication and competition that, in part, motivates Dawkins to push for the gene's-eye-view.

Yet, if this is all Dawkins has to say about the matter, then we need not be persuaded; for competition and replication is also occurring at the level of the individual and its conspecifics (the other members of its species)—a competition that can be rather fierce depending upon the species under consideration. The same holds true for replication; for instance, infants (in some species) replicate the behavior of their parents with a high degree of precision. In terms of competition and replication, there is no obvious reason to favor replicators over interactors. This suggests, then, that Dawkins' argument is off the mark.

Mere competition and replication, however, are not the core aspects of Dawkins' argument. Rather, precision and longevity with respect to replication are the difference makers for Dawkins. As he puts it, "What I am doing is emphasizing the potential near-immortality of a gene, in the form of copies, as its defining property" (Dawkins, 1989, p. 35). From this, he moves to declare, "It is its potential immortality that makes a gene a good candidate as the basic unit of natural selection" (Dawkins, 1989, p. 36). A reconstruction of his argument, call it **The Precision and Longevity Argument**, is as follows:

The Precision and Longevity Argument

P1. The primary unit of natural selection is the entity that can both reproduce with precision and live longer than any other competing entity.

P2. Genetic material can reproduce with precision and live longer than any other competing entity.

C1. Thus, genetic material is the primary unit of natural selection.

replication, etc. I would not only direct the reader to any introductory biology text for these details, but also to Mukherjee's (2016) wonderful historical rendering.

For the moment, let us grant the truth of **P1** and focus on the truth of **P2**. Dawkins' defense of the truth of **P2** of **The Precision and Longevity Argument** can be understood as being tied to the idea that natural selection is mainly about "the differential survival of entities" (Dawkins, 1989, p. 33). In order for entities to survive, they not only must survive, but they must do so in a way that leaves a sort of lasting legacy. Dawkins tells us that this legacy is accomplished by way of "lots of copies, and at least some of the entities must be potentially capable of surviving—in the form of copies—for a significant period of evolutionary time across many generations. Small genetic units have these properties: individuals, groups, and species do not" (Dawkins, 1989, p. 33).

Notice that Dawkins accepts that individuals or groups can function as interactors **(Q1)**, but he does not value such entities relative to the properties possessed by genetic material (see Table 3.3 below). His reasoning is that both the quality of what is copied and the longevity of what is copied are more valuable features than that which aids in ensuring both copying quality and survival longevity; these aiders are, in Dawkins' words, mere "vehicles" (Dawkins, 1989, pp. 253–66). In brief, genes copy and pass along their copies with greater precision and for greater lengths of time than individuals and groups. Thus, this is why Dawkins would confirm the view that the correct answer to **(Q2)**, namely genetic material, *primarily* constitutes the unit of selection.

What about altruism? Well, the title of Dawkins' (in)famous book, *The Selfish Gene*, should give the reader a hint; and Dawkins does not disappoint. His view is that genes have created bodies to interact with the environment to ensure their own success. For added emphasis Dawkins proclaims the following in the preface to the first edition of *The Selfish Gene*: "We are survival machines—robot vehicles blindly programmed to preserve the selfish molecules known as genes" (Dawkins, 1989, p. v).

Dawkins is suggesting a sort of one-way street perspective: genes create phenotypes (observable characteristics of the individual) which interact with the environment so as to ensure the survival of the phenotype-producing genetic material. What this means for the defenders of the gene's-eye-view is that genuine other-regarding altruism is either extremely rare or really nothing more than a clever way of genes finding another vehicle to promote their own survival chances. This appears to be Dawkins' take: "There are special circumstances in which a gene can achieve its own selfish goals best by fostering a limited form of altruism at the level of individual animals" (Dawkins, 1976, pp. 2–3). This claim reveals that Dawkins thinks that the gene is the beneficiary **(Q3)** of its activities and those of the individual. Thus, Dawkins believes that both **(Q2)** and **(Q3)** are of greater importance than **(Q1)** with respect to the unit(s) of selection debate. (As we will see below, Gene's-Eye-View defenders will also attempt to claim that **(Q4)** is more important in terms of vindicating the gene as the primary unit of selection.) The upshot, thinks Dawkins, is that the gene is the primary unit of selection. Table 3.3 offers the following summary of the Dawkins-type gene's-eye-view theory.

QUESTIONS OF IMPORTANCE	UNIT THAT ANSWERS EACH QUESTION	DEGREE OF RELEVANCE
(Q1): What is the *interactor*?	Individuals	Less Relevant
(Q2): What is the *replicator*?	Genes	More Relevant
(Q3): What is the *beneficiary*?	Genes	More Relevant
(Q4): What entity *manifests adaptations*?	Genes	More Relevant (See Below)

Table 3.3 *Gene's-Eye-View and the Unit(s) of Selection*

Is there a way to test Dawkins' rather strong position in defense of the gene being the *primary* unit of selection? That is, it seems that every benefit gained by genes seems to also benefit the individual—or it could, at the very least, be interpreted in this way. For instance, Darwin could argue (if he had the knowledge) that, although genes might very well be the beneficiary of the activity of the individual, such benefits could only accrue as a result of the individual interacting in its environment. So, selection directly encounters the individual interactor, *and* individuals do replicate themselves through their offspring, *and* usually the individual benefits even when genes benefit, *and* adaptations do occur at the level of the behavior of the individual and other phenotypic features as well as at the genetic level. Burian fortifies this sort of reasoning by including the longevity component into the equation. He argues that "once we recognize that we should compare *lineages* of genes with *lineages* of organisms, it is not clear these [i.e., individual organisms and individual genes] are the relevant comparisons; lineages of organisms, like lineages of genes, last until extinction" (Burian, 2010, p. 145; bracketed addition mine). This suggests that the individual organism or individual lineages could be considered the *primary* unit of selection when answering (Q1)–(Q4). What follows, at first glance, is that P2 of **The Precision and Longevity Argument** is false, rendering the entire argument unsound. The implication is that there does not necessarily seem to be a knock-down reason to favor, out rightly, the gene over the individual as the *primary* unit of selection.[5]

One way to save **P2** of **The Precision and Longevity Argument** is to offer another property of genetic material that breaks the tie. Some defenders of the gene's-eye-view do just this as they offer a number of examples designed to show that the gene

---

5 See Moss (2003) for an attempt to advance Darwin's defense of the individual as the *primary* unit of selection by exploring the complexity of cellular, nucleus, and organism activity. Moss' general conclusion is that the view that genetic material is the store-house of all biological information is off the mark.

should be viewed as the *primary* unit of selection. The argument, let us call it **The Distortion Argument**, can be stated as follows:

The Distortion Argument

P1. Either the individual organism or the gene is the *primary* unit of selection.

P2. If there is evidence that there are selective processes functioning at the genetic level that do not make the individual organism more likely to survive and reproduce, then the individual genes are the *primary* unit of selection.

P3. There is evidence, in terms of the activities of "outlaw genes," that there are selective processes functioning at the genetic level that do not make the individual organism more likely to survive and reproduce in the light of the fact that "Outlaw genes" benefit themselves at the expense of the individual organism and the genome to which they belong.

P4. It is not the case that the individual organism is the *primary* unit of selection.

C1. Thus, the gene is the *primary* unit of selection.

As a reminder, **The Distortion Argument** is offered as additional evidence for the truth of **P2** of **The Precision and Longevity Argument**. Keeping this in mind, in the above argument, **P3** of **The Distortion Argument** is true. That is, there are types of genes that appear to benefit themselves with no apparent benefit (and sometimes harm) to the individual or the genome (the set of genes a person has) of which they are a part. This is odd and requires a subtle distinction between *selfish renegade genes* and *selfish genome genes*. The renegade genes, like outlaw genes, are out for only themselves, even at the expense of their fellow genes and the individual in which they reside. As Crow puts it, "There are genes that cheat, perpetuating themselves in the population by tricking the reproductive process to work in their own favor" (Crow, 1979, p. 134). In contrast, selfish genome genes cooperate with their genetic partners and neighbors in order to ensure benefits to both the genome and the individual that houses them. To see this, remember that, normally, half of a parent's DNA *randomly* moves on to the next generation (via the process of meiosis, the process of both separation and recombination of chromosomes from parents). "Outlaw genes," for example segregation distorter genes, are so called because they are able consistently to beat the odds by giving themselves more than a random chance of being present in the next generation.[6]

---

6   Other examples of "Outlaw genes" are sex-ratio distorter genes, "junk" DNA, and the so-called "green beard" gene (Sterelny and Griffiths, 1999).

Putting aside some of the biology details, the point is that these outlaw genes get naturally selected in a way that contributes to their survival and reproductive success even if harm or no benefit accrues to either or both the genome and individual. For example, there are outlaw genes that disturb/harm both tail growth and meiosis in mice and quality of sperm in a species of Drosophila flies (Crow, 1979, p. 134 and p. 136). Importantly, these outlaw genes are able to ensure their own survival into future generations despite their harmful effects. This reveals that natural selection is at work at the level of the gene—even when it is a detriment to the individual. This feature of genetic material, which can be added to **The Precision and Longevity Argument**, points to the truth of **P2** of **The Precision and Longevity Argument**. Thus, given that there are genes that are "renegade selfish" as opposed to being merely "genome-selfish," defenders of the "gene's-eye-view" claim that the gene is the *primary* unit of selection—**The Precision and Longevity Argument** is sound.

The reader should see that **The Precision and Longevity Argument** relies upon giving emphasis to the correct answer to **(Q3)** and **(Q4)**; that is, defenders of **The Distortion Argument** think that distorter genes reveal that benefits accrued to the gene (and not the individual) suggests that the gene is the primary unit of selection (see Table 3.3 above). Additionally, the strategy of disrupting sexual reproduction (i.e., meiosis) to ensure representation in the next generation is an adaptive feature unique to these distorter genes. The manifestation of these adaptive functions gives further credence to the gene, as opposed to the individual, as the *primary* unit of selection (Sterelny and Kitcher, 1988). The implication is that **The Distortion Argument** and **The Precision and Longevity Argument** are sound, apparently vindicating those who champion the gene as the *primary* unit of selection.

A Brief Reply to the Distortion Argument

There are a number of difficulties with **The Distortion Argument**. Let us examine one of the straightforward concerns mentioned at the start of the chapter; namely, that a small set of uncooperative or renegade genes does not constitute a ringing endorsement of the gene's-eye-view. There appears to be a fallacy of composition at play here; that is, some defenders of the gene's-eye-view are inferring a characteristic of the whole from a characteristic of some of the parts of the whole. In this case, they are claiming that the primary unit of selection is the genome from the fact that some small subset of the genome (namely, renegade genes) exhibit unique and anomalous adaptive features. It may very well be the case that these renegade genes cannot be explained at the level of the individual, but that merely shows that there are some gene-level adaptations that cannot be accounted for at the level of the individual. Still, most benefits that accrue to genes also advance individuals (Sterelny and Griffiths, 1999, pp. 59–60).

In terms of **The Distortion Argument, P2** is false; that is, even if it is true that there are outlaw/renegade genes that have adaptive advantages that are not advantages to the individual of which they are a part, it does not follow that the individual is precluded from being the *primary* unit of selection. To think so is to commit the fallacy of composition. Indeed, since most genes (replicators) contribute to the welfare of themselves as well as the individuals (interactors) in which they are housed, it remains unclear whether or not to privilege the gene over the individual as the *primary* unit of selection.

It follows, then, that **The Distortion Argument** is unsound and, therefore, **The Precision and Longevity Argument** is also unsound. To see this, recall that **P2** of **The Precision and Longevity Argument** is thought to be bolstered by **P2** of **The Distortion Argument**. Since **P2** of **The Distortion Argument** is false, the result is that it cannot be used as a difference maker in vindicating **P2** of **The Precision and Longevity Argument**. Another way to put it is that the property of being a distorter is not one that can count as a difference maker in the unit(s) of selection debate. Thus, the gene as the *primary* unit of selection has not been entirely vindicated by these arguments.

V. GOULD'S VISIBILITY AND KALEIDOSCOPE ARGUMENT: A FURTHER REPLY TO DAWKINS

The Argument from Visibility

With the soundness of **The Distortion Argument** and **The Precision and Longevity Argument** in question, it remains to be determined whether the gene or the individual is the *primary* unit of selection. Stephen Jay Gould offers two arguments in defense of the individual. The first argument can be called the visibility argument. In reply to Dawkins's insistence that the unit of selection can only be the gene, Gould replies that it is the phenotype (interactor) that makes a direct contribution to whether or not certain alleles (replicators) are transmitted to the next generation. Even if it is the case that genes "use" bodies to do their dirty work, it is still the body and its features that are being screened by the natural world. Genes cannot be the unit of selection, because nature cannot "see" such inconspicuous entities (Gould, 1980, p. 90). As Gould notes, "If, in favoring a stronger body, selection acted directly upon a gene for strength, then Dawkins might be vindicated" (Gould, 1980, pp. 90–91).[7]

Gould's emphasis on what selection can "see" suggests that he thinks that the answer to **(Q1)**—what is the interactor?—is of highest value (see Table 3.2). Even if one conceded that genes create all phenotypes (this is not the case), it is still the overall individual body plan that maintains the existence of genes. That is, it is the transformation (translation) of genes into morphological, physiological, and behavioral units with which natural selection comes in contact. Since the individual is the interactor and is directly visible to selection, it can be inferred that Gould thinks that it is the individual *qua* body that is the *primary* unit of selection.

---

7   See Brandon (1998, p. 180) for a clever thought experiment that he thinks corroborates Gould's visibility argument.

The Argument from Causal Complexity

For good measure, Gould offers an additional argument to try to vindicate the individual as the *primary* unit of selection. This part of the argument is focused on why the whole individual—and not its parts—must be the unit of selection. In this way, he can block both genes and parts of individuals as the *primary* units of selection. Thinking of the units of selection problem within the framework of *Ockham's razor* may be of help to see why Gould thinks he wins this argument. Ockham's razor is the principle of parsimony, the idea that entities should not be postulated without necessity and that the best explanation is one that posits the fewest number of entities.[8] Since both Dawkins and Gould disagree as to what is the *primary* unit of selection (Dawkins argues for the gene, while Gould argues for the individual), parsimony is not relevant to this part of the discussion. Where parsimony is relevant is in terms of the causal relationship(s) between genes and phenotypes. Dawkins is suggesting that the only entity that need be postulated is the gene with respect to natural selection, because of its unique causal role in creating phenotypes for its own "selfish motives."

Yet, as Gould notes, these various units "are channeled through a kaleidoscopic series of environmental influences: embryonic and postnatal, internal and external. Parts are not translated genes, selection doesn't even work directly on parts. It accepts or rejects *entire organisms* because suites of parts interacting in complex ways, confer advantages" (Gould, 1980, p. 91; my italics).[9] To appreciate the force of Gould's argument, consider the idea of a simple hand movement or the stimulation (i.e., innervation) of some simple behavior. Goldstein (1963, pp. 229–30) offers a whole-body assessment of such simple changes in movements. He makes clear that specific body movements require that all (he probably means a large ensemble or most) body muscles (in addition to the specific muscles of a particular movement) be coordinated or stimulated to ensure that a specific body movement is accomplished. The pattern of movement that emerges is in conjunction with the pressures from the external environment. Goldstein's point is that even a supposed insignificant movement is "not at all insignificant, it rather *enables the organism* to execute the movement correctly" (Goldstein, 1963, p. 230; my italics).

Drawing from Goldstein's account as a way of making sense of Gould's analysis, it is this pattern of movement in response to environmental stress that is the product of natural selection. For example, the overall patterns of swimming motion of fishes or flying patterns in birds are evolved patterns that are crucial to survival and reproductive success. It is these sorts of behavioral patterns that allow individuals to interact in an energetically balanced way with their environments (Ananth, 2007). The point to be gleaned is that Gould postulates, contrary to Dawkins, a

---

8   For an analysis of the history and debate surrounding the proper use of "Ockham's Razor," see Sober's (2015) impressive discussion.

9   Of course, Dawkins is well aware of this causal complexity, but does not give it its proper place within the units of selection debate. Dawkins's "gene's-eye view" rhetoric may be obscuring his considered view.

complex host of causal factors for understanding bodies—this is as parsimonious as one can be with respect to biological entities. With respect to Ockham's razor, Dawkins appears to have cut off more than he is allowed. Gould, on the other hand, thinks he has correctly postulated multiple causal factors, employing Ockham's razor judiciously. From this fact about multiple causal factors in constructing individuals, Gould concludes that causal uniqueness is not present in genes, making it the case that genes cannot be the primary unit of selection. Rather, he thinks that it is the result of such multiple causal factors in constructing individuals (and their parts) that leads to the conclusion that individuals (not even their parts) are the *primary* units of selection. Thus, the combination of the visibility argument and the causal complexity argument reveals that it is quite reasonable to consider the individual as the *primary* unit of selection.[10]

This kaleidoscope causal complexity argument suggests that Gould would be more than happy to concede that the gene is the replicating entity **(Q2)** and has precision and longevity on its side **(The Precision and Longevity Argument)**. He would insist, however, that adaptations are present in both genes and whole individuals **(Q4)**. In the light of this, Gould can go on and defend the view that both genes and whole individuals benefit (the point about altruism) as a result of adaptations occurring in genes and whole individuals **(Q3)**. One implication of this line of reasoning is that the truth of **P2 of The Precision and Longevity Argument** (see p. 95) might very well be true, but can be dismissed with respect to greater value given to the individual as the legitimate interactor and (along with genes) recipient of adaptations. Thus, as is evident from Table 3.2, since the whole individual can be included in answering **(Q1)** and **(Q3–Q4)**, Gould could submit that it is reasonable to view the individual as the *primary* unit of selection. (This whole-individual point will be expanded in the "developmental systems" account to come.)

To summarize, Gould not only rejects the idea that genes are the units of selection, but also that parts of individuals are the units of selection. His defense of this view is that genes cannot be the unit of selection because they are invisible to selection. Moreover, he notes that parts of individuals cannot be the units of selection, because their existence is the product of a host of environmental forces that act on a set of parts "interacting in complex ways." Thus, he thinks that it is reasonable to consider the individual as the *primary* unit of selection.[11]

---

10 It should be noted that Sterelny and Kitcher (1988) have tried to defeat the visibility argument by insisting that phenotypic traits do not "screen off" genotypic traits. They argue, contrary to Brandon (see note 7, p. 100), that the tampering of the allelic environment reveals that fitness varies more with changes in the genotype than the phenotype. See *Philosophy of Science* 59, 1 (1992) for a number of articles dedicated to the issue of Brandon's "screening off" argument. Here is a fun essay topic!

11 This point about parts is important because there are those who defend the idea that genes are the *primary* unit of selection because important complex traits (like the heart) can be linked to a small set of genes. The upshot of this line of reasoning is to reject Gould-type arguments that complex features require complex explanations on the grounds that empirical evidence does not

VI. GENIC PLURALISM AS THE SOLUTION?

It seems that we are at a stalemate. Depending upon which set of questions **(Q1)–(Q4)** is privileged, it is possible to view the gene or the individual as the *primary* unit of selection. Interestingly, drawing upon some of Dawkins' own ideas, Sterelny and Kitcher argue that this impasse is the result of choosing different parts of the gene-to-individual causal structure. For example, should a spider's web, a spider's web-producing behavior, a spider itself, or the genes that contribute to all of this be the primary unit of selection? Sterelny and Kitcher (1988, p. 358) give us the following answer by pointing out that there are alternative explanations one can choose depending upon the particular part of the causal chain of biological events on which one wishes to focus. Importantly, on their version of *pluralism*, the gene-centered explanation is always available. What this means, for example, is that one can always give a selectionist account by focusing on the spider's web, but that the genetic account is *always* an available alternative and is the only causal explanation (of the chain of events) that is always an available alternative. One can always substitute the genetic causal account for any other causal account, but not the other way around. This is why they label their approach *pluralist genic selectionism*. It is a way of privileging the gene's eye view in the light of Gould's kaleidoscope causal complexity account.

Whether or not this assessment by Sterelny and Griffiths constitutes a concession to Gould that the gene may not be the primary unit of selection continues to be debated (e.g., Lloyd, 2005 and Waters, 2005).[12] As we shall see soon enough, *multilevel selection* theorists and *developmental systems* advocates embrace the complexity aspect of Gould's kaleidoscope argument to push for rather different ways of thinking about the unit(s) of selection debate. Before turning to these two alternatives, there is one lingering issue that must be addressed: *should group selection be taken seriously?*

VII. WHAT ABOUT GROUP SELECTION?

Note that one could reply that, at best, the above analysis has only shown that genes may not be the *primary* units of selection and the individual could reasonably be viewed as the *primary* unit of selection. It has not ruled out the possibility that groups can be the *primary* units of selection. In fact, Okasha uses some of Dawkins' own views about genes as part of revealing the plausibility of group selection. He tells

---

appear to support it. See the riveting for-and-against exchange between Sapienza (2010) and Burian (2010) for more details and arguments.

12 In fairness to Sterelny, he has distanced himself from his earlier genic selectionism argument he defended with Kitcher. Notably, in the light of D.S. Wilson and Sober's (1989, 1994) attempts to resuscitate both the parameters and plausibility of group selection as a genuine mechanism of evolutionary change, Sterelny (1996) acknowledges that group selection (at least some version of it) should be taken seriously.

us that Dawkins' own claims appear to work against his obstinate defense of the gene's-eye-view. Specifically, he observes that Dawkins concedes that there was a time when replicators functioned independently of each other, but later assembled together due to the benefits of conglomeration. Yet, as Okasha stresses (2005, p. 1015), such assembly represents the very sort of group selection process that Dawkins rebuffs. If Okasha is to be taken seriously, then as it stands, the primary unit of selection could be either the individual or the group depending upon what evolutionary time period is emphasized. This is a reasonable reply, but some critics continue to resist group selection. These detractors draw upon two arguments that they think help validate the implausibility of group selection as the *primary* (or even possible) unit of selection.

To start, it is important to get clear on what group selection is. *The general idea is that there is a group adaptation, which is a property of the group that benefits the survival and reproduction of the group as a whole.* As Robert Brandon explains, "group selection is natural selection acting at the level of biological groups. And natural selection is the differential reproduction of biological entities that is due to the differential adaptedness of those entities to a common environment" (Brandon, 1998, p. 183). For example, as pointed out in Section I, Wynne-Edwards (1962) argued that some species curtail their reproduction in relation to local food supply. If all individuals in a group reproduced at their maximum capacity, then their offspring would consume all the food supply in the local area. As a result of overconsumption, the group could suffer extinction (see Wiens, 1966). The group could avoid such a fate by collectively restraining their reproduction. This would suggest that there is a group-level property/trait related to population and consumption control. Notice that this rendering of group selection gives over-arching value or relevance to **(Q3)** and **(Q4)** in this debate. With this account of group selection in place, we can look at how it makes sense of **(Q1)**–**(Q4)** in Table 3.4.

QUESTIONS OF IMPORTANCE	UNIT THAT ANSWERS EACH QUESTION	DEGREE OF RELEVANCE
(Q1): What is the *interactor?*	Groups and/or Individuals	Less Relevant
(Q2): What is the *replicator?*	Individuals or Groups	Less Relevant
(Q3): What is the *beneficiary?*	Groups and/or Individuals	More Relevant
(Q4): What entity *manifests adaptations?*	Individuals and/or Groups	More Relevant (See Below)

Table 3.4 *Group Selection and the Unit(s) of Selection*

The Case against Group Selection? The Arguments from Speed and Heritability

Traditionally understood, however, natural selection with respect to individuals does not favor reproductive constraint. An individual that increases its reproduction will automatically have an advantage relative to individuals who have fewer offspring. Within a group, if some individuals produce more offspring than others do, the former will proliferate. Three questions emerge from this individual vs. group scenario:

1. Can individual selection within the group be overcome by selection between groups?

2. Is there a strong likelihood that selection between groups can out-compete selection within groups?

3. Even if group selection is possible, how strong of a process is it?

The answer to the first question is that it is theoretically possible for selection between groups to out-compete selection within groups so long as (i) migration rates are low and (ii) group extinction rates are high. For example, a group of rabbits that controls its reproduction can continue to maintain this behavior so long as (i) migration into the population by those individual rabbits who reproduce at maximum rates is limited (because such reproducers will quickly take over the population in favor of individual selection) and (ii) the group of rabbits dies off quickly so that another group of rabbits can continue the same birth control behaviors. If the group of rabbits survives too long, then invasion from high-level breeding rabbits is inevitable, eventually eliminating the group-level reproduction-constraint property/trait.[13]

The answer to question 2 is that it is not likely that selection between groups can out-compete selection within groups. Moreover, the answer to question 3 is that, even if group selection is possible, it is a rather weak force. There are two arguments—*The Argument from Speed* and *The Argument from Heritability*—that justify these answers to questions 2 and 3. A look at each argument will reveal that **(Q3)** and **(Q4)** of **Table 3.4** cannot be answered in favor of the group. The upshot is that it is difficult to view groups as the *primary* or possible unit of selection.

---

13 See Sober and Wilson (1998) for a defense of group selection along these lines. See also Okasha (2001) for an alternative non-group selection reading of Sober and Wilson's account. I highly recommend this Okasha article for the technical and philosophical details surrounding the group selection debates.

The Argument from Speed

The first criticism is that group selection is a weak force compared to individual selection, because of the slow life cycle of groups with respect to individuals. Individuals reproduce at a much faster rate than groups do. As a result of relatively fast reproduction, individuals can migrate in and out of different groups. The result is that, at any given time, individual adaptations will dominate over any group level trait. In the rabbit-reproduction example, it is very unlikely that the group level property/trait of controlling reproduction could ever survive because groups live too long compared to individuals. Either from new offspring or invasion from outsider rabbits, individuals who reproduce at maximum rates would eventually take over the group before it could die out and form a new group. Thus, in answer to question 2, it is unlikely that between group selection could out-compete with individual selection within groups. Consequently, those who might claim that groups are the *primary* unit of selection would have trouble with this objection.

Reflecting upon all this in terms of altruism vs. selfishness might help. In terms of selfish individuals (e.g., maximum reproducing rabbits) infiltrating a group of altruists (e.g., limited reproducing rabbits), Ridley (1996, p. 326) argues that even if group selection is a process in the natural world, its success would be very short-lived, because "selfish" individuals reproduce much faster than groups of altruists reproduce. Thus, although group selection could be maintained for a few generations of groups, such maintenance is fleeting. What this point reveals is that, given long enough time scales, the reasonable answer to **(Q3)** is that the individual (not the group) is the entity that actually is benefited and that the answer to **(Q4)** is that individual adaptations will eventually trump group-level adaptations. It is this short-lived aspect of group selection that, according to Ridley, renders it a weak force compared to individual selection. So, if Ridley's reply is on the mark, then it is unreasonable to think that the *primary* or genuinely possible unit of selection is the group.

The Argument from Heritability

The second argument against group selection is based on rejecting the idea that there are *heritable features* distinct to groups that are not simply features of individuals that comprise the groups. This line of criticism rejects the relevance of **(Q4)** with respect to this debate or reveals that the individual is the reasonable answer to **(Q4)**. As Brandon makes the point, "In order for differential selection to be group selection (i.e., selection at the group level), there must be some group property (the group phenotype) that screens off all other properties from group reproductive success" (Brandon, 1998, p. 184). The problem is that it is very difficult to make sense of a group having a phenotype in any real sense that can be inherited by another group. Moreover, it has to be the case (as Brandon notes) that the group level property can screen-off all non-group level properties from group reproductive success. Yet this is impossible to do, since the group level phenotype really is the composite phenotype of all the individual phenotypes that comprise the group. Put another way, an altruistic group is nothing more than a bunch of altruistic individuals that comprise the group.

For instance, Brandon (1998, p. 184) argues that it is an empirical question as to whether there are group-level properties beyond individual properties, but he is clear that some effect by the group on the individual would have to be at work. Brandon's point is correct, but not only must there be some effect by the group on the individual, but *the effect must also be heritable at the group level*. In the case of groups, there is no obvious or empirically verifiable reproductive mechanism that is able to pass along a group-level advantageous feature, because groups reproduce by way of individual reproduction within groups. This is why the answer to **(Q4)** ought to be centered around the individual as the primary unit of selection or that **(Q4)** is not relevant to group selection because group-level adaptations do not exist. Okasha (2001, p. 46) makes this point by stressing that parties to this group-selection debate rely upon the phrase "group selection," but their analyses assume that all the reproduction of group-surviving properties are replicated and passed on to future groups by way of individual reproduction. This is what the statistical averages really represent. In other words, group selection advocates posit individual altruists passing on their altruistic behavioral tendencies by way of individual reproduction—no group-level reproduction or property is endorsed. Thus, it is this heritability element (along with whether or not groups actually possess properties) that makes group selection an unlikely candidate as the *primary* unit of selection—if it is even a possible candidate at all.

VIII. TAKING NATURAL SELECTION SERIOUSLY: MULTI-LEVEL SELECTION TO THE RESCUE?

What if natural selection is far more dynamic than the accounts so far have suggested? Defenders of multi-level selection (**MLS**) argue that this is definitely the case. They insist that natural selection can function at distinct levels in isolation or at multiple levels simultaneously (Waters, 2005, p. 313). If **MLS** is on the mark, then one significant implication is that trying to locate "the primary" unit of selection has been a mistake all along!

This idea that selection can work at many levels may sound rather enticing, but its credibility is evident only if there is compelling scientific criteria and evidence. Without all this, multilevel selection comes off as rather fanciful talk. Empirically, then, how can we know that selection is occurring at multiple levels? And is there any empirical evidence to match the relevant criteria?

One of the principle defenders of **MLS**, D.S. Wilson, tells us that a comparison of traits at different levels could help vindicate the truth of **MLS**. Wilson's criterion is basically that *multilevel selection should be taken seriously when individual selection models (or some other selection models) lack predictive sway in particular cases* (Wilson, 1997, S127). Imagine a population of rabbits. On a standard interpretation, Wilson is pointing out that individual fitness prevails within any given population. So, for example, if greater ear curvature proves to be an advantage with respect to predator detection, then those individual rabbits that exhibit this feature are likely to survive and reproduce over other individuals in the group that exhibit less ear curvature. If, however, it is determined that superior ear curvature is beneficial to individuals, but yet it is not the feature that evolves across related rabbit populations (the less-curved-ear feature actually is passed on to subsequent populations), then a more predictive model is required. That model, Wilson suggests, could be a **MLS** one in which selection is working at other levels, presumably simultaneously or discretely. For instance, although superior ear curvature has an advantage to the individual in each group, it might prove to be less advantageous at the genetic level or less beneficial with respect to between-group competition or both simultaneously.

Wilson's criterion above is not without merit, but his own use of "maybe" does suggest that his account is speculative—even if multiple comparisons across levels are performed. This means, for instance, even if it is the case that individual selection models may fail to be predictive in particular cases, this does not by itself validate a multilevel selection alternative approach. It could just as well be the case that a genetic account or an alternative individual selection account is more appropriate. Can more be offered to supplement Wilson's **MLS** suggestion? Put another way, even if Wilson has the criterion for **MLS** correct, evidence or a framework for evidence is still largely missing from his analysis. Without evidentiary support, defenders of the traditional views can continue to retreat to their comfy positions which do have a modicum of empirical corroboration.

In what can be viewed as more empirical support for a Wilson-type **MLS** approach, Okasha believes that thinking in terms of the origin of *biological hierarchy* (i.e., "genes are nested within chromosomes, chromosomes within cells, cells within organisms, organisms within groups, etc." Okasha, 2005, p. 1014) requires a **MLS** perspective. Building upon Wilson's and Okasha's efforts, we can introduce an important aspect of the units of selection debate that has been neglected; namely that how we answer the unit of selection question had better be respectful of the fact that such an answer must help to make sense of the nested hierarchy of biological life pointed out by Okasha. As it stands, the existing alternatives in the units of selection debate—gene selection, individual selection, and group selection, fail to pay fully this respect. Part of showing the requisite deference to the nested hierarchy of life requires recognizing that both the recent past and deep geologic past must be woven into a proper account of the units of selection debate. Okasha attempts such a weaving when he distinguishes between multilevel selection 1 (**MLS1**) from multilevel selection 2 (**MLS2**).

MLS1, according to Okasha, helps make sense of adaptations by individual entities competing against other individual entities, but does not account for group-level transitions in the history of life. How, for example, did individual cells become groups of cells? And how did groups of cells become whole individuals? And how did whole individuals become populations of related individuals (and so on)? For instance, **MLS1** explains why the replicative power of individual RNA molecules[14] gave them a distinct advantage over other related molecules interacting with each other (average individual fitness), but it does not account for "sufficiently cohesive and discrete" groups that possess properties/adaptations for "reducing conflict between their constituent particles" (Okasha, 2005, p. 1019).

In the light of Okasha's contribution to the **MLS** discussion, it is clear that the unit of selection could be either individual units or groups. Which unit natural selection favors would likely depend on the kind of biological transition that is under consideration. This suggests, as is noted in Table 3.5 below, that all the different units of selection alternatives are relevant to **(Q1)–(Q4)**. This is so either in the **MLS1** sense of selection occurring at numerous levels simultaneously or in terms of the **MLS2** sense of natural selection with respect to major transitions in the history of life. In terms of relevance, defenders of **MLS** would have no difficulty allowing all the questions relevant to the units of selection debate. Some questions might be more relevant than others (e.g., **(Q4)** more relevant than **(Q1)** with respect to major transitions), but these differences would be determined empirically to the extent that it is possible.

QUESTIONS OF IMPORTANCE	UNIT THAT ANSWERS EACH QUESTION	DEGREE OF RELEVANCE
(Q1): What is the *interactor*?	Genes, Individuals, and/or Groups	Relevant
(Q2): What is the *replicator*?	Genes, Individuals, and/or Groups	Relevant
(Q3): What is the *beneficiary*?	Genes, Individuals, and/or Groups	Relevant
(Q4): What entity *manifests adaptations*?	Genes, Individuals, and/or Groups	Relevant

Table 3.5 *Multilevel Selection and the Unit(s) of Selection*

A Brief Reply to MLS

Although internal disputes will undoubtedly continue to linger, **MLS** appears to be taken rather seriously these days by most scholars (Waters, 2005). This near-consensus, of course, does not mean that **MLS** is not without some difficulties. For instance, if Okasha is correct that **MLS** had better be sensitive to the major transitions in the history of life as well as accommodate "everyday" population dynamics, then something like **MLS2** would need to be incorporated into the **MLS** toolkit. Yet,

---

14 'RNA' refers to ribonucleic acid. This is a macromolecule (synthesized by DNA) that plays a prominent role in the synthesis (by way of translating, decoding, and transcribing) of proteins.

much like the concern with certain definitions of group selection, **MLS2** assumes that group-level adaptations/properties actually exist. If **MLS** is to endorse the existence of group-level properties (like a property of conflict reduction/resolution), then arguments will have to be put forth with respect to their emergence and source(s) of proliferation that do not rely upon the activities of the individuals of those groups (see group selection discussion). Such arguments would likely venture into various sorts of metaphysical discussions that would either vindicate or reject certain definitions of **MLS2**. This suggests that **MLS** must contend with some of the same critical worries that trouble group selectionism. Beyond the group properties worry, there is one additional assumption made by **MLS** (and the other alternatives discussed thus far) that might be a cause for concern in thinking about the units of selection debate. The next section will explore this potentially problematic assumption by way of the *Developmental Systems* alternative.

IX. A SNOWBALL'S CHANCE? THE DEVELOPMENTAL SYSTEMS ALTERNATIVE

So far, all the participants in the units of selection debate, including those who espouse **MLS**, assume or endorse what can be called the "**conventional interactionist view.**" This is the position that biological features are the product of genetic and environmental factors working in concert with each other. As genes go about the business of producing an individual, this production is influenced by many genetic, embryological, and environmental causes. So, the "information" that the gene unleashes is molded into a growing individual as a result of what Gould calls the kaleidoscope of causal factors (see pp. 100–03). The point is that there appears to be at least two distinct sets of "information entities"—(1) the information in genetic material and (2) the information present in the set of non-genetic/environmental factors—involved in the production of an individual. On the one hand, gene-centered scholars, like Dawkins, give added weight to genetic information in defense of the gene as *the primary* unit of selection. On the other hand, individual-centered scholars, like Gould, favor the individual as *the primary* unit of selection due to the kaleidoscope of causal factors involved in the production of an individual. Further still, some argue in favor of group-level features in defense of the group as *the primary* or possible unit of selection. Distinct from these positions is the multi-level/hierarchical selectionist accounts. Those who defend this line of reasoning insist that natural selection does its pruning either simultaneously or solely at particular levels. From this perspective, there is no *primary* unit of selection. Still, regardless of which unit is or is not considered to be *the primary* unit of selection, all the alternatives thus far assume that the conventional interactionist view is correct.

It is this assumption that captures the second criticism of **MLS**. For instance, R.A. Wilson's resistance to **MLS** is based on his view that the levels of selection are *entwined/fused* and are not separable (Wilson, 2003, p. 532). One implication of Wilson's notion of entwinement is that, if it is on the mark, it tells against the conventional interactionist view. If the causal factors that contribute to the production of life are entwined, then talk of interaction (as opposed to entwinement or fusion) must be rejected. It is the denial of the conventional interactionist view by defenders of developmental systems theory (**DST**) that captures the crux of their "radical" defense (e.g., Lewontin, 1985, Oyama, 2000 and Griffiths and Gray, 1994). As a reply to all the alternative theories in the units of selection debate noted above, the creation of an individual draws upon a great many resources, a variety of information, and numerous forces of nature. The truth of this claim motivates, for example, Lewontin to plead that "Darwinism cannot be carried to completion unless the organism is reintegrated with the inner and outer forces, of which it is both the subject and the object" (Lewontin, 1985, p. 106). Lewontin's point, like Wilson's entwinement above, is that making sense of an individual as an evolutionary entity requires understanding it as an entity that is the product of being intertwined around internal environmental forces and external environmental forces. Being both the subject and object suggests that it is a mistake to distinguish the individual from the very forces that aid in its construction. It is this "individual vs. the environment" or "genes vs. environment" sort of dichotomous thinking that directs Griffiths and Gray to vent that **DST** "rejects the dichotomous approach.... *The role of the genes is no more unique than the role of many other factors*" (Griffiths and Gray, 1994, p. 277; my italics).

But why is it unreasonable to carve the biological world into any set of pairs? **DST** champions, like Griffiths and Gray, answer that there is no way to distinguish different sorts of information-contributors such that one can sort out which information-contributor is more valuable or unique than any other. Again, Wilson makes the same point as Griffiths and Gray when he claims: "Entwinement or fusion of properties or processes implies that it makes no sense to apportion determinate, partial causal responsibility for the resulting evolutionary change" (Wilson, 2003, p. 546).

It is the fusion or coalescence of various forces/information (both internal and external to the developing individual) that could be seen as a part of Griffiths and Gray's argument. With the help of Wilson's notion of entwinement, it may be easier to see that the coming together of different biological forces and different sets of biological information can produce unique biological forces and sets of information that cannot be separated in terms of understanding the nature of an evolving and developing being.[15] The important upshot of the **DST** way of understanding biological entities tells against the importance of the unit(s) of selection problem because, other than the developing system, there is no unit or level that is singled out by natural selection. Their argument, **The Information Argument**, can be reconstructed as follows:

---

15   It is not entirely clear if Griffiths and Gray (and other defenders of **DST**) go so far as to endorse Wilson's entwinement/fusion account. Here is a great paper topic for the curious student! For more on **DST** and its historical and influential predecessors, see Griffiths and Tabery (2013) and Griffiths and Hochman (2015).

The Information Argument

P1. If it is reasonable to view some kinds of biological information (e.g., gene, individual, or group) as the *primary* unit of selection, then it is reasonable to privilege some kinds of biological information over other sorts of biological information.

P2. If it is not reasonable to view some kinds of biological information (e.g., gene, individual, or group) as the *primary* unit of selection, then the developing individual and the various sets of information that aid in its formation tell against the fruitfulness of the unit(s) of selection worry.

P3. It is not the case that it is reasonable to privilege some kinds of biological information over other sorts of biological information.

P4. It is not reasonable to view some kinds of biological information (e.g., gene, individual, or group) as the *primary* unit of selection.

C1. Thus, the developing individual and the various sets of information that aid in its formation tell against the fruitfulness of the unit(s) of selection worry.

The crucial premise of **The Information Argument** is P3. Again, why do defenders of **DST** insist that it is unreasonable to privilege some sorts of causes/information over other sorts? The answer is related to the complex mix of causes that tell against any simple "genes + environment = individual" account.

In what can be viewed as a defense of **P3** above, Griffiths and Hochman (2015, p. 2) explain that there is a legitimate alternative to the conventional interactionist view, call it **"life cycle epigenesis."**[16] The basic idea behind this phrase is that a developing embryo unfolds as a result of both stable and malleable causal factors. Notably, these stable and malleable causal factors are present within the growing embryo and the external environment or niche the embryo will eventually encounter (see Griffiths and Tabery, 2013). By analogy, one could argue that the life cycle epigenesis of an

---

16 'Epigenesis' refers to the set of processes that change the way that genes behave without altering the organization or sequencing of genes. These processes could be internal to the individual or external to it. For the defenders of **DST**, the presence of such processes is a reminder that development is not simply what unfolds from some sort of pre-programmed genetic "information," but that there are robust and sundry non-genetic information processes that directly and indirectly influence the activities of genes (see Shea, 2011, for related discussions). These processes within the individual include, though are not limited to, DNA methylation, histone modification, chromatin remodeling, and hydroxymethalation (see Weinhold, 2006). In addition, there are epigenetic processes external to the developing individual like ambient temperature, chemical signals in its local niche, parental behavioral and auditory expressions, etc. (see Robert, 2004 and Halgrimsson and Hall, 2011).

individual is the product of numerous causes like the growth of a small snowball into a large snowball as it advances downhill. There are numerous causal forces at play that account for why a small snowball develops into a large snowball and the various features it has prior to its final characteristics. These factors might include the initial size, weight, chemical structure, and temperature of the snowball in relation to the local temperature, angle of the downward slope, current wind velocity, quantity of snow and debris on the slope, etc. Notice that the final appearance or shape of the snowball is not determined solely by the initial structure of the small snowball. Information must be gleaned from numerous sources *throughout the development* of the small snowball into the large snowball in order to provide an accurate causal account of its overall development. In much the same way, **DST** defenders of life cycle epigenesis argue that a developing embryo has both an internal environment (that is more than its genetic environment) and an external environment outside the individual that contribute to the construction of the embryo life cycle. This is a version of Gould's causal kaleidoscope argument with the explicit caveat that "A key way in which **DST** tries to improve on the truism that there is an interaction between genes and environment is by rejecting that dichotomy itself as inadequate" (Griffiths and Hochman, 2015, p. 2; my bold). Rather, they claim that there is a dynamic interaction of all relevant causal factors such that "development at each stage builds on the results of development at an earlier stage. The components built by interaction at one stage of development are the components which do the interacting at a later stage" (Griffiths and Hochman, 2015, p. 2).

Notice that if life cycle epigenesis of **DST** is correct, it reasonably should include Wilson's entwinement or fusion view of developing biological systems. In the same way that Wilson thinks that it makes little sense to distinguish greater causal responsibility to some sorts of biological factors that contribute to individual form than other factors, **DST**'s rejection of the conventional interactionist view relies pretty much on this same sort of reasoning. We can see this best by way of counterfactual reasoning. If you modify or eliminate any relevant causal factors that contribute to the building of a life cycle, then it will likely be the case that the life cycle will present differently (however subtle that difference might be). By analogy, if you add more snow to the top of the hill or decrease the overall temperature, then the cumulative effects of these sorts of changes on the small snowball will likely result in a snowball with a different form than it otherwise would have. It is the inclusion of entwinement within **DST**'s reliance upon life cycle epigenesis that provides the empirical "punch" that might vindicate the truth of **P3** of **The Information Argument**; that is, if it is true that there is no reasonable way of privileging some causal information over other sets of causal information, then it would ultimately follow that the worry or benefits of engaging the units of selection debate is greatly overstated.

Not unexpectedly, though, one might reasonably ask the following: why is it not possible to give greater importance or higher value to some causes over other causes? This question can be buttressed with the point that causal complexity and life cycle epigenesis can be granted and still allow for careful parsing of some causes over other "peripheral" causes. For example, one could argue that the fact that DNA is a causal difference maker in the sense that changes to DNA can affect the functioning of RNA, but not the other way around, is reason enough to think that the causal influence of DNA is superior to RNA (Waters, 2007).[17] If this is so, then this suggests that some causes are more valuable than other causes, rendering **DST**'s rejection of the conventional interactionist view dubious and revealing the falsity of **P3** of The Information Argument.[18]

DST's counter reply to the above kind of move is to make a distinction between *causal information* and *intentional information*. 'Causal information' refers to physical information between senders and receivers and the role of channels to help ensure the transfer of information. In contrast, 'intentional information' refers to the relationship between thoughts and things or the "aboutness" of something that is present in the thinking of intelligent beings (Sterelny and Griffths, 1999, pp. 100–05). To illustrate the idea behind causal information, consider that a car radio receives information from a radio station or satellite so that certain kinds of programming (i.e., information) can be heard. There is a channel or passage network that connects the radio station to a car radio. The quality of this channel varies based on weather, quality of both radio station and car equipment, etc. Importantly, there is no distinctive information role played by any of these causal factors that are part of the passage network. The radio signal that arrives at your car is influenced by all the causal factors mentioned (and others not mentioned), all of which jointly contribute to the quality of both the sending and receiving of the radio signal. Of course, humans can value some causes more than others, but this value would not be a feature of the information itself. In the same way, genetic and epigenetic causal factors contribute to the developing organism and there is no distinctive information in these various sorts of causal factors that gives priority of some causes over others. Again, humans can value some types of biological information more than others, but this value would not be a feature of the information itself.

In contrast, consider an example of intentional information. An intelligent being can reflect upon a radio station. This being might ponder its appearance, the quality of its programming schedule, other intelligent beings that work in the radio station, the overall equipment needed in the production of radio signals, etc. All of this envisioning and thinking as it pertains to the radio station captures the "aboutness" of the radio station. It is this kind of "aboutness"/intentional information present in the mind of a thinking being, defenders of **DST** argue, that is present neither in the causal processes related to mechanical contrivances or causal processes in biological

---

17 In his discussion regarding redundant complexity in biology, Nathan (forthcoming) discusses implications regarding redundancy in DNA. Interestingly, he argues that redundancy affords a kind of biological stability that tells against simple forms of reductionism and in favor of a sort of causal complexity that might be favorable to the **DST** project.

18 C.K. Waters (2007) offers a very interesting discussion related to this issue, but his focus is on metaphysical differences in causal processes. He does not directly engage the proponents of **DST**. Still, here is yet another gripping essay topic worthy of pursuit!

phenomena. Ultimately, **DST** poses the following conditional challenge to those who would argue that causal priority can be made transparent in biology: *If it is possible to show causal priority in biology, then it must be possible to show that intentional causal information (i.e., "aboutness") is present in some biological causes as opposed to others*. Defenders of **DST** are confident that this challenge cannot be met. As Sterelny and Griffiths summarize, "it is hard to see how intentional information could be a property of physical systems.... Hence arguments for the special status of genes that rely on attributing intentionality to them face a very serious problem" (Sterelny and Griffiths, 1999, p. 105). Thus, defenders of **DST** can claim that short of invoking some kind of special magic mental/intentional power to some biological causes over other biological causes, it appears that trying to carve-up biological causal priority in developmental processes is a nonstarter. The upshot of this line of reasoning is three-fold: (i) the truth of life cycle epigenesis is legitimized, (ii) the truth of the conventional interactionist view is reasoned to be spurious, and (iii) the truth of **P3** of **The Information Argument** is re-established.[19]

In terms of **(Q1)–(Q4)**, defenders of **DST** would either reject the causal priority framework assumed in the questions and therefore consider the entire set-up as irrelevant or insist that the developing system is the correct answer to each question with the degree of relevance being undifferentiated. Assuming the latter response, here is how **Table 3.6** would look:

QUESTIONS OF IMPORTANCE	UNIT THAT ANSWERS EACH QUESTION	DEGREE OF RELEVANCE
(Q1): What is the *interactor*?	The Developmental System	Relevant
(Q2): What is the *replicator*?	The Developmental System	Relevant
(Q3): What is the *beneficiary*?	The Developmental System	Relevant
(Q4): What entity *manifests adaptations*?	The Developmental System	Relevant

Table 3.6 *The Developmental System and the Unit(s) of Selection*

---

19 It should come as no surprise that this particular debate has not been exhausted by this discussion. For instance, some have tried to deflate the importance of the distinction between causal and intentional information by arguing that intentional information, when properly naturalized, is no different than causal information. For a discussion on the naturalizing intentionality debate, see Stich and Lawrence (1994) and their citations—and yet another thought-provoking essay topic!

A Brief Reply to DST

In some ways, **DST** appears to offer an interesting approach to the units of selection issue. It is sensitive to causal complexity like **MLS**, but moves beyond it by embracing the far-reaching view that there is no legitimate way to carve out causal priority in biological processes. This may very well be due to the entwinement aspect of biological systems defended by Wilson and the lack of intentional value in biological processes. Still, there are two lingering worries about **DST** that might make it less than palatable. First, some critics continue to wonder if the world really is so complex such that it is not possible to weigh some causal factors as more important than other causal factors. Secondly, other critics are skeptical about how committed **DST** is to evolutionary biology. Let us turn to this second concern first.

Again, a particular pressing evolutionary concern is the **MLS2** reminder that a cogent answer to the units of selection issue had better be able to account for the nested hierarchy of life. Specifically, some critics charge that exponents of **DST** take themselves to be promoting a theory that is not in keeping with the overall structure of evolutionary theory as a Darwinian selectionist theory. Rather, **DST** (as the name implies) seems to focus on how development of the organism should be understood, but not on the deep historical changes of species or the deep historical changes of particular characteristics of a specific species.

In response, Griffiths and Gray (2005, p. 419) claim that **DST** is sensitive to the evolutionary hierarchical nature of life by making clear that this hierarchy is itself the product of the very selective forces that assist in the creation and replication of the full gamut of developmental systems. What this means is that both the genetic and the internal and external epigenetic causal forces that contribute to the production and replication of developing systems are present as a result of the force of natural selection in both retaining and weeding-out those variations that are beneficial to developing systems. In terms of evolutionary/geological time scales, this retaining and weeding-out process correspondingly helps make sense of the hierarchical nature of life. From this perspective, making sense of evolution and development goes hand-in-hand.

In brief, as developing systems both modify and are modified by their local environments/niches (which are themselves being modified), they will vary as a result of this "entwined or fused" existence (Robert, Hall, and Olson, 2001, p. 956). Natural selection will then retain or weed-out those variations in developing systems that prove to have a reproductive advantage. It is this entwined picture that supposedly helps make sense of the production and variation of the hierarchical nature of life. Thus, the defenders of the **DST** account can claim that it is not in tension with the **MLS2** requirement of accommodating the hierarchical nature of life.

No doubt, more could be said about the quality of the **DST** response to the hierarchical nature of life requirement.[20] At this point, however, we will shift to

---

20   No doubt, this transition to the second criticism is a bit abrupt. In part, the hope is that the reader will continue the exchange. Do you find the response by **DST** to the hierarchical nature

the second criticism, which is linked to the belief that genuine scientific claims have predictive use. This is the idea that a theory is deemed fruitful to the extent that it shows predictive prowess within its own area of inquiry, and may also reveal unexpectedly surprising insights and predictions in other areas of study. Tooby et al. argue that **DST** should be rejected as a genuine scientific theory because it lacks the predictive element that exemplifies good scientific theories. Indeed, for this reason, they claim that **DST** "is useless as a scientific theory" (Tooby et al., 2003, p. 860).

Griffiths and Hochman (2015) reject this rather pungent criticism and offer the following rebuttal in favor of **DST**:

> The fact that it [i.e., **DST**] makes no *specific* predictions is not a problem. The same could be said of evolutionary developmental biology, or of the 'gene's-eye-view'. However, like those other perspectives it [i.e., **DST**] does make general predictions: that development has no single locus of control, that the impact of any given cause is contingent upon the state of the system as a whole, that the notion of the 'genetic program' is of little heuristic value, that the study of heredity will benefit by an inclusive approach.... (Griffiths and Hochman, 2015, p. 5; bracketed addition mine)

This reply by Griffiths and Hochman is more than a little troubling in that it comes off very fast. First, they claim that it is not a problem that **DST** offers no specific predictions because this lack of predictive power could also be ascribed to both evolutionary developmental biology and the "gene's-eye-view." This is an odd way to try to vindicate one's position. One's opponent could easily retort "then down with evolutionary developmental biology and the gene's-eye-view too!" Additionally, Griffiths and Hochman claim that one *could* submit that evolutionary developmental biology and "the gene's-eye-view" do not put forth specific predictions. Now, it is true that one could make such a claim, but would such a claim be true? The point is that Griffiths and Hochman have not shown that such a claim is true. Again, the concern here is that their reply is rather under-developed, rendering it less than persuasive.

Next, Griffiths and Hochman argue that **DST** is a scientific theory by making a distinction between *general predictions* (**GP**) and *specific predictions* (**SP**). Their point seems to be that a theory that offers only **GP** is no less a scientific theory than a theory that provides **SP**. There are a number of problems with their reply. First, without clear definitions of these two distinct senses of prediction, it is difficult to determine the quality of Griffiths and Hochman's reply.[21] Be that as it may, some preliminary definitions may allow for the beginnings of an analysis. Consider what might be a standard account of **SP** followed by a contrived example:

---

of life requirement reasonable? Methinks that another mind-blowing essay topic lurks here ...

21 One could very well imagine that defenders of both the "gene's eye view" and evolutionary developmental biology would be eager to display why their prognostications are in keeping with the demands put forth by **SP**-type statements.

SP → X is a specific prediction if and only if X is inferred from hypothesis H and X either contributes to the validation or nullification of H by way of particular empirical experiments E.

Example → Plumage color combinations in a specific order contribute differentially to reproductive success in male peacocks. It is inferred that peahens are more sexually attracted to the specifically ordered color combinations yellow, red, and green [color scheme 1] than green, yellow, and red [color scheme 2]. The truth of this specific inference can be determined by genetically manipulating certain alleles such that color scheme 1 is altered to color scheme 2 (or females are restricted to interacting only with color scheme 2 males). If females who showed sexual interest in color scheme 1 show little sexual interest in color scheme 2, then the specific inference is validated, giving support to the general hypothesis about plumage color and sexual selection.

The peacock example above illustrates an instance of **SP** because the prediction regarding a particular color scheme and sexual interest is inferred from the more general hypothesis regarding plumage color patterns and reproductive success. Additionally, the possible experiments to validate the particular prediction also substantiate the general hypothesis.

It is difficult to define **GP**, but we can follow the lead of the examples put forth by Griffiths and Hochman. They are as follows:

1. Development has no single locus of control.

2. The impact of any given cause is contingent upon the state of the system as a whole.

3. The notion of the 'genetic program' is of little heuristic value.

4. The study of heredity will benefit by an inclusive approach.

These examples appear to imply that there is a background theory, namely **DST**, that is thought to be true and, in the light of this true theory, a number of *expectations* can be put forth regarding either related areas of study or theoretical principles governing science. If such expectations prove to be true, then the theory, namely **DST**, is vindicated. The following definition and subsequent example seem to capture Griffiths and Hochman's account:

GP → X is a general prediction if and only if X is an empirical expectation E inferred from theory T and E either validates or nullifies the truth of T.

Example → If **DST** is true, it can be expected that there is no master control center in fetal development.

Now, it is beyond the scope of this chapter to explore the nature of scientific explanation. The distinction between **GP** and **SP** underscores a related difference between an **SP** *inference* and **GP** *expectation*. Primarily, 'inference' refers to statements that can be determined to be true by way of well-defined experiments. In contrast, 'expectation' refers to statements whose truth must be determined by way of examining an entire area of study with no patent criteria. For instance, what would reveal that there is no master control center in development? What would reveal that the notion of 'genetic program' has little heuristic value? Without clear criteria, it is difficult to make sense of the scientific value of Griffiths and Hochman's theory, as suggested by Tooby et al.'s reliance on **SP** as the appropriate standard for what constitutes a proper scientific theory.

If Griffiths and Hochman replied by offering a manifest criteria, then it becomes unclear why their use of **GP** does not collapse into **SP**. For example, imagine that developmental biologists hold the view that there are a set of genes that control body plan development. In fact, these genes are called hox genes or homeotic genes. Griffiths and Hochman could propose the following to developmental biologists: "If disrupting some specific non-hox gene environment of a developing embryo resulted in an embryo that developed in unexpected ways, then it is clear that there is no master control center in fetal development and **DST** is true." Yet, proposing a specific non-hox gene environment is more akin to **SP** than **GP**; that is, a specific experiment can be produced. Thus, either **DST** is not of much value to the bench scientists because of its use of **GP** or it is really offering discreet empirically testable inferences very much like **SP**.

Resolution of this debate requires further inquiry into the nature of scientific explanation and further argumentation from the **DST** camp. At the very least, much more needs to be said in terms of the distinction between **SP** and **GP** and why science should be careful to take seriously the postulations put forth by **DST**'s allegiance to **GP**. At any rate, if **DST** champions can successfully defend themselves against these and other objections, then they will deserve a prominent place amongst the units of selection interlocutors.

X. CONCLUSION

This chapter has offered a brief glimpse into the units of selection debate. In order to make sense of this dispute, the following question took center stage: *Does natural selection primarily range over groups, individuals, genes, or something else?* In order to answer this question, some of the core arguments in defense of selection *primarily* ranging over genes, individuals, and groups, were explicated and critically evaluated within the framework of a particular set of questions, namely **(Q1)–(Q4)**. This analysis revealed that each of these options proved difficult to endorse entirely, so the alternatives theories of multilevel selection and the developmental systems were put forth within the **(Q1)–(Q4)** structure. Although the chapter did not reveal any conclusive resolution to this debate, some strengths and weaknesses of each theoretical framework were made manifest. At the very least, interested readers should be able to continue to engage in this fascinating debate and to participate critically in related discussions that have clearly advanced beyond Darwin's thoughtful reflections.

CHAPTER REVIEW: DISCUSSION AND QUESTIONS

1. In the light of Hamilton's genetics findings, why did Williams and Dawkins reject Wynne-Edwards' group selection hypothesis? [pp. 86–87] [Hint: you should be able to trace this brief history yourself.]
2. What is the "unit(s) of selection" question that governs this chapter? [p. 88]
3. Can you briefly explain **(Q1)–(Q4)** of Table 3.1? Why is this table being used throughout the text to aid in making sense of the unit(s) of selection debate? [pp. 90–91]
4. Why does Darwin think that the individual is the *primary* unit of selection? Why does he make room for group selection in this discussion and why does this inclusion of group selection not take away from his view that the individual is the *primary* unit of selection? [pp. 91–93]
5. Can you use your answers in 4 to help make sense of Table 3.2? [p. 92]
6. Why does Dawkins think that the gene is the *primary* unit of selection? Make sure your answer includes **The Precision and Longevity Argument** and is sensitive to the difference between **interactor** and **replicator**. [pp. 95–96]
7. Why does there seem to be a "tie" between the gene's-eye-view and the individual selection perspective? How do the inclusion of "outlaw genes" and **The Distortion Argument** purport to break the tie? [Make sure your explanation connects **The Distortion Argument** to **The Precision and Longevity Argument**.] [pp. 97–99]
8. How does the fallacy of composition reveal an error in reasoning with respect to **The Distortion Argument**? Explain whether or not you are persuaded by this brief objection to **The Distortion Argument**? [pp. 99–100]
9. Explain how Gould's kaleidoscope argument and his argument from causal complexity tell against the gene as the *primary* unit of selection. How do these arguments constitute a rejection of **The Precision and Longevity Argument** and a vindication of the individual as the *primary* unit of selection? Make sure to connect back your answers to Table 3.3. [pp. 100–02]
10. What is genic pluralism? Does this modified version of the gene's-eye-view tell against or support the gene as the *primary* unit of selection? Does genic pluralism re-establish the soundness of **The Precision and Longevity Argument**? [p. 103]
11. What is group selection? What is Okasha's criticism of Dawkins' rejection of group selection? [pp. 103–04]
12. Does the way in which 'group selection' is defined make a big difference? Make sure your answer includes Okasha's criticism of group selection as the activities of individuals within a group. [pp. 105–07]
13. Explain the argument from speed and the argument from heritability. How do these two arguments supposedly tell against group selection? Make sure that your answer includes

the idea of group-level properties, Brandon's criticism of such properties, and how your answer connects to the summary of Table 3.4. [pp. 106–07]
14. Can you provide the general definition of **MLS**? What is the difference between **MLS1** and **MLS2**? Why, according to Okasha, is **MLS2** needed? [pp. 107–09]
15. What criteria are employed by Wilson in his defense of taking **MLS** seriously? Explain whether or not you are persuaded by his account. [pp. 107–08]
16. With **MLS1** and **MLS2** combined with Wilson's criteria, can you explain each of **(Q1)**–**(Q4)** of Table 3.5? [p. 109] [Hint: connect your answers to questions 14–15 to Table 3.5.]
17. What are the two objections to **MLS**? Make sure you include an explanation of the **conventional interactionist view** assumption with respect to the second criticism. [pp. 109–11]
18. What is the **DST** alternative? How are **Life Cycle Epigenesis** and Wilson's notion of *entwinement* being used to make sense of **DST**? [pp. 110–111]
19. Can you explain The Information Argument in your own words? [p. 112]
20. How do Sterelny and Griffiths use the distinction between *causal information* and *intentional information* as further support for **Life Cycle Epigenesis**, a rejection of the **conventional interactionist view**, and a vindication of the truth of **P3** of The Information Argument? [pp. 114–115]
21. Explain why 'Relevance' is included in every box under the "Degree of Relevance" category in Table 3.6? [p. 115] [Hint: bring together your answers from questions 17–20.]
22. How does **DST** contend with the nested hierarchy of life objection (that was previously discussed in the **MLS** section)? [p. 116]
23. Why do some contend that **DST** is not a real science? How do Griffiths and Hochman use the distinction between **SP** and **GP** to reply to those who think that **DST** is not a real science? Explain whether or not you are persuaded by this reply. [pp. 116–19]
24. In the light of the various discussions of this chapter, explain whether or not you think that the governing question of this chapter (*Does natural selection primarily range over groups, individuals, genes, or something else?*) is actually relevant to the unit(s) of selection issue. [Hint: think!]

REFERENCES

Ananth, Mahesh. *In Defense of an Evolutionary Concept of Health*. Aldershot, UK: Ashgate Press, 2007.

———. "Psychological Altruism vs. Biological Altruism: Narrowing the Gap with the Baldwin Effect." *Acta Biotheoretica* 53 (2005): 217–39.

Brandon, Robert N. "Levels of Selection." In *The Philosophy of Biology*, ed. David L. Hull and Michael Ruse, 180–97. Oxford: Oxford University Press, 1998.

Burian, Richard M. "Selection Does Not Primarily Operate on Genes." In *Contemporary Debates in the Philosophy of Biology*, ed. Francisco J. Ayala and Robert Arp, 141–64. Oxford: Wiley-Blackwell, 2010.

Carroll, Sean B. *Endless Forms Most Beautiful: The New Science of Evo Devo and the Making of the Animal Kingdom.* New York: W.W. Norton & Co., 2005.

Crow, James, F. "Genes That Violate Mendel's Rules." *Scientific American* 240, 2 (1979): 134–43.

Cox, T.M. "Mendel and His Legacy." *Quarterly Journal of Medicine* 92, 4 (1999): 183–86.

Darwin, Charles. *The Origin of Species by Means of Natural Selection or the Preservation of Favoured Races in the Struggle for Life.* 6th ed. New York: Mentor/Penguin, 1958.

———. *Descent of Man, and Selection in Relation to Sex.* London: Murray, 1871.

Dawkins, Richard. *The Blind Watchmaker.* New York: W.W. Norton, 1996.

———. *The Selfish Gene.* New Edition. Oxford: Oxford University Press, 1989.

———. *The Selfish Gene.* Oxford: Oxford University Press, 1976.

Deichmann, Ute. "Gemmules and Elements: On Darwin's and Mendel's Concepts and Methods in Heredity." *Journal for General Philosophy of Science* 41, 1 (2010): 85–112.

Depew, David J., and Bruce Weber. *Darwinism Evolving: Systems Dynamics and the Genealogy of Natural Selection.* Cambridge, MA: MIT Press, 1996.

Ghiselin, Michael T. *The Economy of Nature and the Evolution of Sex.* Berkeley: University of California Press, 1974.

Goldstein, Kurt. *The Organism, A Holistic Approach to Biology Derived from Pathological Data in Man.* Boston: Beacon Press, 1963.

Gould, Stephen Jay. "Caring Groups and Selfish Genes." In *The Panda's Thumb*, 85–92. New York: W.W. Norton, 1980.

Griffiths, Paul E., and R.D. Gray. "Developmental Systems and Evolutionary Explanations." *Journal of Philosophy* 91, 6 (1994): 277–304.

Griffiths, Paul E., and Adam Hochman. "Developmental Systems Theory." *eLS* (October 15, 2015): 1–7. DOI: 10.1002/9780470015902.a0003452.pub2.

Griffiths, Paul E., and James Tabery. "Developmental Systems Theory: What Does It Explain, and How Does It Explain It?" *Advances in Child Development and Behavior* 44 (2013): 65–94.

Hallgrimmson, Benedikt, and Brian K. Hall, eds. *Epigenetics: Linking Genotype and Phenotype in Development and Evolution.* Berkeley: University of California Press, 2011.

Hamilton, W.D. "The Genetical Evolution of Social Behavior." *The Journal of Theoretical Biology* 7, 1 (1964): 1–52.

Hull, David "Introduction to Part II." In *The Philosophy of Biology*, ed. David L. Hull and Michael Ruse, 149–52. Oxford, 1998.

———. "Individuality and Selection." *Annual Review of Ecology and Systematics* 11 (1980): 311–32.

Kaebnick, Gregory E., and Thomas H. Murray, eds. *Synthetic Biology and Morality: Artificial Life and the Bounds of Nature.* Cambridge, MA: MIT Press, 2013.

Lewonton, Richard. "Organism as Subject and Object." In *The Dialectical Biologist*, ed. Richard Levins and Richard Lewontin, 85–106. Cambridge, MA: Harvard University Press, 1985.

Lloyd, Elisabeth A. "Why the Gene Will Not Return." *Philosophy of Science* 72, 2 (2005): 287–310.

———. "Units and Levels of Selection: An Anatomy of the Units of Selection Debates." In *Thinking about Evolution: Historical, Philosophical, and Political Perspectives*, vol. 2, ed. R.M. Singh, C.B. Crimbas, D.B. Paul, and J. Beatty, 267–91. Cambridge: Cambridge University Press, 2001.

Lyons, Sherrie. *Evolution: The Basics*. London and New York: Routledge, 2011.

Maynard Smith, John. "Group Selection." *Quarterly Review of Biology* 51 (1976): 277–83.

Merlin, Francesca. "Evolutionary Chance Mutation: A Defense of the Modern Synthesis' Consensus View." *Philosophy & Theory in Biology* 2 (2010): 1–22. DOI: http://dx.doi.org/10.3998/ptb.6959004.0002.003.

Moss, Lenny. *What Genes Can't Do*. Cambridge, MA: MIT Press, 2003.

Mukherjee, Siddhartha. *The Gene: An Intimate History*. New York: Scribner, 2016.

Nathan, Marco J. "Redundant Causality and Explanatory Robustness." *Midwest Studies in the Philosophy of Science* (forthcoming).

Okasha, Samir "Units and Levels of Selection." In *Routledge Encyclopedia of Philosophy*. London and New York: Routledge, 2009. https://www.rep.routledge.com/articles/units-and-levels-of-selection.

———. "Multilevel Selection and the Major Transitions in Evolution." *Philosophy of Science* 72, 5 (2005): 1013–25.

———. "Why Won't the Group Selection Controversy Go Away?" *British Journal for the Philosophy of Science* 52, 1 (2001): 25–50.

Oyama, Susan. *The Ontogeny of Information: Developmental Systems and Evolution*. 2nd ed. Durham: Duke University Press, 2000.

Ridley, Mark. *Evolution*. 2nd ed. Cambridge: Oxford University Press, 1996.

Robert, Jason S. *Embryology, Epigenesis and Evolution: Taking Development Seriously*. New York: Cambridge University Press, 2004.

Robert, Jason S. Brian K. Hall, and Wendy M. Olson, "Bridging the Gap between Developmental Systems Theory and Evolutionary Developmental Biology." *BioEssays* 23, 10 (2001): 954–62.

Sapienza, Carmen. "Selection Does Operate Primarily on Genes." In *Contemporary Debates in the Philosophy of Biology*, ed. Francisco J. Ayala and Robert Arp, 127–40. Oxford: Wiley-Blackwell, 2010.

Shea, Nicholas. "Developmental Systems Theory Formulated as a Claim about Inherited Representations." *Philosophy of Science* 78, 1 (2011): 60–82.

Sober, Elliott *Ockham's Razors: A User's Manual*. Cambridge: Cambridge University Press, 2015.

———. "The ABCs of Altruism." In *Altruism and Altruistic Love*, ed. S. Post, L. Underwood, J. Schloss, W. Hurlbut, 17–28. Oxford: Oxford University Press, 2002.

———. *Philosophy of Biology*. 2nd ed. Boulder: Westview Press, 2000.

———. *The Nature of Selection*. Cambridge, MA: MIT Press, 1984.

Sober, Elliott, and D.S. Wilson, *Unto Others*. Cambridge, MA: Harvard University Press, 1998.

Sterelny, Kim. "The Return of the Group." *Philosophy of Science* 63, 4 (1996): 562–84.

Sterelny, Kim, and Philip Kitcher. "The Return of the Gene." *The Journal of Philosophy* 85, 7 (1988): 339–61.

Sterelny, Kim, and Paul E. Griffiths. *Sex and Death: Introduction to the Philosophy of Biology*. Chicago: University of Chicago Press, 1999.

Stich, Stephen P., and Stephen Laurence. "Intentionality and Naturalism." *Midwest Studies in Philosophy* 19, 1 (1994): 159–82.

Wade, Michael J. "An Experimental Study of Kin Selection," *Evolution* 34, 5 (1980): 844–55.

———. "Group Selection among Laboratory Populations of *Tribolium*." *Proceedings in the National Academy of Science* 73 (1976): 4604–07.

Waters, C. Kenneth. "Causes That Make a Difference." *Journal of Philosophy* 104, 11 (2007): 551–79.

———. "Why Genic and Multilevel Selection Theories Are Here to Stay." *Philosophy of Science* 72, 2 (2005): 311–33.

Weinhold, Bob. "Epigenetics: The Science of Change." *Environmental Health Perspectives* 114, 3 (2006): A160–67.

Wiens, John A. "On Group Selection and Wynne-Edwards' Hypothesis." *American Scientist* 54, 3 (September 1966): 273–87.

Williams, George C. *Adaptation and Natural Selection*. Princeton: Princeton University Press, 1966.

Wilson, David S. "Reintroducing Group Selection to Human Behavioral Sciences." *Behavioral and Brain Sciences* 17, 4 (1994): 585–654.

———. "Altruism and Organism: Disentangling the Themes of Multilevel Selection Theory." *The American Naturalist* 150, S1 (1997): S122–34.

———. "A General Theory of Group Selection." *Proceedings of the National Academy of Sciences* 72 (1975): 143–46.

Wilson, David S, and Elliott Sober. "Reviving the Superorganism." *Journal of Theoretical Biology* 136, 3 (1989): 337–56.

Wilson, Robert A. "Pluralism, Entwinement, and the Units of Selection." *Philosophy of Science* 70, 3 (2003): 531–52.

Wynne-Edwards, V.C. *Animal Dispersion in Relation to Social Behavior*. Edinburgh: Oliver and Boyd, 1962.

Chapter 4

BIOLOGICAL FUNCTION

[I]f a man were to make a machine for some special purpose, but were to use old wheels, springs, and pulleys, only slightly altered, the whole machine, with all its parts, might be said to be specially contrived for its present purpose. Thus throughout nature almost every part of each living being has probably served, in a slightly modified condition, for diverse purposes, and has acted in the living machinery of many ancient and distinct specific forms.

—Charles Darwin (1984, pp. 283–84; quoted from Lennox, 1983)

I. INTRODUCTION: A FASCINATION WITH KNIVES

At the mischievous age of seven, my father gave me my first knife. It was a simple ivory-handled three-inch penknife with a single blade. With close supervision, I was allowed to tinker with it and play good-guy versus bad-guy as my imagination ran amok. As the years went by, I accumulated quite a collection. It ranged from simple single-blade knives to elaborate multi-blade stilettos. Much like my father's elaborate tool set, I came to appreciate that a simple device, such as the knife, could come to have quite an elaborate set of distinct *functions*. This was quite apparent to me at the age of 10 when I would display to my friends, like a peacock showing off his feathers to watchful peahens or a card magician fanning out his deck of cards to an unsuspecting audience, the many unique devices present in my beautiful blood-red Swiss army-knife. From three distinct blades to a spoon, a fork, scissors, two mini-screw drivers (flathead and Phillips) and a corkscrew, I thought that there wasn't much I couldn't accomplish!

Interestingly, my father (the entomologist) tried to explain that the parts of the knife could be understood like some of the hunting/battle working parts of a giant centipede (I later came to find out that he was referring to the giant Amazon/Peruvian centipede—*Scolopendra gigantea*). Much as the knife has numerous devices for various specific purposes, the centipede had a flat-shield covering for head protection; antennae for tactile and general sensory perception; sharp claws at the front and rear for clamping onto prey or clinging onto certain terrain; venom-injector sites on the claws; and chemoreceptors to assist in assessing its ambient environment. Much as the various tools with distinct functions on my Swiss army knife, the numerous parts of the centipede also had distinct functions.[1] This line of reasoning seemed quite good to me and I was quite keen to employ it while showing off my knife to my friends.

As is plainly demonstrated by my father's centipede analogy and from Aristotle's discussion (ch. 1), explanations about functions can move seamlessly from human artifacts to biological systems. Broadly, of course, there is little dispute about how to make sense of the use of most of the things humans make. For example, a car has a function, again broadly, associated with transportation; scissors have a function of cutting; sunglasses have a function of blocking or reducing both the glare and ultraviolet rays of the sunlight; and my various Swiss army-knife accessories have their distinct functions. Part of how we know what is the function of these artifacts is that we are able, by direct or indirect means, to find out what particular designers intended with respect to the given artifact they designed. For instance, we simply ask the Swiss army-knife maker what purpose, function, or solution he intends for the various accoutrements he has fashioned on this particular kind of knife. Alternatively, a bit of fancy reverse engineering could allow us to infer what the reasonable functions of the knife fixtures might be. As we learned, Aristotle employed this kind of reasoning described by Lewens in his analogy from artifact functions to organism functions. My 10-year old self, with the aid of my father, did the same.

There is, however, a *prima facie* difference between artifacts and biological systems in that the latter display an array of *self-maintenance* activities that seems to be mostly absent in the former.[2] First, and most obviously, as part of sustaining itself, a biological organism can self-replicate itself (entirely or partially) by way of sexual or asexual reproduction. This sense of self-maintenance can be viewed as retention of one's "genetic-self" by means of reproduction. This genetic retention of self does not require a separate designer, like an artifact necessitates in terms of its reproduction. Second, biological systems manifest self-maintenance by way of self-restoration. By degree, of course, biological systems can heal their own wounds or fend off parasites and sometimes regenerate body parts (e.g., star fish, salamander, and worm). In sharp contrast, artifacts do not exhibit this variety or scale of self-restoration. Third, organisms exhibit a range of malleability—both internal to the system and the system as a whole—with respect to external stimuli. From something as simple as moving into the shade on a hot day to clever deception tactics or group predatory strategies, organisms can produce fairly novel behavioral responses that appear quite improvisational. Very little of this sort is put forth by artifacts.[3]

---

1   For those interested in this bat-catching centipede, see Molinari et al. (2005).
2   For a much more in-depth discussion regarding the relationship between living systems and artifacts, see Lewens (2004).
3   No doubt, as advances in technology march onward (e.g., artificial life-robotics, bio-technology, synthetic biology, artificial intelligence computer programs, etc.), we may likely have to re-think, in rather subtle ways, how to distinguish biological and non-biological systems with respect to

Taken together, these aspects of biological systems require a correspondingly sensitive concept of function; a rendering that can legitimately ignore being sensitive to the functionality of artifacts. What this means is that reliance on artifacts to make sense of biological function should be down-played considerably. This suggests that Aristotle's and my father's artifact analogies, Darwin's clever machine-parts metaphor in the blocked quotation at the beginning of this chapter, and my Swiss army-knife example lose some of their explanatory luster. So, if we grant that the self-maintenance differences between biological systems and artifacts are differences that make a difference, then we must search for an alternative way of making sense of organic/biological function that is not strongly tethered to artifact function.[4]

II. FUNCTION TALK AND TALK OF TELEOLOGY

As noted above, then, biological systems present much organization and flexibility by way of a repertoire of self-maintenance attributes. From their internal workings to their gross behaviors, one is hard pressed not to ascribe a level of organization that approaches a sense of goal-directedness or purposefulness to most of these activities. (We saw, in chapter 1, Aristotle attempt to champion something like this with his account of final causation.) Consider the following examples and ways in which this presumed goal-directedness is expressed:

(1) *The function* of the heart is to pump blood.

(2) The liver is *designed to* detoxify blood.

(3) Iduronate sulfatase is an enzyme that is present *for the sake of* metabolizing old connective tissue.

(4) *The goals* of bird feathers are to ensure thermal regulation and flight.

(5) Worker bees *contribute to* the cleaning, nursing, security, and nourishment of their respective bee hives.

(6) The kidneys have *the purpose* of eliminating waste from the blood and regulating fluid levels in the body.

---

    self-maintenance. Some of these issues are discussed in Boden (1990 and 1996) and Pennock (2011).

4 This distancing of biological function from artifact function stands in sharp contrast to Wright's (1973) pivotal work on this topic. For a good collection of articles on the function debate, see Ariew, Cummins, and Perlman (2002) and Buller (1999). Also, two excellent articles on the function debate are offered by Wouters (2003a and 2003b).

'The function,' 'designed to,' 'for the sake of,' 'the goals,' 'contribute to,' 'the purpose' are sample locutions of how we commonly express our understanding of the intricate and organizational patterns that are awash in the biological world around us. The German philosopher, Immanuel Kant, was clearly convinced that biological entities should be understood as functional entities; that is, they are beings that are goal-directed and this goal-directedness is limited to their natural station in life—no other sort of entities, divine or otherwise, is needed to understand their natures. From this perspective, biological organisms have teleological features because such goal-directed features are the product of the natural world. This natural origin of goal-directedness is why Kant thinks that the science of biology can defend a teleological understanding of biological phenomena (Kant, 1908–13, pp. 375–76; drawn from Grene and Depew, 2004, p. 94).

At least one additional reason (beyond organizational and ethological considerations) for the attribution of function *qua* teleology to biological systems is utility; that is, most biological activities of organisms appear to be useful to organisms. Indeed, from the workings at the cellular level to the activities at higher levels, it is plausible to think that biological systems, both parts and wholes, have *functions* in the sense of being goal-directed with respect to bringing about useful survival and reproductive enhancing benefits. The developmental biologist, C.H. Waddington, hints at this connection between utility and goal-directed benefits by noting that the living organism is exceptional in the sense that it displays self-maintenance and continued life by bringing about a "genetic self" via reproduction (Waddington, 1957, p. 2; drawn from Lewens, 2007, p. 529). What is interesting about his claim is that he treats this goal of reproduction as the general goal underlying organism activity. What this suggests is that the concept of biological function must include an over-arching goal of survival-and-reproductive success (or usefulness/benefit if you prefer). For example, it may be correct to claim that the function of the heart is to pump blood, but a complete account of the function of this body-part must include its general goal. In this case, pumping blood is for the sake of survival and reproductive success.[5] Thus, if any part or behavior of an organism is given a specific set of functions, then the additional functions of survival and reproductive success must be tagged on so as to capture the fact of this presumed goal-directed usefulness aspect. Let us call this view the *dual functionalism* account of biological function. We will return to it shortly.

Yet, with respect to explanation, once divine influence from above is discarded and any sort of divine-soul from within is purged, how does dual functionalism gain any credibility? The answer to this question has frequently revolved around explanations of a Darwinian flavor. Biological systems are the product of natural selection, which

---

5   Note that William Harvey, the distinguished and celebrated seventeenth-century English medical doctor, was able to explain the circulation of blood by assuming the Aristotelian teleological doctrine that *nature does nothing in vain* (Harvey, 1928 [1628], p. 129). Although he did not employ a Darwinian-type natural selection teleological account, he still embraced a goal-oriented perspective when thinking about biological/natural processes. See Ribatti (2009) and McMullen (1995) for brief discussions of Harvey's remarkable findings.

retains many features because of their reproductive and life-sustaining effects. The problem, however, is that there is little agreement amongst biologists and philosophers of biology as to what counts as a reasonable concept of biological function. This chapter offers some of the major theories of biological function and their shortcomings. In response, it will be argued that a mixed-evolutionary concept of biological function, which is respectful of dual functionalism, proves to be most compelling.

III. BASIC CRITERIA FOR FUNCTIONAL ANALYSIS

Before turning to specific theories of biological function, let us understand the criteria for what counts as a good theory of biological function. Although the analysis to follow is not likely to be endorsed by all, it gives us a decent framework by which to make sense of the function debate. Reasonably, we may come to find out that some of these criteria may be well worth snubbing. For now, though, a cogent theory of function should be able:

1. *To answer the following questions*: (1) "Why is this feature present in the first place?" [Presence] (2) "Why does the feature persist in the organism?" [Persistence] and (3) "Why does the feature (or some variation of it) continue to be present in successive generations?" [Perpetuation]

It is one thing to be able to explain the possible gene mutation and the subsequent genetics associated with what the kidney does or what chemical interactions are facilitated by a particular enzyme. This would be an answer to (2). It is another thing to be able to convey why the kidney or the enzyme is present *in the first place* in the body of many biological organisms and their successive generations. Answers to questions (1) and (3) require invoking both natural selection and goal-directedness (Pearlman, 2010, pp. 59–60). As Aristotle and Kant have made clear, entities/species and their parts, which are the product of nature, have purposes that facilitate their survival and reproduction. To this end, determining X's function requires X's specific function and X's overall contribution to survival and reproduction—that is, dual functionalism. The point is that there is a mechanistic component to making sense of biological functions *qua* biological traits and a developmental aspect. The former is about explaining why a trait is currently present in an organism. This will require a clear genetic account. The developmental part reminds the reader that a function is not merely present in organism X. Rather, slight variations on such a function can be found in other related future generations and an account of biological function must be given to make sense of such continued fruitfulness, integrity, and replicability.

2. *To distinguish genuine functions from side effects.*

Not only is it the case that the heart pumps blood, but it also makes a thumping sound when it is pulsating. Even if we grant that the thumping sound can be diagnostically useful to a doctor, it is reasonable to think that the pumping of blood is a genuine function of the heart. The pulsating/thumping sound, in sharp contrast, is a mere side effect. A persuasive concept of biological function should be sensitive enough to rule out mere side effects from genuine effects (Lewens, 2007, p. 531). It is these genuine effects that get to be called "functions."

3. *To account for malfunctioning.*

Since an organism and its parts produce effects that have an advantage for an organism, a cogent concept of biological function should allow for the fact that the organism and its parts can fall short of producing the requisite effects. The reasons for failure include injury, genetic abnormality, disease, etc. (see Millikan, 1989 and Neander, 1995). To illustrate, there may be scenarios in which the heart does not pump enough blood through the body or the foraging bee does not perform its food location dance upon its return to the hive. In either case, *the part or the organism is not doing what it is supposed to do.* In short, there is a *malfunction* of the part and the organism. As Perlman stresses, "Any successful theory of the source of function needs to account for this possibility of malfunction (Perlman, 2010, p. 54)."[6]

4. *To exclude conscious design—divine or otherwise.*

Once it is granted that life on Earth is the product of hundreds of millions of years of evolutionary forces, it is then those forces alone that should take center stage in our explanations. Of course, this leaves open the possibility that a biological account of function is compatible with a conscious design concept of function, but the former does not require the latter (Wright, 1973, pp. 142–43). Indeed, as we saw in ch. 2, in his *Origin*, Darwin was at pains to show that natural selection is a rather plausible force of nature in terms of making sense of the unfolding of life *without* the need to invoke divine conscious influence. Let it be granted, then, that a plausible concept of biological function should *not* include conscious design. Notice that a further implication of this conclusion is that human contrivances, namely artifacts, must also be excluded in making sense of biological function.

See Table 4.1 on the next page for a brief table summary of this section.

IV. SYSTEMIC FUNCTIONALISM

Because of their resistance to embracing evolutionary analysis as a focal point, there is one important family of naturalistic attempts to make sense of biological function—let us call this set *systemic functionalism*—that requires our attention (this will also help make sense of the above four-fold criteria). The prominent defender of this anti-evolutionary naturalistic concept of function is Robert Cummins.[7] He explicitly points out that "functional analysis can properly be carried on in biology quite

---

6   For further support of the normative view of functions, see Neander (1995). For a dissenting view as to why evolutionary/historical view of function cannot actually satisfy the malfunction criterion, see Davies (2000).
7   For more on systemic functionalism, see Cummins and Roth (2010), Boorse (2002), and Davies (2001).

CRITERIA	EXPLANATION
Presence, Persistence, and Perpetuation	A persuasive concept of biological function should be able to make sense of why a given trait (i) is present at all in an organism, (ii) continues to persist in an organism, and (iii) manages to perpetuate itself in future generations.
Genuine Functions vs. Side Effects	A persuasive concept of biological function should be able to distinguish genuine functions from side effects or accidents.
Malfunctioning-Normativity	A persuasive concept of biological function needs to accommodate malfunctioning by acknowledging that biological functions are normative; that is, there is a set of activities that are *supposed to be* performed by the biological functions.
Conscious Design	A persuasive concept of biological function offers an analysis of function that stays clear of endorsing conscious design—be it divine or otherwise.

Table 4.1 *Four-Fold Criteria for Biological Function*

independently of evolutionary considerations: a complex capacity of an organism (or one of its parts or systems) may be explained by appeal to a functional analysis regardless of how it relates to the organism's capacity to maintain the species" (Cummins, 1975, p. 756). On Cummins' view, if a component X causally contributes to system S's performance of Z, then X's contribution constitutes a function of S's performance of Z (Cummins, 1975, p. 765). The heart, for example, is a feature of the circulatory system. This system, which includes the lungs and blood vessels, works to keep about five liters of blood continuously moving through the body in a constant exchange of waste-filled blood for oxygen-rich blood. So, the heart has the function of pumping and exchanging good and bad blood through the body because it causally contributes to the circulatory system's overall task of moving blood throughout the body. We can formally make sense of systemic functionalism as follows:

Systemic Functionalism: A feature X has a function in system S if and only if activity Y of X causally contributes to S's overall capacity/performance of Z.[8]

So, the foraging bee's (X) food location dance is a function of the bee hive (S) if and only if the food location dance (Y) of the foraging bee (X) causally contributes to the hive's overall performance of maintaining the well-being (Z) of the bee hive community. Or, bird feathers (X) have a function of flight in the life of a bird (S) if

---

8   See Cummins (1975, p. 72) for his own formal account.

and only if thrusting and lifting (Y) of feathers (X) causally contributes to the bird's overall capacity for life-sustaining movements (Z).

Objections to Systemic Functionalism

With respect to our criteria, what are we to make of systemic functionalism? First, although systemic functionalism can likely account for why a biological feature is currently present in a system, it cannot account for why it originally came into existence and why the feature persists in successive generations. Again, to the extent that dual-functionalism should be taken seriously, Cummins' account addresses *the what-X-does* aspect (mechanistic aspect) of biological function, but disregards the *why-X-does-what-it-does* aspect (teleological/developmental aspect) of biological function. The upshot here is that systemic functionalism cannot account for why a feature of an organism originally was retained and perpetuated in successive generations.

Not surprisingly, Cummins and Roth (2010) have a reply to this sort of concern in which they distinguish teleology from neo-teleology. Basically, Cummins and Roth approve of function by natural selection so long as the feature under consideration is not already thought of as a functional trait prior to a selection history. It is this attribution of function to a feature that has not gone through the rigors of natural selection to which Cummins and Roth consider neo-teleology objectionable. For instance, given Cummins and Roth's distinction, we should not attribute flight as a function of early or proto-wings of birds. Rather, such an attribution should be given only after the flight mechanism has proven itself to be a survival and reproductive benefit for many generations/life cycles.

This appears to be a considerable concession on the part of Cummins and Roth. So long as evolutionary function attributions are appropriately sensitive to the relevant selection history, they are willing to consider those features as genuine functions. (This is an important distinction made by Cummins and Roth and, as we will see, proves to be a problem for the propensity interpretation of function. This will be discussed below.) It is clear, then, that Cummins and Roth are not entirely hostile to an evolutionary perspective on biological function depending upon how natural selection is employed by the Darwinian. To this extent, their analysis can handle the first criterion pertaining to the origin and successive presence of a biological trait.

Surprisingly, though, Cummins and Roth then proceed to offer an appeal to "common sense" over an appeal to evolutionary selectionist accounts. Their argument relies, in part, upon a knowledge problem; namely, that it is difficult to know what features were under selection millions of years ago. So, rather than wait on speculative answers from evolutionary biologists, Cummins and Roth submit that drawing upon what seems obvious to be a function should guide our judgments (Cummins and Roth, 2010, pp. 77–78).

Granting that it can be a serious chore to determine selected effects from non-selected effects, does it follow, however, that we should then rely upon our common sense function attributions? Cummins and Roth give the impression that it is relatively obvious and transparent to determine the function attribution of some traits and it would be silly to wait for biology to confirm our intuitions. As they put it, "It seems obvious that wings are for flight. But it is not at all obvious that, for example, sparrows, or birds in general, have wings because they enabled flight in an ancestral population" (p. 77). So, they think that we should stick with systemic functions which are "ubiquitous, important, and deeply entrenched" (p. 77).

Now, Cummins and Roth make a good point here, but over-state their position. It is true that it can be difficult to determine whether or not a particular trait is present as a result of selective advantage or is merely fortunate enough to be linked genetically to some other trait that has a selective advantage. What this means is that we may have to advert to DNA-sequence research, phylogenetic analysis, and nifty reverse engineering coupled with genetic tinkering to determine whether or not a particular trait is the product of natural selection.[9] What this suggests is that we may have to wait for the relevant information from biologists rather than rely on common sense impressions about function. For the sake of this first criterion, then, Cummins and Roth appear to waffle back-and-forth about whether or not an evolutionary selectionist account should be the favored methodology for determining biological functions. The effect of this vacillation is that it is not clear that they have offered an account of function that can make sense of the origin and successive presence of a trait.

Importantly, at this juncture of the discussion, we should be a little cautious with respect to endorsing Cummins and Roth's reply. They move from the obviousness of certain common sense function attributions to a rejection of evolutionary function attributions, on the grounds that the findings of the latter's methodology, is somewhat suspect. This gives the impression of an uncharitable handling of evolutionary functional analysis—especially since they just conceded the cogency of certain evolutionary function ascriptions! First, science is not always transparent or obvious in either its methodology or its findings. Cummins and Roth may be correct about the obviousness of the function of bird wings, but not all character traits admit of such "transparent" assessment. In fact, the findings of science are frequently difficult to understand and, at times, flat-out counter-intuitive. Indeed, biological findings have made common sense judgments look rather puerile, since natural selection is forced to work with and modify existing structures. The result is that a feature of an organism that appears to have an "obvious" function may not have a function at all! This is exactly the case of the cave salamander's eyes. One's common sense response to seeing this species' eyes would be to claim that they are for seeing. It turns out, however, that they are vestigial organs and no longer have a function.[10] Thus, even if Cummins and Roth are correct about the tentativeness of some function determinations in evolutionary biology, it does not follow that we should be quick to embrace

---

9   See Rosenberg (2006, chapters 7 and 8) for a summary and citations of recent work done in DNA-sequence research.

10 For an accessible discussion of this species of salamander, see Dawkins (2009, pp. 351–55). Also see Jeffery (2005) for a more technical developmental biology discussion of this same sort of vestigial-eye example as it pertains to the Mexican cavefish (*Astyanax mexicanus*).

those common sense function ascriptions that are "ubiquitous, important, and deeply entrenched." At the very least, Cummins and Roth would need more than this sort of reply to tell decisively against an evolutionary selectionist account of function.

Let us turn to our second criterion of being able to distinguish genuine effects from side effects. In general, a lot of criticism has been levied against systemic functionalism's handling of this concern. The problem, say the critics, is that systems are bombarded with all sorts of influences that have an impact on a system *to some degree*. The implication is that those who defend systemic functionalism are forced to accept that all sorts of side effects of a system, which do not contribute to the system's overall capacity/performance or do so rather insignificantly, are genuine functions. For example, the heart's thumping sound is a capacity of the heart, but the system to which the heart contributes, namely the circulatory system, performs the task or has the capacity for blood circulation. The heart-thumping sound does not contribute to this overall performance, so this capacity should not be considered a function. Yet, the systemic functionalists *appear* to be forced into accepting *the thumping sound* as a genuine function of the heart.

Cummins and Roth counter this criticism by simultaneously (1) modifying the systemic functionalism account and (2) happily endorsing this implication of not being able to locate *the function* of an entity (Cummins and Roth, 2010, p. 82). First, they relativize/contextualize the system under consideration so that if medical diagnostics is the relevant system, then the heart thumping sound has a function for that system. In contrast, if the circulatory system is the relevant system, then heart thumping does not have a function. From this perspective, then, a feature and its effects can be either genuine functions or genuine side effects depending upon the relativized/contextualized system under consideration. Second, since containing systems can have numerous properties/capacities, it is possible for related sub-systems to contribute to numerous properties/capacities. Reasonably, then, these sub-systems will have an assorted set of functions. It is this multi-function aspect of systems that drives Cummins and Roth to reject the possibility that a given feature of a system has a single and unique function. Thus, Cummins and Roth can claim that they have responded appropriately to this criticism of genuine functions and side-effects by rejecting it as a legitimate criterion for functional analysis.

Notice that, with respect to this criticism related to genuine functions and side effects, Cummins and Roth *now* wish to distance themselves from intuitions. In fact, they concede that biological findings can be rather counter-intuitive and not track common-sense notions about biological function very well (Cummins and Roth, 2010, p. 83). So, when the criticism is about the origin of functions and their passage from one group to the next, Cummins and Roth are quick to endorse the obvious or common sense notions of function. When, however, the discussion shifts to the criticism regarding genuine functions and side effects, they want to distance themselves from common sense intuitions regarding function. Clearly, Cummins and Roth conveniently help themselves to common intuitions when they agree with their account and conveniently denounce common intuitions (in favor of a kind of counter-intuitiveness akin to some of the findings of science) when they hinder their account. They appear to endorse openly this approach when they claim that selectionist accounts are "misleading and not well-poised to interface with the rest of science and common sense" (Cummins and Roth, 2010, p. 80).

It is difficult to see how Cummins and Roth can have it both ways. The problem is not the multiple-functions account. As we will see, this is a plus with respect to their analysis. Rather, they draw on the insights of science or abandon those very same insights based on how they correspond to their intuitions regarding the nature of biological functions. Yet, their concession that science can be rather counter-intuitive clearly tells against giving too much weight to intuitions regarding biological function.

Additionally, the fact that a feature can have numerous functions does not forestall the possibility of a given feature having no function. Even an evolutionary account can accommodate a multi-functional body part and still claim that a specific activity of the very same multi-functional body part is a side-effect. In contrast, it is not clear that the systems account can do this because it does not offer any substantive restraints regarding its relativism/contextualism. For example, a function of the thumb can be to strike the spacebar of a computer or to give a vote of confidence (as in "thumbs up") when one has successfully completed a task or is about to embark upon a task. Either of these thumb activities can count as genuine functions, for systemic functionalists like Cummins and Roth, so long as one can locate the relevant system. Yet, it would be better to say that not all uses of a feature get to count as functions because not every set of circumstances should be considered a legitimate system. Thus, in their excitement to capture the importance of multi-functional entities, Cummins and Roth have opened the flood gates to all sorts of inappropriate function ascriptions rather than observing that some features are side effects. As it stands, this is a failing of systemic functionalism. So, although systemic functionalists, like Cummins and Roth, have tried to circumvent the genuine functions/side effects objection, our discussion has revealed that it is not one that they are able to evade very well. The further upshot is that systemic functionalism is not able to address persuasively or accommodate this second criterion.

We can now move to the third criterion: malfunction. Recall that the concern is that a serious account regarding the nature of functions must be able to accommodate the demand that genuine functions are normative; that is, to be a function is for X to do that which is supposed to be done by X. And when it occurs that X does not do what it is supposed to do, X is malfunctioning. As Davies illustrates, "A token mammalian heart that is unable to pump blood in the normal way as a result of some sort of defect nevertheless is 'supposed' to do so. There is a specific standard or norm, imposed by the selective efficacy of ancestral tokens, that applies to this token despite its current capacities" (Davies, 2000, p. 20).

Can systemic functionalism accommodate the normative dimension of malfunctioning? Perlman answers this question as follows: "[M]any find failure to account for malfunction to be a glaring shortcoming of systemic functions.... [W]e often seek to explain the behavior of a biological system in terms of success or failure, and failure often turns on whether its systems function properly or malfunction" (Perlman, 2010, p. 63). Perlman's point can be expressed by the following argument:

> **P1.** If an integral part of X's successfully doing Y is partly understood in terms of X's possible failure to do Y, then malfunctioning is an integral aspect of the proper function of X.
>
> **P2.** An integral part of X's successfully doing Y is partly understood in terms of X's possible failure to do Y.

> **C1.** Thus, malfunction is an integral aspect of the proper function of X.

If we assume the truth of **P1**, how are we to make sense of **P2**? To start, lurking behind the language of success or failure is a more fundamental point pertaining to the retention of a property even when the physical instantiation of it cannot be successfully actualized. *The question is: can a functional property continue to be present in, for example, the heart if the heart's physical task of pumping has been compromised?* Those who defend malfunction answer "yes" and those who reject the importance of malfunction say "no." Consider one of the proponents of systemic functionalism, Christopher Boorse. Boorse thinks that this worry about malfunction is a red herring (that is, he would think that **P2** above is false). He demonstrates this by way of the following heart example:

> If Carla's heart cannot pump blood, then pumping blood is not, in fact, the function of her heart; it has no function. Since blood-pumping is the normal function of a human heart, it would be the function of Carla's heart if Carla's heart pumped blood normally; but it does not, so it is not. (Boorse, 2002, p. 89)

Boorse uses the Carla example to show that malfunctioning reveals a loss of function as opposed to it revealing an integral part of the nature of function. Carla's defective heart, according to Boorse, does not have a function and is not supposed to do anything. From this perspective, we can infer that Boorse rejects the notion that a property can be retained after its corresponding physical task is lost.

The first response to the heart example is to point out that success admits of degree. The heart can pump blood optimally or it can fail to pump blood at all. These are the limiting cases of which a spectrum exists. If this is granted—and it is hard to see why one would object to this point—then it is also reasonable that the function of blood pumping is still present in cases of sub-optimal blood pumping. Only when cases under consideration reveal a degree of blood pumping that compromises the organism's overall fitness can it be claimed legitimately that the function of blood pumping is no longer present. What this suggests is that Boorse's use of "normal" is too coarse of a description of how to think about the heart's blood pumping function; for even sub-normal blood pumping may be good enough with respect to overall organism fitness. In these sub-normal cases, the heart is best understood as malfunctioning as opposed to non-functioning. Thus, it is not a stretch to claim (contrary to Boorse) that the functional property of blood pumping is lost only when the pumping of blood is significantly compromised. The overall implication is that this "normal function" aspect of Boorse's systemic functionalism is less than desirable in the light of his rejection of malfunction.

Does the above reply to Boorse reveal the truth of **P2**? I do not think it does; for, in a sense, the above account actually shows that there is some range of malfunction that constitutes non-function. This is all that systemic functionalists require to argue that it is the case that functional properties are lost when a certain range of sub-normal performance is present. If this is correct, then systemic functionalists can conclude that either the "supposed to" norm attached to functions persists only to the extent that its corresponding physical task persists well enough or they can conclude that the "supposed to" norm is merely a figment of the imagination of those functionalists that simply cannot let go of a recondite teleology (see ch. 1). Either route appears to suggest that **P2** is false. So, it seems that systemic functionalists have the upper hand regarding normativity and biological function.

In response, Perlman (2010, p. 67) uses the evolution of eyes to defend normativity and malfunction. He remarks that eyes—even a blind person's eyes—are *supposed to* aid in visual perception.[11] This task is what they were selected to do and account for why they are currently present. It is the Darwinian selection history and not some arcane teleological magic that accounts for this. To see this, think of Perlman's argument as a kind inference to best explanation. He is pointing out that it would be very odd, in terms of energy and resource expenditure, for eyes and their accompanying complex optic nerves and visual centers to be present and not be there for some purpose—even if that purpose cannot be actualized. We could infer that these visual parts are not supposed to do anything, but that does not seem like a reasonable inference to make in the light of the costs to have such features. Thus, contrary to both Boorse and Cummins and Roth, Perlman concludes that the malfunction aspect—not doing what X is supposed to do—of biological function cannot be dismissed on both Darwinian and complexity grounds. The implication, thinks Perlman, is that systemic functionalism does less well than selectionist accounts at making sense of malfunction. It should be clear that Perlman is pushing hard for the truth of **P2**!

One final point should help to illustrate why Perlman may have the edge in this debate and why **P2** is likely true. Consider the human appendix and the eyes of a blind person. In the case of the appendix, it is a feature that once had a function, but no longer has a function. This vestigial organ, which once aided in the digestion of certain types of grasses, is present as a mere remnant of an organ that once had a

---

11   Perlman is specifically responding to Cummins and Roth (2010).

function. So, in the case of the appendix, it does not have a function because it not only cannot perform the task it once could, but it also cannot realize that ancient function given its developmental history. There is nothing that the appendix is supposed to do in its present form, because its present physiological state does not map onto its functional historical form at all. In contrast, a blind person's eyes cannot perform their function presently, but their developmental history reveals that they are not a remnant of a once functioning feature. Rather, the eyes of a blind person reveal a defect of a feature that would otherwise realize its use if not for the defect. This is not so in the case of the appendix. From this perspective, we can understand Perlman to be suggesting that a worthy concept of biological function should be able to accommodate why a vestigial organ does not have a function and why a malfunctioning organ still retains its developmental (i.e., what it is supposed to do) function. This can only be accomplished by giving malfunction its due as part of the concept of biological function. It does not seem to be the case that systemic functionalists can make this accommodation very well. In fact, it must treat the appendix and the eyes of the blind person in the same way. Thus, the force of this distinction between vestigial organs and non-functioning organs should propel us to endorse the truth of **P2**, understanding that we would do well to weave malfunctioning/normativity into our concept of biological function.[12]

Let us complete our assessment of systemic functionalism by looking at the last criterion; namely, that the concept of biological function should exclude conscious design or intentions of any sort. Can the systems approach handle this? It is not clear that it can or wants to, at least on Cummins and Roth's account. Recall, since locating a legitimate functional entity is merely a matter of relativizing/contextualizing to whatever system is invoked, systemic functionalism must allow the religious system and any mechanical system (be it human or otherwise) as legitimate systems from which function ascriptions can be made. If Cummins and Roth can allow (as we saw above) the diagnostic system of medicine to be a legitimate system, then it seems *prima facie* rather difficult to put constraints on what counts as a legitimate or illegitimate system. Yet, as Darwin tried so hard in the *Origin* to make clear (see ch. 2), a naturalistic concept of X is one that necessarily excludes divine or intentional content. This is precisely what systemic functionalism cannot exclude and is deficient to this extent.

In summary, systemic functionalism appears to struggle in its attempt to address the criteria of biological function. At first glance, in the light of the analysis above, systemic functionalism cannot (1) account for why a feature has a function in the first place or the origin of a feature, (2) distinguish genuine functions from side effects, (3) account for malfunctioning, and (4) provide the conceptual constraints needed

---

12   There should be enough analysis here to give the reader a glimpse into the malfunction part of the biological function debate. Much more can and should be said, but stopping at this point hopefully will motivate instructors and students to push the analysis further. A good place to keep going would be Neander (1995).

to cease the inclusion of divine and/or intentional design as part of its concept of function. After examining Table 4.2 as a reminder of this section, we will turn to these alternatives posthaste.

CRITERIA	EXPLANATION
Presence, Persistence, and Perpetuation	Although systemic functionalism can account for why a trait continues to persist in an organism, it is neither able to account for why the given trait is present at all in the organism nor why the trait manages to perpetuate itself in future generations.
Genuine Functions vs. Side Effects	Although some defenders of systemic functionalism take pride in not meeting this demand, nevertheless, systemic functionalism is not able to distinguish genuine functions from side effects.
Malfunctioning/ Normativity	Systemic functionalism's defenders take pride in not endorsing the normative "supposed-to" component to biological malfunctioning, while the critics openly assert that this resistance to malfunction is a serious limitation to this theory.
Conscious Design	Since systemic functionalism relies upon a perspectival/ contextualist methodology, it is in no position to dismiss or block conscious design—divine or otherwise—as a legitimate context from which to make function ascriptions.

Table 4.2 *Four-Fold Criteria and Systemic Functionalism*

It is worth reminding the reader that the above table represents the conclusions that have been drawn from the brief arguments offered in this section. This table does *not* represent a knock-down final assessment of systemic functionalism. Rather, it should be viewed as both an opportunity for interested parties (hopefully our students!) to continue the exchanges initiated herein and an invitation to explore some of the other biological/natural accounts of biological function in the light of the above *prima facie* objections to systemic functionalism.

V. VARIETIES OF EVOLUTIONARY FUNCTIONALISM

Given that functions need only range over biological systems (as opposed to including artifacts or conscious design more generally), which are the product of millions of years of evolution, evolutionary approaches to the concept of function are worth pursuing. To this end, a general description of *Evolutionary Functionalism* will prove useful. Ernst Mayr's position is a good place to begin. To illustrate, consider the heart. According to Mayr, it is not only important to understand the details of *how* the heart is able to perform the pumping that it does, but a complete causal understanding of such a phenomenon requires that a persuasive defense of *why* it is the case that the heart makes manifest such behavior be provided (Mayr, 1974, p. 108). With respect to the heart, the answer a Darwinian like Mayr would give is that the blood-pumping activity of the heart is a behavior that is retained by natural selection not only because it circulates blood, but also because it ensures the survival and reproduction of the organism and the species to which it belongs (goal-directed element). Notice that Mayr is defending a dual functionalism account (see beginning of this chapter). Such an account, thinks Mayr, completes a causal explanation of why a feature is currently present in an organism and the reason why such a feature can correctly be thought of as a function for the organism. The formal definition of Evolutionary Functionalism is as follows:

> **Evolutionary Functionalism**: A feature **X** has a function in an organism **O** if and only if activity **Y** of **X** produces effect **E** because **Y** and **E** were naturally selected (over some other causes and effects) to bring about the goals **G** of survival and reproductive success of **O**.

For example, the liver has a function of blood detoxification in mammals, because the activity of converting ammonia into the less toxic compound urea produces the effect of detoxified blood. Moreover, this activity and effect were naturally selected for the sake of survival and reproduction.[13]

Note that there are more subtleties to Mayr's account than are captured in the above definition. Specifically, how to understand 'selection' in the above definition is not made clear. Are features currently selected? Were they only selected in the past? Will they be selected in the future? All of these questions will be addressed in the discussion that is to follow. Specifically, two philosophical approaches to understanding Evolutionary Functionalism will be distinguished: (1) the backward-looking *Etiological Evolutionary Functionalism* and (2) the forward-looking *Propensity Evolutionary Functionalism*. Additionally, this section will argue that neither of these versions of biological function is adequate, but that a worthy alternative, (3) *Mixed Evolutionary Functionalism*, is available.

Etiological Evolutionary Functionalism

On this interpretation, a feature performs a function in a system of an organism if and only if (1) the feature's presence in a system was useful with respect to the organism's reproductive success in previous generations and that (2) it is the result of evolutionary selection forces. Ruth Millikan stands as one of the prominent defenders of Etiological Evolutionary Functionalism. She argues that a trait is functional because its presence

---

[13] For those interested, there is a growing literature regarding the inclusion of evolutionary thinking into the fields of medicine and public health. See Ananth (2016) for this discussion.

is due to its ability to produce, in a self-sustaining fashion, a beneficial difference that related traits were unable to produce (1993, p. 38).[14] The result is that, in contrast to side effects or lucky features, what counts as a functional trait is one that can re-cycle itself as a result of delivering a reproductive advantage to the organisms of which it is a part. Similarly, Valerie Hardcastle articulates this version of Etiological Evolutionary Functionalism by noting that it is a kind of "backward-looking" functional explanation "for why something is there in natural systems by picking out the property of that thing most valuable to previous generations plus some selecting mechanism" (1999, p. 32; see also Wright, 1976, p. 84). The formal definition looks like this:

Etiological Evolutionary Functionalism: A feature X currently has a function in an organism O if and only if activity Y of X produces effect E because Y and E were naturally selected (over some other causes and effects) to bring about the goals G of survival and reproductive success of O.

With this general description in place, a specific example is in order. In humans, iduronate sulfatase is the lysosomal enzyme that is designed to breakdown mucopolysaccharides (a gel-like substance found in the body of cells). For example, connective tissue outside of cells needs to be replaced on occasion. When this replacement occurs, iduronate sulfatase metabolizes the old connective tissue. On occasion, in males only, a genetic error occurs such that not enough iduronate sulfatase is present to breakdown the mucopolysaccharides that build-up from the remaining old connective tissue. The result of the build-up of mucopolysaccharides (in lysome cells) is the following multi-system collapse: hyperactivity, aggressive behavior, coarse facial features, enlargement of internal organs, dwarfism, stiffening of joints, progressive deafness, and severe mental retardation. This genetic disease is known as Hunter syndrome.[15]

From an Etiological Evolutionary Functionalism perspective, the function of these iduronate sulfatase enzymes is to metabolize mucopolysaccharides, because, ancestrally, there was selection pressure in favor of them doing just this to ensure survival and reproductive success. In severe cases, human males who either lack iduronate sulfatase enzymes or do not produce enough of them have multi-system dysfunction, rendering them physically unfit.

A quick glance at the etiological account might move one to consider this version of Evolutionary Functionalism credible. For, as part of its content, it appears to include both the necessary and sufficient conditions for what it means for X to have a function. As John Bigelow and Robert Pargetter affirm, "The big plus for the etiological theory is that it makes biological functions genuinely explanatory, and explanatory in a way most comfortable with the modern biological sciences" (1987, p.

---

14   For a similar etiological account, see Neander (1991a and 1991b).
15   Note that there are two forms of Hunter syndrome. There is a severe form that occurs in juveniles (between ages 2–4) and a mild form that occurs in early adolescence (between ages 5–10). A more detailed account of Hunter syndrome can be found in Contran et al. (1989, pp. 149–51) and the following internet website: http://www.nlm.nih.gov/medlineplus/ency/article/001203.htm.

187). Moreover, it provides a general framework for distinguishing genuine functions from mere accidents, because a feature of a biological system is a function if and only if it is the product of natural selection.

REPLIES TO ETIOLOGICAL EVOLUTIONARY FUNCTIONALISM

Against Etiological Evolutionary Functionalism, scholars have offered four objections. First, some argue that the etiological account presupposes a rather unduly simplistic view of the evolution of traits. Specifically, as Hardcastle (1999, p. 32) has pointed out, the etiological account cannot guard against Panglossian-like "just-so-stories" about how a given feature *qua* trait X is retained as the fittest in a population as a result of the advantage conferred on O as a result of the fitness advantage had in an ancestral population.[16] There are two points related to Hardcastle's criticism that require our attention. The first point is related to a fear about letting "Mother Goose" in the house of biology. Specifically, given the combination of deep historical time and complexity, it is no easy task figuring out the potentially very odd and complex ways bodies have been put together. We can tell various tales from where and about how eyes and noses came, but linking these stories with the potentially tortuous meanderings of "real" genes, phenotypes, and behavioral repertoires is a tall order. Moreover, she points out that there are other ways (pleiotropy, random drift, etc.) that features could have been perfected to their local environments such that natural selection of a feature cannot be taken for granted.[17] From this perspective, Hardcastle is reminding us that we need to set aside our puerile imaginations and distance ourselves from taking the facile story-telling approach regarding the complexity and fitness of biological systems.[18]

---

16 'Panglossian' refers to the character, Dr. Pangloss, in Voltaire's satirical novella *Candide*. Dr. Pangloss would contrive all sorts of positive interpretations, which would strain credulity. For example, he claimed that the nose has the design that it does have for the purpose of wearing eyeglasses. It is just this sort of absurd story that we do not want in our explanation of biological features. To offer such an account about the nose, for example, would be to advance a "Panglossian just-so-story." For a philosophical discussion of all this, see Radner and Radner (1998).

17 'Pleiotropy' refers to distinct phenotypic effects produced by a single gene. For example, some chickens exhibit frizzle feather trait. This condition is the curling and outward and upward production of feathers (the feathers do not lie flat on the body). Additional pleiotropic effects of this condition, caused by a single deletion in the genomic region coding for *a-keratin*, include delayed sexual maturity, excessive food consumption, and increased metabolism. 'Random drift' refers to the presence of more genes left behind by some individuals in a population than other individuals—just a chance event. The result is that the next generation of genes will include some genes that are present, not as a result of more fit individuals, but by luck, from comparatively less fit individuals.

18 Note that, in his critique of the Wright-Millikin-Neander etiological concepts of function, Buller (1998) argues this point of Hardcastle in detail. Briefly put another way, 'random drift' refers to the loss of variation that can occur by chance or by unpredictable changes of alleles. See Sober (1984, p. 34, fn. 11) for more on random drift. 'Pleiotropy,' again, is a term used to describe the fact that a single gene can have multiple phenotypic effects or a cluster of genes may have several effects on a given phenotype. See Rosenberg (1985, p. 237) for more on this

Part of Hardcastle's criticism is uncharitable to the etiological account, which is not necessarily committed to the optimality interpretation of natural selection. The etiological account recognizes that not all traits are selected because they are optimal for a particular environment; rather, many traits are selected because they are superior or "just better" than other traits in contending with the existing environment. So, with respect to her claim that the etiological account assumes "a natural selection of the fittest," Hardcastle overstates her criticism.

Moreover, it is important to note that the *adaptationist program* need not be abandoned (although the optimality condition as a general requirement should be cautiously embraced) in the light of epistemic limitations related to the complexity of historical entities like organisms. To illustrate, Lewontin (1978, p. 230) stresses that it is safe to assume that most features of organisms are adaptive functions, even if particular features may not be adaptive. The reason is that there is no way to test whether or not a feature is the product of chance, but it is possible to test whether or not a feature is an adaptive function.[19] So, on the one hand, Hardcastle claims that epistemic constraints make it the case that the adaptationist program should be abandoned. On the other hand, Lewontin submits that epistemic constraints only give further support in favor of the adaptationist program.

No doubt, Hardcastle's concern about story telling in biology should be taken seriously, but is not a knockdown argument against the adaptationist program. Rather, it should be seen as a warning that the assumption that a given feature is an adaptive trait is tentative until empirical justification is forthcoming. (This is the same sort of response that was given to Cummins and Roth earlier in this chapter.) Thus, Hardcastle's first criticism is not as successful as she insists and does not refute the etiological account of function, which relies on the adaptive history of traits.

Second, some have argued that the reasoning behind Etiological Evolutionary Functionalism is circular.[20] In general, a charge of circularity is the suggestion that one is attempting to defend or explain a particular conclusion by employing the very conclusion to be defended as a premise. Again, Hardcastle (1999, p. 33) notes the circularity criticism of the etiological account. She summarizes that the proponents of Etiological Evolutionary Functionalism include the historical function of X as part of their account for the historical selection for X. Restated, the reason why X is currently present is because X had a function at some time in the past that was selected because of the function it had. For example, to argue that feathers currently function to help ensure thermal regulation in birds, because thermal regulation was

---

technical term. Finally, 'exaptation' is the term used to suggest that it is possible for phenotypes to be co-opted from their present function in order to serve some other distinct function. See Gould and Vrba (1982, pp. 4–15).

19 For an instructive discussion of the criteria (e.g., correlation between character and environment, results from altering a character, comparison of naturally occurring variants, etc.) needed to determine whether or not a feature is an adaptation, see West-Eberhard (1998, pp. 8–14).

20 This criticism has been offered by Bigelow and Pargetter (1987, p. 190) and both Hardcastle (1999, p. 33) and Bechtel (1985, p. 150).

selected for this function in the past, already builds into the analysis the idea that feathers had this function, for which they were then selected. The conclusion is that those who favor Etiological Evolutionary Functionalism are smuggling into their defense the very thing they are trying to defend.

Etiological Evolutionary Functionalism can avoid this difficulty by making it clear that physiological feature X is a function of organism O if and only if X is selected not because it has a function, but because it is useful to the survival and reproduction of O and X is retained and passed on (genetically or behaviorally) to successive generations. On this version of Etiological Evolutionary Functionalism, the initial retention of X is not present because it already has a function, but because it is useful to the organism. It is only after X has been retained and transmitted to many successive generations that it acquires a function. For example, feathers acquire the function of flight only after it becomes present in successive generations as a result of survival and reproductive advantage. Beth Preston stresses this historical point when she notes that Etiological Evolutionary Functionalism "is a strong one, in the sense that a protracted period of time is required during which the performance of the trait is tested against alternatives and found successful, thus ensuring its own reproduction" (Preston, 1998, p. 227). Thus, the combination of the usefulness of a feature and its successive propagation through reproductively established families tells against Hardcastle's charge of circularity.

Third, some argue that Etiological Evolutionary Functionalism is erroneously committed to the view that advantageous acclimations cannot be functions, because it allows only those activities that have a particular evolutionary history to be genuine functions (McLaughlin, 2001, p. 90). In reply to those who subscribe to Etiological Evolutionary Functionalism, imagine a possible world in which the human lungs successfully acclimate to an atmosphere that has suddenly changed from being oxygen-rich to being almost entirely carbon dioxide-rich. The function of lungs before the atmospheric shift was to inhale oxygen and exhale carbon dioxide, but in the new environment the lungs function to inhale and exhale carbon dioxide. According to Etiological Evolutionary Functionalism, this new activity of the lungs cannot be a genuine function of the lungs because it is not the product of natural selection. The point of this counterfactual is that the etiological account is forced to accept the counterintuitive view that many beneficial acclimations that help to keep an organism alive are not functions.[21]

This third criticism seems to be the most damaging, but it is not clear that it is. What seems to be assumed within this criticism is that this counterintuitive implication of the etiological account renders it implausible. As McLaughlin claims, "The evolutionary etiological view asserts—counterintuitively—that newly advantageous traits have no functions" (McLaughlin, 2001, p. 134). Yet, intuitions do not resolve this issue as we saw with the systemic functionalism discussion in the previous section.

---

21 This example is a variation on one offered by Margolis (1976, pp. 238–55) and this point about confusing function and evolution is offered by Bechtel (1985, pp. 131–70). This concern is also captured by the standard "swampman" counterfactual (see Davidson, 1987).

For example, imagine the first few instances in which birds used feathers for flight. The evolutionary etiologist would argue that these early instances of flight do not confer the function of flight on feathers, because such features have not been screened by natural selection—that is, they are not yet heritable. This sort of reasoning would apply to the claim that the thumb has the function of pressing the space bar on a keyboard. This activity is beneficial, but it is not the function of the thumb, because such an activity is not part of a natural selection history. Only after such an accidental beneficial feature has been part of many cycles of generations and confers a fitness advantage is it the case that it becomes a genuine function. In this way, the etiologist can claim that this account best explains how to distinguish genuine functions from mere advantageous acclimations or side effects. It is able to do this based on its reliance on evolutionary theory, not intuitions about what appears to be a function and what does not appear to be a function. Thus, the etiological account can simply "bite the bullet" that mere advantageous acclimations, which do not have an evolutionary selective history, are not functions.

The fourth criticism is that proponents of Etiological Evolutionary Functionalism must attribute functions to those features that no longer have functions. For example, the appendix, which is part of the digestive organ known as the caecum, once had the function in early mammalian evolution to aid in digestion of certain plants with low nutritional value.[22] In hominid evolution, however, the appendix is still present but no longer has this digestive function; that is, it has not been maintained by natural selection. The problem for the etiological account is that it must accept that the function of the appendix still is to digest plants with low nutritional value, because that is what its evolutionary causal history designed it to do. Yet, it is clear that the human appendix is an evolutionary remnant that no longer has a function. The point is that, although Etiological Evolutionary Functionalism can distinguish genuine functions from accidents, it cannot distinguish between features that have genuine functions and those that once had, but no longer possess, genuine functions.

This fourth criticism is one that the etiological account cannot evade. By relying solely on evolution, the etiological account must ascribe functions to those features that no longer have functions. As Nissen correctly remarks, "Since history is forever, if functions are determined by their history, functions are forever. New functions can be added, but old ones never die. That means that vestigial organs still have their original functions" (Nissen, 1997, p. 185). Thus, Etiological Evolutionary Functionalism is triumphant in distinguishing genuine functions from accidents because of its reliance on evolutionary causal history, but such an achievement proves to be a somewhat pyrrhic victory.

There is one final concern that needs to be noted. Some argue that Etiological Evolutionary Functionalism is committed to defining functions in terms of actual

---

22   The caecum is a pouch-like portion of the digestive tract (near the end of the large intestine) that connects the small and large intestines in humans and many other animals. For more on the function of the appendix, see Nesse and Williams (1994, pp. 129–30).

reproductive success (Bigelow and Pargetter, 1987, p. 190). As it is described in this section, it is ambiguous whether or not such a commitment is entailed.[23]

Clearly, this may be a serious problem with versions of the etiological account that are committed to the view that Y is a function of X if and only if Y contributes to *the actual* reproductive success of O. This leads to the absurd implication that only those things that actually perform their activity are functions. Imagine a case in which bananas are dangled above the heads of two monkeys. Assume that the hands of monkeys have evolved to grasp things; that is, grasping is an activity that contributes to reproductive success. Further imagine that as one monkey goes to grasp the dangling bananas, a lightning bolt strikes its hands, rendering them physically incapable of grasping. The other monkey, however, moves forward and grasps the bananas and proceeds to eat them. On the etiological account, the lightning-struck monkey's hand does not have the function to grasp, while the other monkey's hand does have the function to grasp, because the former monkey's (and not the latter's) hand can no longer actually contribute to the reproductive success of the monkey. In this case, it is obvious there is no difference between the two monkeys except for this accidental event, which is not in any way related to the adapted nature of the monkey's hand. The point is that it is absurd to confer a function on an organism or take away a function from an organism based on lucky or unlucky anomalous environmental perturbations, which are not part of the normal environment in which the organism has evolved.

It is this sort of example that lead Bigelow and Pargetter to claim that the "etiological theory is mistaken in defining functions purely retrospectively, in terms of actual survival" (Bigelow and Pargetter, 1987, p. 191).[24] So, if it is the case that the etiological account is committed to actual reproductive success, then this criticism is quite relevant. The point that can be taken away from this criticism is that it is better to say that Y is a function of X if and only if Y has *the capacity* its ancestors had to contribute to the reproductive success of O. By substituting 'capacity' for 'actuality,' the problem of environmental anomalies disappears. This point is explicitly captured in the propensity interpretation of function (see next section below).

In terms of the four criteria, Etiological Evolutionary Functionalism is able to account for why a functional trait is present in the first place. This is done by establishing a legitimate comparative selective history and revealing the likely selective advantage the specific trait provided for the system of which it is a part. Furthermore, Etiological Evolutionary Functionalism is able to distinguish genuine functions from side-effects because of its reliance on natural selection. Additionally, it openly endorses normativity regarding how a functional trait *should* perform its task in the light of the historical advantage of performing its task. Moreover, by drawing on Darwinian tenets, Etiological Evolutionary Functionalism neither endorses nor includes any sort

---

23 A version of this criticism can be found in Smart (1963, p. 59).
24 Note that Bigelow and Pargetter are aware that this criticism is not relevant to every version of the etiological account, but might be appropriate for some versions.

of divine or intentional causal factors in its account. Of course, one could claim that "divine causation or divine grace" is lurking behind all of this stuff,[25] but that is no fault of Etiological Evolutionary Functionalism's strict reliance on non-intentional natural processes. Table 4.3 offers a summary of all this.

CRITERIA	EXPLANATION
Presence, Persistence, and Perpetuation	By drawing on natural selection, Etiological Evolutionary Functionalism is able to offer plausible answers to the presence, persistence, and perpetuation of functional traits.
Genuine Functions vs. Side Effects	Because of its reliance on evolutionary history and natural selection, Etiological Evolutionary Functionalism is able to distinguish genuine functions from side effects.
Malfunctioning/ Normativity	Since an evolved feature captures what it means for a trait to do what it is "supposed to do," Etiological Evolutionary Functionalism is able to accommodate the normative dimension of malfunctioning.
Conscious Design	By giving causal priority to evolutionary forces, Etiological Evolutionary Functionalism offers a naturalistic concept of biological function that eschews conscious design.

Table 4.3 *Four-Fold Criteria and Etiological Evolutionary Functionalism*

Etiological Evolutionary Functionalism appears to be a good contender for making sense of biological function. It does a fine job of incorporating the four-fold criteria and handling most of the other objections noted above. Still, recall that Etiological Evolutionary Functionalism cannot handle fully the objection that once a feature has an evolved function it cannot lose the function. This is an odd implication, but one that weighs on Etiological Evolutionary Functionalism's account because of its sole reliance on evolutionary history. Moreover, versions of Etiological Evolutionary Functionalism that rely on actual reproductive success are forced to make erroneous function ascriptions. Additionally, because of its reliance on evolutionary history, there is also the worry that Etiological Evolutionary Functionalism's answers to the four-fold criteria are tentative at best. Of course, the quality of the biological data employed would go a long way in mitigating these worries. At any rate, there is enough concern regarding Etiological Evolutionary Functionalism to move forward and determine how well the propensity account holds up under our philosophical microscope.

---

25   This maneuver has been attempted by Pope (2008).

Propensity Evolutionary Functionalism

As a way of both maintaining Evolutionary Functionalism and contending with some of the criticisms put forth against Etiological Evolutionary Functionalism, Bigelow and Pargetter advance what can be called *Propensity Evolutionary Functionalism* (1987, pp. 191–94).[26] This account draws on the need to move away from the actual contributions to reproductive success toward the propensity or capacity to contribute to reproductive success (1987, p. 192). For example, the function of iduronate sulfatase enzymes is to metabolize mucopolysaccharides and not some other substance found in the body of cells, because creatures whose mucopolysaccharides are broken down by iduronate sulfatase enzymes have a greater disposition of surviving and reproducing than creatures whose mucopolysaccharides cannot be metabolized.

As Bigelow and Pargetter note, a propensity or disposition "is a subjunctive property: it specifies what will happen or what is likely to happen in the right circumstances, just as fragility is specified in terms of breaking or being likely to break in the right circumstances" (1987, p. 190). For instance, the ear was selected in the past because it enhanced both predator/prey detection and the sense of body equilibrium, both of which confer a survival-enhancing propensity on the organism. So long as similar (not exact) environmental pressures are present, one can be confident that, for example, the ear will continue to confer its survival-enhancing propensity upon tokens of the species at any time in the life history of the tokens. The formal characterization of Propensity Evolutionary Functionalism looks like this:

> **Propensity Evolutionary Functionalism**: A kind of Evolutionary Functionalism explanation that argues that a feature X has a function in an organism O by performing activity Y if and only if Y produces effect E because Y and E confer and will continue to confer a propensity P (within a certain range of environmental pressures) to bring about the goals G of survival and reproductive success of O.

For example, the reason iduronate sulfatase enzymes have a function in the human body is because their activity of metabolizing mucopolysaccharides confers a survival-enhancing propensity on the human body within a certain range of environmental pressures. Moreover, it will continue to confer a survival-enhancing propensity on the human body so long as the same range of environmental pressures is present.

It is this ability *qua* propensity of a feature to actualize a specific task(s) in the presence of a certain range of environmental perturbations that leads Bigelow and Pargetter to consider such a propensity as a survival-enhancing propensity and a biological function. Thus, Bigelow and Pargetter conclude that the function of a feature "generates propensities that are survival-enhancing in the creature's natural habitat" (1987, p. 192). For example, the gene that causes sickle-cell anemia is a highly adaptive feature

---

[26] An earlier version of the propensity view of function can be found in Mills and Beatty (1979).

of the immune system of people, who live in low altitudes, that defends them against malaria.[27] Notice that the natural habitat must include a low altitude, otherwise the harmful sickle-cell anemia effects "kick in." Natural habitat, then, is that environment in which adapted traits *were* selected (within a certain range of variation).

REPLIES TO PROPENSITY EVOLUTIONARY FUNCTIONALISM

Propensity Evolutionary Functionalism faces a number of objections. The first concern is noted by Godfrey-Smith, who claims that the propensity interpretation of function "draws on the historical facts it sought to avoid" (1994, p. 352). The point is that the propensity interpretation is designed to resolve the problem faced by the etiological account of relying on biological history to make sense of what is and is not a function. Indeed, after arguing that it would be a mistake to insist that the existence of functions must be based on the contingent truth of evolution (1987, pp. 188–89), Bigelow and Pargetter claim that their propensity account includes evolutionary history because it is sensitive to a concept of biological function that does not ignore the survival-enhancing benefits related to a past habitat (1987, p. 196). This is why the propensity account is included with the evolutionary functionalism family.

It is this sort of reply that can make sense of Godfrey-Smith's claim that, for Bigelow and Pargetter, it appears that natural habitat "is understood historically" (Godfrey-Smith, 1994, p. 352). Primarily, Bigelow and Pargetter want to make sure they can keep functions and mere accidents (i.e., acclimations) distinct. Yet, they are clearly committed to the view that a feature already has a function *qua* propensity independent of *historical* notions of survival.

Rather than thinking that Bigelow and Pargetter are committed to the historical element they hoped to avoid, it is better to claim that their account is ambiguous with respect to how they understand 'past habitat'; that is, they want to distance themselves from including evolutionary selection mechanisms as part of their propensity interpretation of function. They also *appear* to concede, however, that evolutionary selection mechanisms are part of their propensity account when they claim that some of their judgments about function do not conflict with the judgments made by proponents of Etiological Evolutionary Functionalism. It is this claim that drives Godfrey-Smith to insist that Bigelow and Pargetter are committed to an evolutionary sense of 'past habitat.' Yet, this is not the case. Bigelow and Pargetter are only committed to the claim that their function ascriptions are compatible (i.e., do not conflict) with those function ascriptions that are rendered from an Etiological Evolutionary Functionalism perspective. To avoid this confusion, they need only make clear that 'past habitat' refers to the recent past habitat in which the survival-enhancing propensity is relevant. In fact, their propensity account makes clear that a propensity is understood within the context of a certain range of environmental fluctuations. Radical environmental changes simply mean that a particular trait does

---

27 The details of sickle-cell anemia will be discussed in a later section of this chapter.

not have a survival-enhancing propensity in the existing radical environment. This reply is quite in keeping with their overall account. Thus, although Bigelow and Pargetter are ambiguous in how they understand the term 'past,' it is clear that they can overcome this objection by Godfrey-Smith and embrace (as they would like to do) an evolutionary history-free propensity interpretation of function. The implication is that, contrary to Godfrey-Smith, Bigelow and Pargetter are not committed to the evolutionary history they are trying to avoid. Of course, then one has to wonder how seriously evolutionary history is being taken by defenders of the Propensity account.

The second criticism is raised by Hardcastle, who argues that propensity functional naturalism is open to the very circularity problem that it is designed to avoid. Much like Etiological Evolutionary Functionalism, Propensity Evolutionary Functionalism assumes that evolved functions will continue to be selected for now and in the future. Since, for Bigelow and Pargetter, 'function' means 'propensity to enhance survival,' they are claiming that, for example, the function of the ear will proceed to be selected because it will go on to be selected. As Hardcastle warns, "if we define functions in terms of future selection of [trait] T, then we cannot use functions to account for the future selection of T for that would be using the future selection of T to explain the future selection of T" (1999, p. 34).[28]

Hardcastle's point is that Propensity Evolutionary Functionalism defends a conclusion by incorporating the conclusion within the defense. The difficulty with this criticism is that it assumes that, according to Propensity Evolutionary Functionalism, functions are defined in terms of future selection. This, however, is not the case. The propensity theory is committed to the view that feature X is a function only if X has a survival-enhancing propensity. As Bigelow and Pargetter make clear, "On our theory, the character already has a function, and by bad luck it might not survive, but with luck it may survive, and it may survive because it has a function" (1987, p. 195). Their point is that function *qua* propensity is independent of any sort of selection mechanisms. So, for Hardcastle to insist that the propensity interpretation is committed to defining functions in terms of the future selection of some character is to ignore the priority that Bigelow and Pargetter give to function over selection. Thus, Propensity Evolutionary Functionalism does not fall prey to the charge of circularity levied by Hardcastle.

The third problem with Propensity Evolutionary Functionalism is that it takes function as ontologically prior to selection—a move that raises the question of how they are able to determine what is and is not a function. As Bigelow and Pargetter note, "On our theory, *the character already has the function*, and by bad luck it might not survive, but with luck it may survive, and it may survive *because* it has a function" (1987, p. 192; my italics). Similarly, in response to the etiological account, Bechtel claims, "The correct order is to claim that *those things that are functional will evolve*, rather than to claim that those things that evolve are functional" (Bechtel, 1985, p. 150; my italics).

---

28 For further criticisms of the propensity theory approach, see Mitchell (1993).

The above claims by Bigelow, Pargetter, and Bechtel are problematic. The obvious problem with giving priority to function over selection is that it raises the question of *why* it is the case that X has the function in the first place. X has the function to do Y, because X yields a survival-enhancing propensity on O. But why does X have the function *qua* survival enhancing propensity that it has? Surely, they cannot rely on propensity here, because propensity and function are one and the same once selection is no longer part of the concept of function. That is, if Bigelow, Pargetter, and Bechtel presume (as they do) that character X "already has the function" prior to selection, then this means that X already has a propensity prior to selection. If not selection, then what confers "having a propensity" that makes it the case that X is a function? They could respond by claiming that a propensity is a property or capacity of a trait to do X. Yet, this leads to a regress problem. For now it can be asked, how is it the case that a property or capacity "already" exists in a creature without introducing some sort of causal history to account for the capacity? As it stands, Bigelow, Pargetter, and Bechtel have no answer, because a capacity or property is an unexplained metaphysical element of their analyses. Natural selection, on the other hand, is a physical force or process (like gravity). Are propensities thought to be the same? This seems unlikely, because Bigelow and Pargetter have already ruled out the possibility that propensity relies on contingent natural phenomena. In terms of the function criteria, Bigelow, Pargetter, and Bechtel cannot account for the presence of a trait (see Criterion #1, p. 131). The upshot of this overall objection is that Bigelow, Pargetter, and Bechtel have not offered a persuasive account of what a function is. Thus, it is not at all clear that giving priority to propensity over selection is preferable.

Bigelow, Pargetter, and Bechtel might object to the above criticism by claiming that "What is a function?" is different from the question "What causes functions to be present?" For example, if Bigelow and Pargetter find a watch, they can claim to know it's a watch first and then ask who made it. Obviously, they know that they have to give a causal history to account for its function, but that is different from defining what a function is. That is, they could argue that they do not have to answer the second question in order to answer the first. This reply, however, is vulnerable to the next criticism.

This final objection to Propensity Evolutionary Functionalism is that it cannot distinguish genuine functions from mere side effects. For example, imagine that the metabolic activities of iduronate sulfatase enzymes not only breakdown mucopolysaccharides, but they also have the accidental benefit of improving the sense of smell. On the propensity interpretation, both the metabolic activities and the improved sense of smell would have to be considered genuine functions because the former (directly) confers a survival-enhancing propensity and the latter (accidentally) confers a survival-enhancing propensity.[29] This implication leads McLaughlin to conclude correctly that those who embrace the history-free propensity interpretation of fitness are "forced to

---

29 Of course, this is not to suggest that an improved sense of smell could not be a genuine function. For example, in part, this could be determined by showing that survival and reproductive success (i.e., fitness) decreased as a result of the absence of the improved sense of smell *while* the metabolic waste removal benefits of iduronate sulfatase remained present.

attribute a function to more or less everything" (McLaughlin, 2001, p. 126). Indeed, this criticism reveals why they must address the question "What causes functions to be present?" Thus, Propensity Evolutionary Functionalism should be rejected on the grounds that it cannot distinguish genuine functions from mere side effects.

What about the four-fold criteria? As we just saw, Propensity Evolutionary Functionalism cannot distinguish genuine functions from side effects. Moreover, because a trait has a function prior to selection, the question regarding "why is the trait present in the first place?" still requires an answer. The difference is that the question can be reformulated as follows: "Why is the function present in the first place?" Propensity Evolutionary Functionalism does not offer an answer to this question. Maybe this sort of ontological concern is not important to the Propensity Evolutionary Functionalism project, but it is a question that Etiological Evolutionary Functionalism can answer. Still, because Propensity Evolutionary Functionalism takes selection, environment, and reproductive success seriously, it can account for why a functional trait is able to persist and perpetuate in future generations. In terms of malfunctioning, Propensity Evolutionary Functionalism could claim that any functional trait that performs below its "supposed-to" selected propensity would be malfunctioning and this would be a matter of degree—the lower the propensity, the greater the malfunction. So, it seems that Propensity Evolutionary Functionalism could accommodate the normative dimension to malfunctioning. Lastly, because Propensity Evolutionary Functionalism gives ontological priority to function over selection, it must concede that the presence of a functional trait could be the product of conscious design. The problem is that function is taken as a primitive starting point, rendering it difficult for those who defend Propensity Evolutionary Functionalism to put constraints on the origin of the functional trait. This is not what one would have hoped for in a naturalistic concept of function. All of this is summarized in Table 4.4 on the next page.

Mixed Evolutionary Functionalism

Thus far, the general conclusion is that neither Etiological Evolutionary Functionalism nor Propensity Evolutionary Functionalism is able to emerge unscathed en route to its respective concept of biological function. Etiological Evolutionary Functionalism suffers from focusing on actual reproductive success and not being able to allow an entity to lose its function, whereas Propensity Evolutionary Functionalism cannot distinguish genuine functions from fortuitous accidents and it cannot justify giving priority to function over selection.

The more defensible alternative combines these two accounts. Propensity Evolutionary Functionalism has the advantage that it is not committed to the actual reproductive success of a trait, but only to the disposition of such a trait to enhance reproductive success. The advantage of Etiological Evolutionary Functionalism is that it is able to distinguish genuine functions from mere side effects. Moreover, it gives priority to selection over propensity in order to determine what is a genuine

CRITERIA	EXPLANATION
Presence, Persistence, and Perpetuation	Propensity Evolutionary Functionalism presumes that a trait has a function prior to selection, but relies on selection and reproductive success as part of its account. So, it is able to make sense of the persistence and perpetuation of a functional trait. It is not clear, however, that it can make sense of the presence of a trait since it presumes that the trait has a function prior to selection.
Genuine Functions vs. Side Effects	Because of its reliance on propensity without any sort of constraints, Propensity Evolutionary Functionalism cannot distinguish genuine functions from, for example, beneficial side effects.
Malfunctioning/ Normativity	Propensity Evolutionary Functionalism could incorporate malfunction to the extent that propensities admit of degree. If a trait is functioning below its "normal" propensity, then it is malfunctioning to that extent.
Conscious Design	By giving priority to function over selection, Propensity Evolutionary Functionalism must allow for the presence of a functional trait to be the product of conscious design.

Table 4.4 *Four-Fold Criteria and Propensity Evolutionary Functionalism*

function because it takes causal history into account. In the spirit of unification, the appropriate account of function will give priority to natural selection, but claim that selection ranges over propensities to survive and reproduce. In full, a feature of an organism is a function if and only if it confers a propensity to produce a specific set of activities/effects and corresponding specific benefits *and* to enhance the goal of survival and reproductive success on an organism *and* that such a propensity-set is established through natural selection. This is a rendering of *dual-functionalism* noted in the early stages of this chapter. The formal characterization of this dual-functional mixed account is as follows:

> **Mixed Evolutionary Functionalism**: A kind of Evolutionary Functional explanation that maintains that a feature X has a function in an organism O by performing an activity Y if and only if Y produces effect E and both Y and E confer a survival enhancing propensity P on O (within a certain range of environmental pressures) and will continue to confer P on O (so long as a certain range of environmental pressures is present). And, moreover, P is currently present, because, ancestrally, there was natural selection in favor of retaining P to bring about the goals G of survival and reproduction.

A return to the enzyme example will help to explain the above account. Recall that iduronate sulfatase is the lysosomal enzyme that is designed to breakdown mucopolysaccharides. With respect to the mixed account above, iduronate sulfatase is a function of the human organism (and other species), because of its ability to produce the specific effect/activity of metabolizing mucopolysaccharides; and this effect correspondingly confers a survival-enhancing propensity on the human organism (and other species). Importantly, the reason why it currently confers such a propensity is because, ancestrally, there was natural selection in favor of retaining such a propensity (over a range of environmental pressures) for the sake of survival and reproduction.[30]

Turning to the four-fold criteria, Mixed Evolutionary Functionalism appears to hold up pretty well. First, because it relies on selection of propensity, it can account for the presence, persistence, and perpetuation of a trait. For instance, the kidney has the functions of blood purification and fluid regulation in the first place because of the selected survival enhancing propensity these activities conferred on organisms. Notably, these effects/activities became functions only after enough life cycles allowed for the relevant propensity range to be stabilized. Also, the kidney continues to have these functions in organisms because selection pressure ranges over a gamut of beneficial propensities and so long as the environment remains fairly stable, this propensity scope will continue to be present. Finally, the survival enhancing propensity of the kidneys will aid in the likelihood of the survival and reproduction of the organisms, ensuring the perpetuation of the species. Thus, even though the actual biology behind determining these propensity details may be both daunting and tedious, Mixed Evolutionary Functionalism stands as a reasonable way to address the presence, persistence, and perpetuation of a trait.

Second, given that natural selection aids in the presence, persistence, and perpetuation of a trait, Mixed Evolutionary Functionalism is able to distinguish genuine functions from side effects. For instance, the blood pumping of the heart confers a survival-enhancing propensity such that it is reasonable to consider this activity a function. In contrast, the heart's thumping sound offers no or very little survival enhancing propensity to the organism. Unlike Propensity Evolutionary Functionalism, any old benefits and effects and corresponding propensities will not confer a function ascription. Only those propensities that have withstood the test of natural selection over a great many life cycles will be justifiably deemed genuine biological functions. Thus, the outcome is that Mixed Evolutionary Functionalism is able to distinguish genuine functions from side-effects.

---

30   Godfrey-Smith has suggested that 'ancestrally' must refer not to the geologic past, but to the recent past. His justification is that a feature may have evolved in the recent past by natural selection to have a different function than it had in the geologic past (see Godfrey-Smith, 1994). In the light of this reasonable suggestion, 'ancestrally' refers to both or either historical past—the deep geologic past or the more recent evolutionary past—to accommodate a feature having a new function or multiple functions. Note, however, that contrary to Godfrey-Smith, this account does not dismiss the geologic past in favor of the recent evolutionary past, because this would leave out the possibility that the geologic function and the more recent evolutionary function could co-exist.

Third, again, Mixed Evolutionary Functionalism's reliance on natural selection reveals that it is able to accommodate malfunctioning. For instance, a blind person's eyes *are still supposed* to assist in visual perception even if they cannot due to defect or injury. The survival enhancing propensity is still present because that is what the eyes were selected to do (unlike vestigial organs). Remember, it would be very odd and downright implausible, in terms of complexity and energy expenditure, for the various optic systems to be present and not be there in order to do something. To this extent, Mixed Evolutionary Functionalism aligns itself with Etiological Evolutionary Functionalism and possibly Propensity Evolutionary Functionalism (and distances itself from systemic functionalism's snubbing of malfunction) by wholeheartedly underwriting the malfunction/normative dimension of biological functions.

Fourth, because Mixed Evolutionary Functionalism gives special weight to evolution and natural selection, it is able to reject the need to incorporate conscious functions as part of its account. Given that an account of biological function should be "natural" and free from having to make sense of the functional nature of artifacts, this feature of Mixed Evolutionary Functionalism is a definite plus. With respect to the other accounts of function discussed in this chapter, only Etiological Evolutionary Functionalism can meet the demands of this criterion. Thus, if the rejection of conscious design is accepted as an important criterion for making sense of biological function, then it should be clear that Mixed Evolutionary Functionalism is in good standing in its unwavering endorsement of this criterion. Table 4.5 summarizes Mixed Evolutionary Functionalism and the four-fold criteria.

CRITERIA	EXPLANATION
Presence, Persistence, and Perpetuation	Mixed Evolutionary Functionalism is able to make sense of the presence, persistence, and perpetuation of a trait because of its reliance on natural selection ranging over propensity.
Genuine Functions vs. Side Effects	Because of its reliance on natural selection, Mixed Evolutionary Functionalism is able to distinguish genuine functions from side effects.
Malfunctioning/ Normativity	Mixed Evolutionary Functionalism can openly endorse the malfunction/normative dimension of biological function because, from an evolutionary perspective, traits retain their "what they are supposed to do" characteristic.
Conscious Design	By giving priority to natural selection, Mixed Evolutionary Functionalism is able to distance itself from being forced to accept that the presence of a functional trait is the product of conscious design.

Table 4.5 *Four-Fold Criteria and Mixed Evolutionary Functionalism*

Table 4.5 reveals that Mixed Evolutionary Functionalism is a position worth taking seriously. From a philosophical perspective, this means that, at the very least, it is a rendering of a theory of biological function that is worthy of a critical response. Not only does Mixed Evolutionary Functionalism appear to satisfy the four-fold criteria, but it does so in a way that endorses *dual functionalism*; that is, not only are propensities selected in favor of a particular effect they produce, but also because of their overall contribution to survival and reproductive success.

Table 4.6 compares the various accounts of function with respect to the four-fold criteria. "Yes" indicates that the theory can accommodate the specific criterion; "No" indicates that the theory cannot accommodate the specific criterion; and "Unclear/Perhaps" indicates that the theory may or may not be able to accommodate the criterion or can only do so to a certain extent.

CRITERIA	THEORIES OF FUNCTION
Presence, Persistence, and Perpetuation	Systemic Functionalism: **Unclear/Perhaps** Etiological Evolutionary Functionalism: **Yes** Propensity Evolutionary Functionalism: **Unclear/Perhaps** Mixed Evolutionary Functionalism: **Yes**
Genuine Functions vs. Side Effects	Systemic Functionalism: **No** Etiological Evolutionary Functionalism: **Yes** Propensity Evolutionary Functionalism: **No** Mixed Evolutionary Functionalism: **Yes**
Malfunctioning/ Normativity	Systemic Functionalism: **No** Etiological Evolutionary Functionalism: **Yes** Propensity Evolutionary Functionalism: **Unclear/Perhaps** Mixed Evolutionary Functionalism: **Yes**
Conscious Design	Systemic Functionalism: **No** Etiological Evolutionary Functionalism: **Yes** Propensity Evolutionary Functionalism: **No** Mixed Evolutionary Functionalism: **Yes**

Table 4.6 *Comparison of Various Theories of Biological Function*

VI. CONCLUSION

Upon introducing the topic of biological function and explaining the relevant criteria for this discussion, this chapter has put three standard accounts of function on display: (1) Systemic Functionalism, (2) Etiological Evolutionary Functionalism, and (3) Propensity Evolutionary Functionalism. The accompanying analysis has revealed that, although there are aspects to each of these theories that make them somewhat attractive, all of them fall short of satisfying the demands of the four-fold criteria. The implication is that none of these accounts is worthy of a full endorsement, even in the light of Etiological Evolutionary Functionalism's ability to accommodate each of the four criteria. As an alternative, this chapter also offered a fourth theory of biological function: (4) Mixed Evolutionary Functionalism. It is argued that this permutation does better than its competitors at taking seriously that biological systems are best understood as a bundle of evolutionary compromises and that these are best understood in terms of evolved propensities to produce specific effects and to ensure survival and reproductive success. If this assessment is on the mark, then Mixed Evolutionary Functionalism is an account of biological function that should receive further critical attention—maybe even a critical interest that parallels my youthful fascination with knives!

CHAPTER REVIEW: DISCUSSION AND QUESTIONS

1. From Section I, you should be able to discuss the specific self-maintenance activities in defense of why an understanding of the nature of artifact functions and biological functions should be kept distinct. Are you persuaded by the argument offered? [pp. 126–27]
2. Section II starts with various phrases used to capture "function talk" and teleology. Can you make sense of *dual-functionalism* in this section? Use an example to show your understanding. [pp. 127–28]
3. What are the four criteria for the function debate (Section III)? At first glance, which of these criteria (and the arguments given) seem less appealing than the others? [pp. 129–30]
4. Using an example (use one from Section II), can you explain Cummins' version of Systemic Functionalism? [pp. 130–31]
5. You should be able to discuss the objections to Systemic Functionalism and how these objections are used to make sense of the four-fold criteria with respect to Systemic Functionalism. [pp. 132–39] [This is a lengthy discussion and should be broken-up into parts. See also question 6 below.]
6. Notably, the normative/malfunction discussion is lengthy and more time should be spent discussing it. First what is the malfunction/normativism issue? How does Systemic Functionalism deal with it? Are you persuaded by the reply to systemic functionalists on the normativism/malfunction issue? [pp. 135–38]
7. Again, using an example, you should be able to explain Etiological Evolutionary Functionalism. [pp. 140–41]
8. Why does the fourth criticism related to "losing a function" appear to be the most damaging to Etiological Evolutionary Functionalism? Or is the third objection related to "advantageous-traits-are-not-necessarily-functions" a more serious objection? [pp. 144–46]
9. Yet again, you should be able to use an example to explain Propensity Evolutionary Functionalism. [pp. 148–49]
10. What is the problem with Propensity Evolutionary Functionalism giving ontological priority to function over selection? [pp. 150–51]

11. How does an answer to the above question relate to how Propensity Evolutionary Functionalism can handle the four-fold criteria? [pp. 152–53]
12. Use an example to explain Mixed Evolutionary Functionalism. [pp. 152–54]
13. How does Mixed Evolutionary Functionalism draw upon both insights of Etiological Evolutionary Functionalism and Propensity Evolutionary Functionalism? What pitfalls does it supposedly avoid that Etiological Evolutionary Functionalism and Propensity Evolutionary Functionalism are unable to avoid? [p. 153]
14. Is Mixed Evolutionary Functionalism a legitimate alternative to Systemic Functionalism, Etiological Evolutionary Functionalism, and Propensity Evolutionary Functionalism? [pp. 154–156]
15. What objections might be offered to Mixed Evolutionary Functionalism if you think the answer to question 14 is "no"? [Hint: Is there a metaphysical concern regarding having natural selection selecting for a propensity set?]

REFERENCES

Ananth, Mahesh. "Human Organisms from an Evolutionary Perspective: Its Significance for Medicine." In *Handbook of the Philosophy of Medicine*, ed. Thomas Schramme and Steven Edwards, 1–29. Dordrecht: Springer, 2016.

———. *In Defense of an Evolutionary Concept of Health: Nature, Norms, and Human Biology*. Aldershot, UK: Ashgate Press, 2008.

Ariew, André, Robert Cummins, and Mark Perlman, eds. *Functions: New Essays in the Philosophy of Psychology and Biology*. Oxford: Oxford University Press, 2002.

Bechtel, William. "In Defense of a Naturalistic Concept of Health." In *Biomedical Ethics Review 1985*, ed. J.M. Humber and R.F. Almeder, 131–70. Clifton: Humana Press, 1985.

Bigelow, John, and Robert Pargetter. "Functions." *Journal of Philosophy* 84, 4 (April 1987): 181–96.

Boden, Margaret. *The Philosophy of Artificial Life*. Oxford: Oxford University Press, 1996.

———. *The Philosophy of Artificial Intelligence*. Oxford: Oxford University Press, 1990.

Boorse, Christopher. "A Rebuttal on Functions." In *Functions: New Essays in the Philosophy of Psychology and Biology*, ed. André Ariew, Robert Cummins, and Mark Perlman, 63–112. Oxford: Oxford University Press, 2002.

Buller, David J., ed. *Function, Selection, and Design*. Albany: SUNY Press, 1999.

———. "Etiological Theories of Function: A Geographical Survey." *Biology and Philosophy* 13, 4 (1998): 505–27.

Contran, Ramzi S., et al. *Robbins Pathological Basis of Disease*. 4th ed. Philadelphia: W.B. Saunders Company, 1989.

Cummins, Robert. "Functional Analysis." *Journal of Philosophy* 72, 1 (1975): 741–64.

Cummins, Robert, and Martin Roth. "Traits Have Not Evolved to Function the Way They Do Because of a Past Advantage." In *Contemporary Debates in the Philosophy of Biology*, ed. Francisco J. Ayala and Robert Arp, 72–85. Hoboken, NJ: Wiley-Blackwell, 2010.

Darwin, Charles. *The Various Contrivances by Which Orchids Are Fertilised by Insects.* 2nd ed. Revised with a New Foreword by Michael Ghiselin. Chicago: University of Chicago Press, 1984.

Davidson, Donald. "Knowing One's Own Mind." *Proceedings and Addresses of the American Philosophical Association* 60, 30 (1987): 441–58.

Davies, Paul. *Norms of Nature: Naturalism and the Nature of Functions.* Cambridge, MA: MIT Press, 2001.

———. "Malfunction." *Biology and Philosophy* 15, 1 (2000): 19–38.

Dawkins, Richard. *The Greatest Show on Earth: The Evidence for Evolution.* New York: Free Press, 2009.

Godfrey-Smith, Peter. "A Modern History Theory of Functions." *Noûs* 28, 3 (1994): 344–62.

Gould, Stephen Jay, and Elizabeth Vrba. "Exaptation: A Missing Term in the Science of Form." *Paleobiology* 8, 1 (1982): 4–15.

Grene, Marjorie, and David Depew. *The Philosophy of Biology: An Episodic History.* Cambridge: Cambridge University Press, 2004.

Hardcastle, Valerie. "Understanding Functions: A Pragmatic Approach." In *Where Biology Meets Psychology*, ed. Valerie Hardcastle, 27–43. Cambridge, MA: The MIT Press, 1999.

Harvey, William. *Exercitatio Anatomica d Motu Cordis et Sanguinis in Animalibus* [*Anatomical Studies on the Motion of the Heart and Blood*]. Springfield: Charles C. Thomas, 1928 [1628].

Jeffery, W.R. "Adaptive Evolution of Eye Degeneration in the Mexican Blind Cave Fish." *Journal of Heredity* 96, 3 (2005): 185–96.

Kant, Immanuel. *Kants gesammelte Schriften.* Berlin: George Reimer, 1908–13.

Lennox, James G. "Darwin *Was* a Teleologist." *Biology and Philosophy* 8, 4 (1993): 409–21.

Lewens, Tim. "Functions." In *Handbook of the Philosophy of Science: Philosophy of Biology*, ed. Mohan Matthen and Christopher Stephens, 525–47. Amsterdam: Elsevier, 2007.

———. *Organisms and Artifacts: Design in Nature and Elsewhere.* Cambridge, MA: MIT Press, 2004.

Lewontin, Richard. "Adaptation." *Scientific American* 239, 3 (1978): 212–30.

Margolis, Joseph. "The Concept of Disease." *Journal of Medicine and Philosophy* 1, 3 (Sept. 1976): 238–55.

Mayr, Ernst. "Teleological and Teleonomic, a New Analysis." *Boston Studies in the Philosophy of Science* 14 (1974): 91–117.

McLaughlin, Peter. *What Functions Explain.* Cambridge: Cambridge University Press, 2001.

McMullin, Emerson Thomas. "Anatomy of a Physiological Discovery: William Harvey and the Circulation of the Blood." *Journal of the Royal Society of Medicine* 88, 9 (1995): 491–98.

Millikan, Ruth. *White Queen Psychology and Other Essays for Alice.* Cambridge, MA: MIT Press, 1993.

———. "Defense of Proper Function." *Philosophy of Science* 56 (1989): 288–302.

Mills, S.K., and J.H. Beatty. "The Propensity Interpretation of Fitness." *Philosophy of Science* 46, 2 (1979): 263–86.

Mitchell, Sandra. "Dispositions or Etiologies? A Comment on Bigelow and Pargetter." *Journal of Philosophy* 90 (1993): 249–59.

Molinari, Jésus, et al. "Predation by Giant Centipedes, *Scolopendra gigantea*, on Three Species of Bats in a Venezuelan Cave." *Caribbean Journal of Science* 41, 2 (2005): 340–46.

Neander, Karen. "Misrepresenting & Malfunctioning." *Philosophical Studies* 79, 2 (1995): 109–41.

———. "Functions as Selected Effects: The Conceptual Analyst's Defense." *Philosophy of Science* 58 (1991a): 168–84.

———. "The Teleological Notion of 'Function'." *Australasian Journal of Philosophy* 69 (1991b): 454–68.

Nesse, Randolph M., and George C. Williams. *Why We Get Sick*. New York: Times Books, 1994.

Nissen, Lowell. *Teleological Language in the Life Sciences*. Lanham: Rowman & Littlefield Publishers, 1997.

Pennock, Robert. "Negotiating Boundaries in the Definition of Life: Wittgensteinian and Darwinian Insights on Resolving Conceptual Border Conflicts." *Synthese* 185, 1 (2011): 5–23.

Perlman, Mark. "Traits Have Evolved Because of a Past Advantage." In *Contemporary Debates in the Philosophy of Biology*, ed. Francisco J. Ayala and Robert Arp, 53–71. Oxford: Wiley-Blackwell, 2010.

Pope, Stephen J. *Human Evolution and Christian Ethics*. Cambridge: Cambridge University Press, 2008.

Preston, Beth. "Why Is a Wing Like a Spoon? A Pluralist Theory of Function." *The Journal of Philosophy* 95, 5 (May 1998): 215–54.

Radner, Daisie, and Michael Radner. "Optimality in Biology: Pangloss or Leibniz?" *Monist* 81, 4 (1998): 669–86.

Ribatti, Domenico. "William Harvey and the Discovery of the Circulation of Blood." *Journal of Angiogenesis Research* 1, 3 (2009): 1–2.

Rosenberg, Alexander. *Darwinian Reductionism: Or, How to Stop Worrying and Love Molecular Biology*. Chicago: Chicago University Press, 2006.

———. *The Structure of Biological Science*. Cambridge: Cambridge University Press, 1985.

Smart, J.J.C. *Philosophy and Scientific Realism*. New York: Random House, 1963.

Sober, Elliott. *The Nature of Selection*. Cambridge, MA: The MIT Press, 1984.

West-Eberhard, Mary Jane. "Adaptation: Current Usages." In *Philosophy of Biology*, ed. David L. Hull and Michael Ruse, 8–14. Oxford: Oxford University Press, 1998.

Wouters, Arno. "Philosophers on Function." *Acta Biotheoretica* 51, 3 (2003a): 223–35.

———. "Four Notions of Biological Function." *Studies in the History and Philosophy of Biology and Biomedical Science* 34, 4 (2003b): 633–68.

Wright, Larry. *Teleological Explanations: An Etiological Analysis of Goals and Functions*. Berkeley: University of California Press, 1976.

———. "Functions." *Philosophical Review* 82, 1 (1973): 139–68.

Chapter 5

THE SPECIES DEBATE

It is really laughable to see what different ideas are prominent in various naturalists minds, when they speak of "species" [...] It all comes, I believe, from trying to define the undefinable.

—*Charles Darwin (in Burkhardt and Sydney, 1991, p. 309; my emphasis)*

I. INTRODUCTION: A CONFUSING DAY AT THE ZOO

As I am eating popcorn and strolling through the local zoo (at the curious age of eight) with my mom, I find that our meandering has led us to the "big cats" locale. In a span of a few hundred yards, I am astounded by the size of the tigers, the rugged regality of the lions, the sleek motions of the onyx-hued jaguars, and the rather domestic-looking pumas. Upon listening to the litany of details offered by the tour guide about the various cats, I ponder to myself that it is incredible that these are all different *kinds* of "cat." Interestingly, although there are many differences between these *types*, I see that they share many resemblances such that it is reasonable to accept why the tour guide would group them together as being part of a larger unified group—that is, "cat." I recall that my reasoning was that in the same way that my father had different *kinds* of screwdrivers in his tool box, all of these items still belonged to the *class* of screwdriver.

Yet, when we made our way to the marine life section of the zoo, I was surprised to be told that dolphins, whales, sharks, and tuna are not all "fish." In fact, I am told that dolphins and whales are similar *kinds*, but sharks and tuna belong to separate *types*. At this point, I was rather perplexed. All the "cats" looked alike (with clear differences) and they were put into the same group; so should the different kinds of "fish"! I yelled to myself. For it was clear to me that all the different birds belonged together, the different monkeys belonged together, and that the giraffes did not belong in the same group as either birds or monkeys. The natural world seemed pretty well carved up to me. I could not help but blurt out to the tour guide: "I am confused. Why are the cats, birds, and monkeys put into their own groups, but the "fish" are grouped differently? I know that giraffes and spiders are different kinds of animals and should not be grouped together, but the fish are pretty much the same (like the cats were pretty much the same with some obvious differences)." It was clear that the tour guide understood my concerns. After looking at me for a minute and all of the other visitors doing the same (all the while my mom begging me to keep quiet), he let out an almost Santa-like laugh and simply said, "this is all complicated business, son. We had better leave it to the biologists to figure out." Everyone laughed, but I recall that I was not very amused.

No doubt, most of us are quite aware that there are many sorts of cats, including lions, tigers, and good-old Tom the domestic cat (and his various cartoon-depicted schemes of trying to catch Jerry!). We also know that these cats are mammals, which are also animals. Still, we might not be fully aware of how complicated biological classification is. The following cat example is a sample of a traditional classification, which does not even go into the more complex divisions.

Consider Tom, our friendly house cat. The biological term for house-cat is *Felis catus*. The first word names the genus, and the second specifies the species within that genus. Returning to some of our high school biology (and our domestic cat example), you might recall that animals are broadly categorized as follows:

Domain: Eukarya (organisms with cells that have an enclosed nucleus).

→ *Felis catus* belongs to the domain of *eukarya*.

Kingdom: the six major kingdoms include: plantae, animalia, fungi, protista, eubacteria (monera), and archaebacteria.

→ *Felis catus* belongs to the *animalia* kingdom.

Phylum: the five major phyla (plural for phylum) include: cnidarian (invertebrates), chordata (vertebrates), arthropods, molluscs, and echinoderms.

→ *Felis catus* belongs to *chordata*.

Class: This is the further division of the five major phyla above. For example, the phylum chordata is further divided into the following vertebrate classes based on distinct shared characteristics: amphibians, birds, mammals, reptiles, and fish.

→ *Felis catus* belongs to the class *mammalia* (distinct characteristics: warm blooded, hairy, and milk-producing glands).

Order: Classes are further divided based on certain common characteristics, which result in numerous sub-orders based on more refined shared characteristics.

→ *Felis catus* belongs to the order of carnivora (this includes over 280 species of primarily meat-eating placental mammals, ranging from the tiny weasel to the massive polar bear).

Family: At this level of classification, animals appear much more similar to each other since they are being grouped based on more specific common characteristics. Also, sub-families (e.g., pantherinae) are also used to distinguish further common features.

→ *Felis catus* belongs to the family of felidae (36 different cats comprise this family, including both domestic cats and wild cats—e.g., lion, tiger, jaguar, leopard, etc. Distinguishing characteristics include carnivorous, rounded heads, short muzzles, and retractable claws. The big cats in this grouping can roar, but not purr, while the small cats in this group can purr, but cannot roar).

Genus: Much like the family taxon, the genus taxon relies on a narrower set of common characteristics to group organisms.

→ *Felis catus* belongs to the genus of felis. This grouping is based on smallness of body size and the inability to roar. So, both domestic cats and small wild cats (e.g., mountain lion) are usually placed in this category.

Species: Although this classification level is the focus of this chapter, a distinguishing feature of a species is that it constitutes a group of individuals that actually or potentially interbreed in nature.

→ *Felis catus* belongs to the species catus. This cluster group includes over 40 pure breeds, and more breeds are added as they are developed.

So, our domestic cat, Tom, has the overall classification of:

eukarya → animalia → chordata → mammalia → carnivora → felidae → felis → catus

Now, consider just one of these many divisions: Felinae and Pantherinae, which are sub-families of the Felidae family. How are we to make sense of these divisions? In general there are two sorts of related but distinct concerns wrapped around the species discussion. First, in terms of a *categorizing issue*, why put something into one of these sub-families, and not the other? Felinae (e.g., house-cats) are in general smaller than Pantherinae (e.g., lions); but cougars (= pumas) can be quite large, despite being classified as Felinae. Felinae can't roar, but Pantherinae can—except for couple of kinds of leopards, which can't. So, why can't some of these non-roaring leopards be placed in the felinae sub-family rather than the pantherinae sub-family? Put another way, what kinds of characteristics are—or should be—used to make this division? And why? There is a good deal of controversy among biologists and philosophers of biology about matters here. Second, in terms of a *metaphysical issue*, a second question is: when some sort of agreement about what's relevant to classification is reached, what reason is there to think that this division constitutes anything significant in nature? Maybe it's just a way we prefer to classify things, rather than something out there. Questions like these can be raised about any grouping of biological kinds, not just the grouping into species; but most of the controversy in the literature talks about species groups, and "the species problem." This chapter will address both of these issues to varying degrees. Specifically, we want to describe and evaluate the method employed by a few prominent species concepts and then briefly assess the metaphysical implications of these accounts.

As we shall come to see, determining a satisfactory concept of species has proven to be quite vexing on both biological and philosophical grounds. As Mayr, one of the founders of modern biology and key players in the species debate, forcefully states: "There is perhaps no other problem in biology on which there is as much dissension as the species problem" (Mayr, 2004, p. 171). For instance, within the field of biology, Luckow points out that her research reveals that "botanists are routinely applying a number of species concepts." Luckow recommends that (1) criteria on restricting species should be made clear, (2) the species concept should be stated clearly up front, (3) the use of a particular species concept should not produce internal inconsistencies, and (4) the use of a particular species concept should not be based on the organism under study, but on careful reflection on the concept of species itself (Luckow, 1995, p. 600).

Luckow's suggestion that (1)–(4) should be part of the conceptual tool-kit of card-carrying biologists reveals that there may very well be a bit of in-house tension amongst biologists regarding their overall handling of the species issues. To some extent, this conceptual tension—even within biology—is understandable. *Not only must one provide a cogent account of the nature of species, but one must also do so with one eye focused on both taxonomy and other evolutionary considerations and another eye attentive to a bundle of philosophical worries that may appear orthogonal but are quite central to the issues at hand* (Brandon and Mishler, 1996, p. 111).

Foundationally, the tension is associated with both the complexity and variability of organisms in the natural world and the philosophical penchant for offering a unified definition that ranges over this vast variation. To make matters worse, according to some critics, the existing Linnean genus-species method (i.e., the two Latin names that accompany the vernacular name; for example, lion = *Panthera leo*) appears to be ill-equipped to accommodate this diversity.[1] Maybe all this complicated stuff cannot be left entirely to biologists to resolve. After all, they are busy enough with their experiments and field observations!

---

1   This is not to cast disparagement on Linnaeus' impressive efforts. For more discussion and citations, see Paterlini (2007).

So, as a point of emphasis, the problem of making sense of the nature of species only seems to get worse when philosophical concerns are thrown into the mix (Ghiselin, 1974, p. 541). For instance, the philosopher of biology, Stamos, elaborates on this philosophical issue of universals as follows: "In a sense, the species problem is quite simple. Are biological species real, and, if real, what is the nature of their reality ... do species words refer to entities in the objective world with a real existence independent of science?" (Stamos, 2003, p. 1).[2] Although Stamos is correct in locating the issue/question at hand, the answer is anything but simple. As it turns out, providing a cogent concept of species requires navigating through difficult terrain—a terrain whose topography is both biological and philosophical. Although all of the issues in the species debate cannot be handled in this chapter, upon laying the preliminary groundwork, the following theories will be discussed: (1) Species as Natural Kinds, (2) Pheneticism, (3) Biological Species Concept, (4) Species as Individuals, (5) Phylogenetic-Cladistic Species Concept, and (6) Relational Biosimilarity Species Concept. This analysis will reveal that, with some reservation, (6) does the best job of navigating both the biological and philosophical demands that constitute the species debate.

II. WHO CARES? WHY CARE?

Before diving into the details of the species debate, it is worth noting a few of the reasons for engaging this line of inquiry in the first place. First, although this may come off as cliché, humans are naturally curious creatures. Not only do we seem to have a desire to categorize and organize all kinds of stuff (e.g., baseball cards to biological systems), we have a penchant also to have our curiosity satisfied. To the extent that it is possible in our busy lives, we are eager to move beyond mere wonder to a richer satisfaction of those things that catch our interest. (Recall that Aristotle was eager to remind us of this desire to know.) Thus, at a basic level of curiosity and as knowledge-seeking beings, we do and should take an interest in how the natural world is organized.

Second, think of some activity that is of great importance to you. Are you: an avid deep-sea fisher? a skilled poker player? a basketball maniac? a tennis champion? Whatever happens to catch your fancy, imagine that you were told that you did not really understand your special area(s) of interest. Not only might this come as a surprise to you, but you might find this to be a source of frustration, if it proves to be true; or you might attempt to defend yourself against such a charge of ignorance; or you might endeavor to improve your knowledge upon realizing the gaps in your understanding. The point is that those who take seriously the details of their areas of interest will be highly motivated to make sure they understand the conceptual issues surrounding

---

2   Note that this text by Stamos should be viewed as the "species Bible." I highly recommend it to those who wish to engage the many intricacies of this debate that cannot be covered in this chapter.

these details. Biology is no exception. So, those who have a special interest in the field of biology will and should be eager to understand the conceptual issues that pertain to this discipline—the species debate is a classic example.

Third, some of the details of how the concept of species has been employed reveal the social implications of applying definitions. For example, both race and IQ have been understood differently based on particular definitions of species. Specifically, there was a period in which some "races" were thought to have a higher IQ than other "races." One of the major assumptions of this view is that there are four or five distinct human races—that is, species, with inherently different attributes. The result was that many people—for example, Native Americans, Africans, Jews, Hispanics, and African Americans—were either treated as less than human or as an inferior "species." Sadly, as many of us know, this treatment extended to the level of policy decisions that imposed unjust treatment and immoral burdens on a great many people.[3] Today, in the light of a better understanding of human evolution and a correspondingly more careful view about the concept of species, the idea of different human species/races is looked upon with much suspicion. The point is that it is important to have a clear understanding of biological concepts because history reveals that abuse looms large when such concepts are applied erroneously or maliciously.

Fourth, putting to one side the obvious need for biologists to be able to communicate effectively with each other, the nature of species has clear practical implications with respect to conservation biology.[4] As Rojas (1992, pp. 174–76) stresses, a careful examination of the species problem is crucial to addressing how to manipulate local environments—either by modifying the existing environment or adding organisms to the existing environment or removing organisms from the existing environment—for the sake of maintaining life's diversity.[5] Thus, it is clear that taking seriously the species concept has important practical ramifications. In fact, the further upshot of all this is that all of us—from curious people, to policy decision-makers, to field biologists, and to philosophers—have reasons to care about the species concept.

III. JUST A BUNCH OF INDIVIDUALS?
THE NOMINALIST CHALLENGE

Within the history and philosophy of biology literature there is a so-called metaphysical problem regarding the reality of the classification schemes used to make sense of organisms. Philosopher of biology, David Stamos, elaborates on this philosophical

---

3   Think of the unfair medical care ("a study in nature") given to those syphilis-suffering African Americans in what is now infamously known as "The Tuskegee Studies" (Brand, 1978) and the numerous medical abuses perpetrated against Native Americans (Hodge, 2012).
4   Stamos (2003, p. 3) provides an instructive illustration of potential communication problems and their effects on research by way of John Maynard Smith's summary regarding his research on the various flight adaptations of birds.
5   A similar point can be found in Sterelny and Griffiths (1999, p. 183) and Richards (2008, p. 161).

issue as follows: "In a sense, the species problem is quite simple. Are biological species real, and, if real, what is the nature of their reality ... do species words refer to entities in the objective world with a real existence independent of science?"[6] Stamos' 'independent of science' phrase indicates that the concern is not primarily an epistemic one (i.e., a knowledge concern), but predominantly an ontological one (i.e., an existence concern). As we saw, Darwin struggled with how to deal with the nature of species and his worries have had a lasting impression on the subsequent debates on this topic. For instance, the biologist, Michael Ghiselin, tells us that the "species problem has to do with biology, but it is fundamentally a philosophical problem—a matter for *the theory of universals*."[7]

What does Ghiselin mean by 'theory of universals'? In general, the answer to this question is captured by the distinction between *universals* and *particulars*—also understood as the debate between *realists/universalists* and *nominalists/individualists*. Broadly, although controversy abounds, the distinction between these two camps is usually drawn by way of spatiotemporal terms. For instance, some philosophers claim that particulars are separated by their location in space and time, while others maintain that a given universal is bound by neither space nor time and can exist in entirely distinct places at the same time. For example, consider the case of a particular black jaguar, Jake. Assume further that Jake lives in a particular place at a particular time. It seems eminently reasonable to claim further that Jake cannot also live in another specific place and time.[8] Furthermore, it correspondingly follows that there cannot be another black jaguar, who is identical to Jake, that occupies the same space and time as Jake. So, via spatiotemporal terms, it is possible to individuate distinct particulars/individuals.

In clear contrast, consider the universal property of *blackness* that our black jaguar, Jake, possesses. Some philosophers would insist that this property of blackness not only resides in the specific space and time occupied by Jake, but also that the exact same property of blackness can simultaneously be present at a different location in space—for instance another black-hued jaguar, James, in an entirely different locale. Indeed, these same philosophers may even claim that numerous universals (e.g., carnivore, nocturnal, warm-blooded, etc.) relevant to Jake may be present in a particular area of space and time it occupies and also in a distinct space and time occupied by James. In general, the point is that a universal can be present in two different places at the

---

6   David N. Stamos, *The Species Problem* (Lanham: Lexington Books, 2003), 1.
7   Michael Ghiselin, "A Radical Solution to the Species Problem," *Systematic Biology* 23 (1974): 541 (my italics).
8   For the sake of simplicity, it will be assumed that time-travel is not possible. If time-travel were possible and no personal identity issues are relevant, then it would have to be conceded that Jake could travel back to a time and place in which he formally resided. The upshot is that individuating particulars by way of spatiotemporal terms would not be entirely adequate. At any rate, let's not muddy the waters any more than needed!

same time. In point of fact, those who take universals seriously think that they are best understood like the abstract entities of mathematics (e.g., numbers, sets, classes, etc.). As Lowe (p. 2002, p. 350) makes the point, "many philosophers who believe in existence of universals ... consider that all universals are themselves abstract entities and consequently do not exist in space or time."

With the distinction between universals and particulars in hand, the pressing question is whether or not the former should be taken as seriously or more seriously to exist than the latter or if the former exists in a different way than the latter. Those philosophers who defend the existence of universals are known as *realists*. Realists insist that universals exist and that they are not bound by space and time. What this means is that realists not only believe that universals exist, but they do so *in a way* that is distinct from particulars (since the latter are bound by space and time).

In contrast, nominalists come in two flavors and both renditions reject the notion of universals endorsed by realists. First, some nominalists, let us call them *inclusive nominalists*, reject the existence of universals (as understood by realists) in favor of only the existence of particulars. What makes this version of nominalism inclusive is that it includes both individual concrete things as particulars (e.g., individual jaguars) and insists that particular abstract entities (e.g., particular instances of a set, property, feature, etc.) also exist. Second, there are those nominalists, let us refer to them as *exclusive nominalists*, who argue that 'particular' refers only to concrete particular things bound by space and time and both realists and inclusive nominalists are misguided.

So what is at stake here? Broadly, within the metaphysics literature, there is a concern pertaining to the ontological status of abstract entities/objects like sets or classes. Using sets as an example, Lowe (2002, p. 377) summarizes the debate as follows:

> [B]oth the claim that sets are abstract objects and, more radically, the claim that sets exist may be challenged. On the one hand, it might conceivably be urged that sets all of whose members are concrete objects are concrete objects themselves. On the other hand, it might be conceivably urged that talk about 'sets' is just a convenient grammatical device for talk about members.

Lowe suggests that talk of sets as abstract entities can be understood in two ways. One way, for example, that abstract entities could be understood is that the set of jaguars could not only refer to the existence of individual jaguars, but also to the existence of an individual jaguar set. This is basically the position taken by the inclusive nominalist noted above. Alternatively, one could conceive of the phrase 'set of jaguars' as a mere linguistic device and a sort of short-hand way of talking about only the existence of particular jaguars. As suggested above, this is the route taken by exclusive nominalists. In what follows, two arguments put forth in defense of exclusive nominalism will be explicated with respect to the species debate. It is this camp that captures the nominalist challenge in the species debate.

Argument from Causal Efficacy

To see this discussion played out in the species debate, Ghiselin (1997) initially goes on to explain this problem in terms of nominalism and classes without entirely taking one side of the debate or the other:

> According to one very popular philosophical notion, nominalism, individuals are "real" but classes are not. This makes a certain amount of sense: a nominalist would say that this chair is real, whereas 'chair' in general or in the abstract is not. Nominalists generally go somewhat further than that, and might even say that chairs share nothing at all except the name 'chair'—hence the etymology. One does not have to be a nominalist to admit that "chair" and the one on which I sit do not have the same ontological status. (1997, pp. 13–14)

Ghiselin begins by reminding the reader that nominalism is the view that individuals and not classes (i.e., sets or groups) are real entities in the world. More specifically, he points out that the name (e.g., 'chair') used to group individuals shares nothing with respect to the actual individuals who are the referents of such terms. Indeed, although Ghiselin is a bit guarded, he concedes that class/set terms do not have the same ontological status as actual individual organisms that comprise such labeling. Still, many nominalists more strongly urge that such labels do not exist in any way similar to that of individual entities. This interpretation of nominalism is in keeping with the exclusive nominalism position noted on the previous page. It appears that this is the version of nominalism that is on Ghiselin's mind when he moves to the species debate:

> In the controversies that surround the species problem, one position that has been taken is the so-called nominalistic species concept. According to this view, species are classes, classes are not real, therefore species are not real. Consequently, perhaps, they are mere conventions and have no role to play in biological thinking. The traditional response to this kind of nominalism was to deny nominalism itself, and take a "realistic" view of species. Species are classes, and they are real, so there was no problem. But the nominalistic argument made a certain amount of sense: if species are classes, how could they evolve or become extinct? It would be like sitting on chair in the abstract. (1997, p. 14)

As Ghiselin points out above, the nominalist challenge in biology is that grouping categories, such as class, should not be given much ontological weight since such categories cannot produce the sorts of behaviors, like sexual activity, indicative of individual organisms. From this perspective, the nominalist is claiming that classes or any other sort of grouping label should not be taken as having any sort of "real" existence out in the world because they are incapable of having any sort of causal efficacy in the world.

Argument from Arbitrariness

To see this exclusive nominalist perspective forcefully expressed from another angle, Burma draws upon the long evolutionary trail of descent with modification as follows:

> When we try to deal with larger aggregates of individuals, our categories become more and more abstract and empty of any real meaning ... a species, be it a plant or animal, is a fiction, a mental construct without objective existence. To set up species in this continuous line of descent, we must chop it into units, and in any such process the divisions are purely arbitrary.... Species and subspecies are the units with which the taxonomist deals, but they are merely convenient labels for arbitrary groupings....[9]

So, from Burma's perspective (and Darwin to some degree),[10] "species" and "subspecies" do not capture any unique divisions out in the world. The reason, he argues, is that there will always be a demarcating problem when individuals are grouped together. Now, this may not be obvious when we go about separating Siamese cats from African lions, but it proves to be deeply problematic when we try to separate, for instance, the Mexican soil-dwelling salamander (*Lineatriton*) from its near identical looking Mexican mountain-dwelling salamander. In fact, such cases of identical-appearance-but-distinct-species have been found among birds, snakes, whales, fungi, and flowering plants. This has led some biologists to claim that bio-diversity has been grossly under-reported (David Wake, 2006 and 2009). The implication, thinks Burma, is that there is a degree of arbitrariness regarding how to distinguish some species from others. It further follows, he thinks, that the various labels used by biologists to distinguish species are useful but are not part of the fabric of real biological systems that are part of one long evolutionary history. Again, much like Ghiselin, we see Burma offering a defense of exclusive nominalism.

Drawing on the arguments of both Ghiselin and Burma, the **nominalist challenge** is clear. Any theory about the nature of species had better be able to address the exclusive nominalism position or had better not run afoul of it. Put another way, the argument from causal efficacy and the argument from arbitrariness reveal, *prima facie*, that not only is the realist position implausible, but also that inclusive nominalism is far-fetched.[11]

---

9   Benjamin Burma, "The Species Concept: A Semantic View," *Evolution* 3 (1949): 369–70.
10 Stamos has argued that Darwin's considered view, which includes relying on some of Darwin's personal letters, reveals that he is not a nominalist. For further discussion, see Stamos (1996, pp. 127–44) and Ereshefsky (2010b, pp. 405–25).
11 It may very well be the case that many contemporary philosophers of biology and biologists are not worried about this ontological concern as it pertains to the realism vs. nominalism debate. Regardless, it has had some historical and philosophical pull and is worth pointing out that it has been a genuine worry in the debate regarding species independent of essentialist concerns. Still, given the history of some of these philosophical terms, it is important to point out that inclusive nominalism mentioned above, also known as philosophical nominalism, accepts the

CRITERIA	EXPLANATION
1. Ranges over all life forms	A feasible species concept should be able to make sense, in a uniform way, of animals, plants, and other living organisms.
2. Incorporates organisms as evolutionary entities	An accurate species concept should acknowledge life forms as a product of evolutionary forces.
3. Incorporates real organism characteristics	A cogent species concept should be able to incorporate the actual physical features that aid in *uniquely* distinguishing some set of life forms from other sets of life forms.
4. Contends with the nominalist challenge	Either directly or indirectly, a persuasive species concept should be able to address why the nominalist challenge is either misguided or complementary.

Table 5.1 *Four-Fold Criteria for Species Designation*

IV. CRITERIA FOR SPECIES ASCRIPTION

Along with the strong nominalism challenge, a successful species concept must satisfy some general requirements in order to have a fighting chance (These are summarized in Table 5.1 above). First, in order to claim that one has a schema or theory that allows for the proper organization of organisms, the assumption is that such a theory can be used to organize *all* life on Earth. Presumably, this means that a good species concept will be able to organize life across bacteria, protists, animals, fungi, and plants. If not all life, then a persuasive species concept should be able to distinguish and categorize, at the very least, the varieties of individual animals and plants. Clearly, it would be strange to offer a species concept that could not help categorize major forms of life on our planet (Grene and Depew, 2004, p. 294).

Second, a reasonable theory of species must acknowledge that biological systems are the product of evolutionary forces. This view reveals a world of complicated

---

reality of particular objects only, and rejects the reality of *any* type of category of things—not just species, but shoes, planets, silver things, rivers as well. It allows for the existence of red things, but denies the existence of redness that somehow functions to realize red things as a real category. We will not worry about this version of inclusive/philosophical nominalism, but will keep our eye on how exclusive nominalism is addressed. Additionally, we will focus on the idea that while many distinctions into kinds of things reflect real differences, perhaps even differences that make a difference causally in the physical world, some supposed distinctions are arbitrary, or vague, or based only on social ideology or perception, not what is perceived—in other words, merely verbal. This sort of argument has successfully cast doubt on the reality of the category "witch," for example. Many scholars see this type of concern as being the primary focus in the species debate.

overlapping and integrated relationships between and amongst organisms and their environments. This nested hierarchy view of life is the gift bequeathed to us by the efforts of Darwin and those that followed in his footsteps. As Mishler stresses, "Evolutionary biology will be richer and much more accurate in its models of the world if this Darwinian hierarchical perspective is accepted. Evolutionary and ecological processes are occurring at many nested levels" (Mishler, 2010, p. 120).

Third, there must be some set of features out in the world that is near-indispensable with respect to demarcating which individuals belong to which group. The point of this criterion is that one must actually do some field biology—as opposed to only armchair speculations—that yields some set of material differences that make a difference.

As we will come to find out in this chapter, it is no trivial task to offer a species concept that incorporates the set of criteria provided above. Primarily, the Darwinian view of organisms is that of entities that exhibit organized variability, interrelatedness, and malleability. What this means is that there appears to be no unique essence upon which to anchor a unified species concept. This suggests that endorsing criterion #2 requires rejecting criterion #1. Moreover, the complex interrelatedness of life implies that embracing criterion #2 also requires rejecting criterion #3. Alas, as already noted, the nominalist challenge of criterion #4 puts even further pressure on those who take the species category seriously. It would seem to follow, then, that the theoretical maturation of biology has correspondingly produced an impending death knell with respect to a unified species concept. Let us determine, by examining a few valiant efforts, to what extent this Darwinian deathblow should be taken seriously.

V. ORGANIZING THE SPECIES DEBATE

The concern over the nature of species stretches back at least to Aristotle's efforts. Not surprisingly, many subtleties have emerged in the ensuing years, especially the plethora of biological insights that must be countenanced in the wake of Darwin's ruminations. Yet, organizing such a large and nuanced body of literature is no trivial task. Still, as a way of gaining access into this discussion, the various accounts of the nature of species can be divided into non-historical, historical, and mixed proposals. Drawing partly on Stamos' pragmatic lead (2003, pp. 21–29), Table 5.2 on the next page displays a manageable way of organizing some of the prominent species concepts, while also keeping in mind how they address criteria 1–4.

The first point to stress is that the categories in Table 5.2 can house modified variations of these specific species concepts. For example, there are numerous natural kind/essentialism species concepts (e.g., Aristotle, 1984; Copi, 1954; Quine, 1965; Kitts and Kitts, 1979; and Elder, 2008); a variety of phenetic species concepts (e.g., Sneath and Sokal, 1973; Sokal and Crovello, 1970; and Michener, 1970); a bundle of biological/reproduction species concepts (e.g., Mayr, 2004; Patterson, 1982 and 1984); an assortment of species-as-individuals accounts (e.g., Ghiselin, 1974; Hull, 1976 and 2006; and Eldredge, 1985); plenty of phylogenetic-cladistic species concepts

SPECIES CRITERIA ------------------- SPECIFIC SPECIES CONCEPTS	CRITERION #1: ranges over all life	CRITERION #2: sensitive to evolutionary history	CRITERION #3: accommodates particular features of life	CRITERION #4: Can fight off nominalist challenge
NON-HISTORICAL EMPHASIS:				
Species as Essential-Natural Kinds	Yes	No	No	No
Pheneticism	Yes	No	Yes	? Unclear
Biological Species Concept	No	No	Yes	? Unclear
HISTORICAL EMPHASIS:				
Species as Individuals	Yes	Yes	No (or not easily)	No (or not easily)
Phylogenetic-Cladistic Species Concept	No	Yes	Yes	No
MIXED SPECIES ACCOUNTS:				
Relational Biosimilarity Species Concept	Yes	No	Yes	No

Table 5.2 *Working Species Concepts and Four-Fold Criteria*

(e.g., Hennig, 1979; Wiley, 1978; Cracraft, 1989; de Queiroz and Donoghue, 1988; Baum and Donoghue, 1995; Shaw, 1988; and Wheeler and Platnick, 2000); and more than a handful of mixed species accounts (e.g., Griffiths, 1999; Stamos, 2003; and Depew and Grene, 2004). Again, so as to acquire a flavor for this debate, the non-historical renderings, which will be discussed in this chapter, include: (1) Essentialism/Natural Kind, (2) Pheneticism, and (3) The Biological Species Concept. The historical alternatives that will be examined are: (4) Species as Individuals and (5) The Phylogenetic-Cladistic Species Concept. Finally, a version of the mixed account, (6) The Relational Biosimilarity Species Concept, will be reviewed.[12] Upon completing each species account, a smaller reminder table (that includes the four criteria) will be presented to assist you in digesting what you have read.

---

12 *And* these various accounts barely scratch the surface of many alternative species concepts!

VI. ENGAGING THE SPECIES DEBATE: ESSENTIALISM AND NATURAL KIND

Species as Essentialist-Natural Kind

As my opening zoo anecdote reveals, it is still commonplace to hear people speak of there being distinct *kinds*, *classes*, or *types* of organisms out in the world. On one interpretation, this means there are distinct unchangeable kinds of organisms in the natural world. For example, before engaging this topic of species, many of my students routinely are sympathetic with the view that there are horses, butterflies, fish, lemons, red roses, etc. and these living entities are unique to the extent that they have a core set of properties that cannot and do not change. In fact, many students insist that core properties do not change even in the light of the sundry biological and environmental changes related to Earth's history. This view is associated with the concepts of *essentialism* and *natural kind*. Essentialism has two major elements:

(1) There is a cluster of indispensable properties that belongs to all and only those entities that possess the set of indispensable properties. If an entity possesses the set of indispensable properties, then it is a member of a distinct kind. [**Indispensability Condition**]

(2) The cluster of indispensable properties is causally efficacious with respect to other properties that usually belong to members of the kind.[13] [**Causal Efficacy Condition**]

For example, a material object belongs to those things gold, if and only if, the material object has an atomic number 79. This means that atomic number 79 (i.e., the number of protons found in the nucleus of the atom) is the indispensable property that distinguishes the kind gold. Moreover, as it turns out, gold is usually a yellow malleable metal that is also a good conductor of heat and electricity. These additional characteristics owe their existence to the indispensable characteristic of atomic number 79. As the word suggests, in terms of gold, atomic number 79 is the essence—the essential, that is, indispensable property—of the kind gold.

'Natural kind' refers to those entities that have a set of indispensable properties, but the concept also includes the following two additional conditions:

(i) The cluster of indispensable properties that designate an individual belonging to a particular kind cannot change and continue to designate that same individual to that same particular kind. Additionally, an individual

---

[13] These two points are variations of the two found in Ereshefsky (2010a, p. 256).

cannot lose the cluster of properties that designate that individual belonging to a particular kind and still be that particular kind. So, if at any time X is the cluster of properties indispensable for being of a kind Y, then X is always the cluster of properties indispensable for being of a kind Y. [**The Static Condition**]

(ii) In any possible world **W**, so long as individual X has cluster properties Y, X belongs to those things **Z** in any possible world **W***. [**The Universal Condition**]

First, the static condition has two parts: (i) it is not possible for the atomic number 79 to change and the entity still be part of those things gold and (ii) those things gold cannot lose the indispensable property of atomic number 79. Second, the universal condition means that, in any possible world, so long as a material object has an atomic number of 79, that material object belongs to those things gold in any possible world. The universal condition can also be understood to mean that there are no members of those things gold that do not have atomic number 79 and all things that have atomic number 79 belong to those things gold.

We can now bring together the concepts of essentialism and natural kind in the species debate. First, on the essentialist view, there exists a set of properties that are possessed by all and only those members of that species that explains why they are as they are (Sober, 1980, pp. 353–54).[14] On the natural kind view, what is required to be part of an individual of a particular class does not change in time and deviations from this unchanging essence are nothing more than imperfections (Mayr, 2004, pp. 174–75).[15] With the ideas of essentialism and natural kind in hand, the essential-natural kind view of species, the first non-historical species account, can be defined as follows:

Species as Essentialist-Natural Kind → An entity E belongs to species X if and only if E has a set of characteristics that all and only members of X have in all possible worlds, and that are causally responsible for other usually present characteristics of Xs.[16]

---

14   Hull (1994, p. 313) offers a similar definition.
15   Mayr gives similar accounts, with varying degrees of detail, in his other major works (1982 and 1988).
16   It should be noted that there could be theoretical accounts of species in which organisms are grouped by a specific set of DNA/genetic markers and a set of corresponding characteristics. Whether or not these accounts should be considered "essentialist" depends on the details of the specific species theory being offered. If, for instance, the theory allows for these DNA markers to be non-static, then it does not satisfy the essentialist/natural kind account offered here.

REPLY TO SPECIES AS ESSENTIALIST-NATURAL KIND

There is much consensus that **Species as Essentialist-Natural Kind** is rendered otiose in the light of Darwinism. A critical look at each of the above conditions will help to see why this is the case. First, from a Darwinian perspective, there is no set of indispensable properties. In fact, populations evolve and branch off to create new populations. Basically, natural selection functions in the light of phenotypic variation from our gene-environment complex. As stated earlier in the book, heritable variation that proves to be beneficial can be retained and passed on to future generations. The result is that these future generations can appear and behave quite differently from their ancestral populations. The reasons for these kinds of changes are numerous, but some of the standard biological factors are: genetic recombination by way of sexual reproduction, heterochrony (changes in the timing of development), behavioral adjustments, mutation events, genetic drift, and neutral selection (see chapter 1 for a reminder of these definitions). These are some of the central sources of new heritable variations—some of which prove to be advantageous to local environmental stressors (e.g., new predator)—of which selection can take advantage in the long run. Given the scope of variation and change in the light of Darwinism, both the tenets of essentialism and natural kind are unsupportable (Okasha, 2002, p. 197), because the static and indispensability conditions require that organisms and their offspring remain part of the same species. Thus, not only is allegiance to the static condition of **Species as Essentialist-Natural Kind** unwarranted, but also a willingness to embrace the notion of an indispensable property or characteristic is also suspect.

Still, what about the causal efficacy condition? Well, if Darwinism is correct that biological systems do not have any indispensable properties, then whether or not such properties are causally efficacious with respect to other regularly occurring properties is a moot concern. Nevertheless, even if it were granted that an indispensable set of genes is present in each population, this still does not allow for a stable set of phenotypic features. The reason for this is that phenotypes are a product of genes plus local environmental factors. Enough of a change in the environment can accommodate unexpected phenotypic and behavioral changes. For example, in the wake of some devastating El Niño hurricanes, researchers observed so-called distinct species of Finches (small and large birds) reproducing with each other—a behavior brought on only by the diminution of options resulting from the El Niño effects. (We can speculate that "others" really start to look sexy when partner choices become greatly constrained!)[17] The example illustrates that sexual behavior, which is generally species-specific, can be drastically modified in the wake of increased

---

Alternatively, if the account treats the DNA markers as static and also satisfies the causal efficacy condition and the universal condition, then it would clearly satisfy the essentialist-natural kind account offered here. No doubt, there may very well be "essentialist" accounts that do not fit this analysis, but unfortunately not every variation on an essentialist concept of species can be tackled in this discussion.

17 The details of this account are wonderfully presented by Weiner (1994).

environmental stressors. Thus, the causal efficacy condition of essentialism cannot be taken seriously from a Darwinian perspective. Yet in fact, given the numerous sources of biological variation noted above, it is unlikely that there is any stable set of genetic and environmental conditions that will yield a regular unchanging set of phenotypic characteristics. The implication is that the causal efficacy condition must be jettisoned as well. The further upshot is that both conditions of essentialism—the indispensability condition and the causal condition—are rendered unserviceable when we are forced to take Darwinism seriously.

This leaves only the universal condition—the second element of the concept of natural kind. On this view, there are finite beings known as *particulars*—individual organisms that live and die—that are in possession of properties that do not change. What is the ontological status of these unalterable properties? Notice that, much like universals, these properties can exist simultaneously in different locations and can also exist in different time periods. For instance, the cluster of properties that characterizes jaguars (e.g., being carnivorous, being nocturnal hunters, having black-colored hair, etc.) are the very properties that can simultaneously exist at different times and locations with respect to different jaguars. Defenders of **Species as Essentialist-Natural Kind** must be committed to existence of these sorts of properties in order to be able to claim that organism X belongs only to species Y at any given time or place. The implication is that endorsing **Species as Essentialist-Natural Kind** includes embracing either the universalist or inclusive nominalism positions. Yet, keeping in mind the exclusive nominalist challenge, if the **Argument from Causal Efficacy** and the **Argument from Arbitrariness** are on the mark, then the acceptance of either the universalist or inclusive nominalist positions further weakens the **Species as Essentialist-Natural Kind** account.[18]

Additionally, the universal condition of **Species as Essentialist-Natural Kind** is not appropriately sensitive to the foundational assumption that the process of evolution includes the fact of shared common descent. On this view, it makes no sense to talk of organisms sharing a unique set of properties. For example, both dolphins and bats are able to echolocate, but it would be a mistake to say that they belong to the same species. This just happens to be a case of *parallel evolution*, the idea the similar traits have evolved separately in different species. Ereshefsky forcefully states the problem in terms of common descent with two points—one about common descent and the second regarding developmental constraints. First, as Ereshefsky explicates, due to a common evolutionary history, there is enough sharing of genetic material for common features to emerge in different species. He uses the example of birds and bats. Both these flying wonders have wings, but the origin of them comes from different evolutionary routes. The point is that the same feature or

---

[18] It is unclear whether or not champions of **Species as Essentialist-Natural Kind** are committed to the existence of abstract entities like sets. If they are, then this further tells against their position in the light of the exclusive nominalist challenge.

set of indispensable features can be *multiply realized* in different species by distinct causal factors.[19] So, imagine two organisms having all the same features, but those features are the product of distinct causal evolutionary routes. Then, further imagine that flight and all other features of bats and birds are the same set of indispensable features of what it means to be a bird or a bat, it follows that one would be forced to include birds and bats in the same species (Ereshefsky, 2010a). This implication, however, reveals that **Species as Essentialist-Natural Kind** is not a very cogent account of what constitutes a species.

Second, Ereshefsky makes it clear that organisms are also constrained by their biology. This means that there is a range of physical features that can most likely emerge in the light of the evolution of a given species and its local environmental challenges. For example, given the advantages of size that have proven successful to the ostrich in its local environment, its corresponding weight constrains its possibility of flight. The same holds true for the Emu. This example illustrates the idea of the body as a bundle of evolutionary compromises in the light of local environmental challenges (Ananth, 2008, ch. 7). Although it might be ideal for the Ostrich and the Emu to be able to fly and run at high speeds, their corresponding body weight (relative to their wing strength) and the lack of a keel makes flight a virtual impossibility.[20] The absence of the keel is the major developmental constraint here. Yet, both species are able to run at high speeds. Note, however, that high speed running is the product of developmental constraints and local environmental pressures. Both of these distinct species lack a specific characteristic and possess a specific characteristic, leading one to think that they belong to the same species (assuming that high speed running and all of their other indispensable characteristics are the same and are indispensable attributes). Again, this grouping would be a mistake. Thus, contrary to those who accept the universal condition associated with **Species as Essentialist-Natural Kind**, Darwinism reveals that it should be abandoned—along with **Species as Essentialist-Natural Kind** (see Dennett, 1995, pp. 91–100).

Finally, there is the idea of "all possible worlds." Recall that the concept of natural kind includes the notion that all and only those organisms which possess the set of indispensable attributes associated with a particular species necessarily belong to that species—in all possible worlds. Imagine that on some distant possible planet, call it Terra, there are life forms very similar to those on Earth. Consider further that there happens to be a particular species on Terra that looks identical to the Lion species on Earth. Notably, Terra-Lion came to exist by the same forces that are associated with the evolution of Earth-Lion. For example, assume that it is the case that Terra-Lion evolved by way of a DNA molecule just like Earth-Lion; every conceivable phenotypic and behavioral trait of Terra-Lion is identical to that

---

19 The plausibility of multiple realizability will be explored further toward the end of this chapter. Nevertheless, for an interesting discussion, see Shapiro (2000 and 2008).

20 A keel is an extension of the sternum (breastbone) that acts as anchor for the wing muscles of birds. It is the keel that is crucial to providing the kind of leverage needed for bird flight.

of Earth-Lion. It hunts, roars, yawns, copulates, sleeps, etc. just like Earth-Lion. Would we be willing to say that Terra-Lion really is a species of lion like those on Earth? On the natural kind view, I believe that the answer is "yes," because this satisfies the natural kind conditions set forth by **Species as Essentialist-Natural Kind**. From a Darwinian view, however, Terra-Lion did not evolve on Earth. Specifically, Terra-Lion does not share the same phylogenetic history as Earth-Lion. The point is that, according to a certain reading of Darwinism (as we saw in the chapter on biological function), biological evolution on Earth is bounded by the history of the biological processes on Earth. If this is correct, then it makes no sense to talk about Terra-Lion being a species of Earth-Lion. Thus, in the light of Darwinism, **Species as Essentialist-Natural Kind** should also be rejected for its "all possible worlds" feature.

This section has explored the **Species as Essentialist-Natural Kind** account of species and a number of criticisms against it. Importantly, given a Darwinian perspective, there is little support for (1) the indispensability condition, (2) the causal efficacy condition, (3) the static condition, and (4) the universal condition. Additionally, given the species criteria discussed earlier, it is reasonable that **Species as Essentialist-Natural Kind** ranges over all biological systems (criterion #1), but it does so in a way that ignores these systems as evolutionary entities (criterion #2). Furthermore, it incorporates immutable essences as part of its ontology and requires drawing upon unwarranted features of organisms (criterion #3). Finally, in the light of its support of unchanging universal essences, **Species as Essentialist-Natural Kind** stands in opposition to nominalism (criterion #4). Put another way, **Species as Essentialist-Natural Kind** needs to provide evidence that warrants making a reasonable inference that there is a cluster of properties that are anything more than individual properties unique to individual organisms. Unfortunately, **Species as Essentialist-Natural Kind** has very little to offer as a response to this nominalist challenge. The conclusion of this section, as is clear in the table below, is that **Species as Essentialist-Natural Kind** is not a reasonable account of the nature of species.

Table 5.3 reminds us of all of these concerns.

	CRITERION #1	CRITERION #2	CRITERION #3	CRITERION #4
Species as Essentialist-Natural Kind	Yes	No	No	No

Table 5.3 *Species as Essentialist-Natural Kind and The Four-Fold Species Criteria*

Phenetic Species Concept

One aspect of **Species as Essentialist-Natural Kind** that one may find appealing is that organisms appear to be organized according to their appearances. It seems eminently reasonable, then, that species designation should still focus on groupings based on shared physical and behavioral characteristics. From this perspective, abandoning the essentialism and natural kind tenets of **Species as Essentialist-Natural Kind** and still retaining shared physical and behavioral features may be the best way to make sense of the nature of species. Recall my little excursion to the zoo. I (like many people I suspect), was quick to organize animal life by way of features that seemed eminently reasonable to me: dolphins and sharks are part of the same set due to their overall similar external features; lions, tigers, and other ferocious felines were all cats due to their common characteristics, and all the primates were part of the same "monkey" group due to their common traits. Although these were mere ruminations of a confused young boy, there is a theory that is in-keeping with this sort of classificatory schema. It is known as the *Phenetic Species Concept*. According to the **Phenetic Species Concept**, a species is a collection of organisms that looks similar to each other and correspondingly distinct from other groups of organisms. The **Phenetic Species Concept** can be defined as follows:

Phenetic Species Concept → X is a member of species Y, if and only if, X has a set of features F that only members of Y possess.

The similarities and dissimilarities related to F in the above definition usually range over (i) morphology, (ii) behavior, and (iii) genetics. Part of the reason for relying on (i)–(iii) is that a great deal of data on the historical relatedness of organisms is still unavailable and constantly under revision as information is updated. Basically, because our knowledge of deep geologic history is inherently limited, defenders of the **Phenetic Species Concept** do not want to endorse a concept of species that relies on such ignorance. Rather, they insist that careful measurement of observable features related to (i)–(iii) can produce a serviceable concept of species. As Ridley points out, the overall goal of the pheneticists is "to measure so many characters that the idiosyncrasies of particular samples should then disappear" (Ridley, 1996, p. 378).

Primarily, the **Phenetic Species Concept** relies upon various statistical procedures and assumptions that can produce what they claim are on-the-mark groupings of organisms. For example, assume that tongue length and neck girth are the two characteristics that assist the pheneticist in determining which organisms should and should not be clustered together. For example, if frogs A, B, C, etc. share a similar tongue length and neck girth, then they will be grouped together as being part of the same species. If another set of frogs share a different set of tongue length and neck girth measurements, then they will be grouped in a different species set. Now, these groupings can be corroborated by simply adding on a great many more features (e.g., mating call differences, jumping ability, skin and eye pigmentation, sexual behavior, etc.) and doing the requisite number crunching. With enough features determined, it should be possible to produce conclusive sets of frogs that are uniquely distinct from other sets. These sets of frogs, according to pheneticists, will be understood as unique frog species.

REPLY TO PHENETIC SPECIES CONCEPT

Unfortunately, much like my young self's categorizing of the natural world, the **Phenetic Species Concept** defenders have greatly over-estimated the power of their approach. Although the **Phenetic Species Concept** might very well do a decent job of distinguishing many related varieties of animals, its overall methodology would also produce many mistakes. First, some scholars rightly worry about arbitrariness related to choice of measurement methods. For example, one statistical method might group frogs A and C together and another method might group B and C together (see Ridley, 1996, pp. 375–80). How is the pheneticist able to determine which statistical method most closely captures "the way the world really is"? Unfortunately, the **Phenetic Species Concept** does not offer a resolution to this concern except to rely on large pools of data. Rather, whatever statistical method is employed, it merely allows for the grouping of similar features. This may produce a bit of epistemic satisfaction, but not necessarily ontological accuracy (Sterelny and Griffiths, 1999, p. 184). One need only take a casual glance at the variations present in beetle populations; indeed, both the similarity and variation of stag beetles could give the **Phenetic Species Concept** fits! The result is that, since the natural world is replete with tremendous in-population variation, the **Phenetic Species Concept** could very well produce contradictory species accounts or may simply generate mistaken species groupings.

In terms of our criteria, the **Phenetic Species Concept** is able to range over all life forms (criterion #1). It might need to adjust its set of features for non-sexually reproducing organisms as part of its statistical methodology, but such an accommodation would not seem to be problematic. Also, it is able to focus on actual organism features to demarcate species (criterion #3). Of course, as noted above, it is not clear if it is able to do this in a non-arbitrary way. Additionally, the **Phenetic Species Concept** is *not* able to incorporate the evolutionary dynamics that are an integral part of biological systems (criterion #2). Finally, since the **Phenetic Species Concept** is only interested in producing a statistical composite image of an organism, what counts as a species is a collection of similar-looking individuals. If this is all that the **Phenetic Species Concept** hopes to achieve, then it could be compatible with exclusive nominalism. If, however, the **Phenetic Species Concept** is committed to the existence of a unique cluster of properties or features (or even the existence of sets *qua* abstract entities) that is in keeping with its constructed composite image, then it needs to provide some account of why these additional features (beyond mere individual features and individual organisms) should be taken seriously. It might be the case that different statistical methods would require different ontological commitments, but the **Phenetic Species Concept** does not directly address all of this. As noted earlier, it is a failing of any species concept that is not sensitive to this metaphysical concern. Thus, it is unclear where the **Phenetic Species Concept** stands with respect to the exclusive nominalist challenge (criterion #4) and this lack of clarity is a mark against this account. At any rate, since erroneous groupings would be almost inevitable as a result of **Phenetic Species Concept**'s methodology, the noted conceptual and practical difficulties have resulted in its abandonment by most participants in these debates. Table 5.4 offers a summary.

	CRITERION #1	CRITERION #2	CRITERION #3	CRITERION #4
Phenetic Species Concept	Yes	No	Yes	? Unclear

Table 5.4 *Phenetic Species Concept and The Four-Fold Species Criteria*

Biological Species Concept

It should be clear that both **Species as Essentialist-Natural Kind** and the **Phenetic Species Concept** will not do. Although both approaches are sensitive to drawing upon the real features of organisms, they do so at the expense of ignoring the shared history of life and the quirky ways that organisms are and are not related to each other in the light of this history. This brings us to the last non-historical account of species—one that has had much appeal. This is the **Biological Species Concept**, which was founded and ardently defended by Ernst Mayr. According to Mayr (2004, p. 178), the basic idea behind the **Biological Species Concept** is that a population of individuals is a species, if and only if, the individuals within the population breed only with members within the population—and not with other local populations (Mayr, 2004, p. 179).

Mayr expands on his interbreeding criterion in response to the question, "Why are there species?" He argues that the reason why there are species is that there is a set of individuals who have very similar, well-balanced, and harmonious genotypes. This is evident from the selection pressure in favor of different kinds of isolating mechanisms that either ensure mating within one's population or prevent successful mating outside of one's population (Mayr, 2004, pp. 178–79 and Shaw, 2001, p. 880). For example, mating seasons vary across populations such that a chance sexual rendezvous with another outside of one's population will illicit little or no sexual stimulation. Also, sexual behavior cues can also be population specific, rendering a chance encounter behaviorally uninviting.[21] Thus, what counts as a species are those individuals who mate exclusively within their population, resulting in a well-balanced and harmonious genotype.

The further implication of this argument is that the **Biological Species Concept** makes it clear that if it can be established that a given population has little or no gene flow with respect to other local populations, then that population can be deemed a species. We can describe formally the **Biological Species Concept** as follows:

> **Biological Species Concept** → Population X constitutes a species Y, if and only if, members **M** of X are reproductively restricted to mate with only other **M**s of X by various isolating mechanisms that secure genotype harmony.

---

[21] For further details on isolating mechanisms, see Mallet (1998, pp. 379–80) and Templeton (1989, p. 6).

REPLY TO BIOLOGICAL SPECIES CONCEPT

There have been numerous criticisms of the **Biological Species Concept**. First, Mayr's **Biological Species Concept** may be far too restrictive in terms of species membership. Although there are surely plenty of scenarios in which populations are isolated enough to block gene-flow from neighboring related populations, it is unclear if such scenarios capture the common gene-flow dynamics of neighboring sub-populations. In effect, either Mayr's **Biological Species Concept** applies only to a subset of *sympatric* cluster groups (i.e., conspecific groups that do not hybridize despite their close proximity to each other) or only reflects a kind of "snap-shot" or "freeze-frame" diagnosis of conspecific gene-flow dynamics. In either case, the **Biological Species Concept** appears to portray a dubious picture of population dynamics or applies to only a small set of cases (Sterelny and Griffiths, 1999, p. 189). The upshot of this worry is two-fold. Either the **Biological Species Concept** does not really capture the dynamism present in real-world conspecific interactions or it unnecessarily applies only to a sub-set of populations.

Still, defenders of the **Biological Species Concept** could insist that the species-designation strictures imposed by the **Biological Species Concept** should be accepted. For example, imagine there are ten cluster groups of rabbits in a particular environment. If there is any level of gene-flow amongst these populations, then the **Biological Species Concept** would have to remain agnostic regarding species anointment to any of these groups or insist that there is no appropriate species-level population in this given set of rabbit clusters. Rather, one would have to wait for a kind of isolating mechanism to prevent gene-flow. For instance, a powerful enough earthquake might create a separation in the earth that would prevent one of the ten rabbit populations from interacting with the other nine. Now, given enough time, this isolated population could occupy and create a new niche such that it can continue to thrive and reproduce independently of the other nine (or any other) rabbit populations. In this way, a unique "genotype harmony" could be sustained. In a biological flash, then, a new species would be created (according to the **Biological Species Concept**). Thus, rather than seeing the Biological Species concept as being unnecessarily too restrictive regarding which sub-populations are genuine species, its defenders could insist that this conservatism is warranted in the light of real-world dynamics. So, it is conceivable that the **Biological Species Concept** could distance itself from the criticisms noted above.[22]

Yet, even if the **Biological Species Concept** is a reasonable account of sexually reproducing organisms, it does not appear to apply to a wide range of asexually reproducing life forms. For example, many life forms reproduce without the fusion of gametes such that an offspring arises from a single parent. Just to illustrate, organisms that reproduce by *fission* (e.g., bacteria, protists, and fungi), *budding* (e.g., yeast, hydra, certain parasites and worms), *vegetative propogation* (e.g., strawberries and tulips), *sporulation* (e.g., certain fungi and algae), and *fragmentation* (e.g., annelid worms, liverworts, and echinoderms) would be excluded from species analysis by the **Biological Species**

---

22 It is not clear how persuasive this sort of response is. I direct the reader to Coyne and Orr's (2010) analysis for further discussion.

Concept (Ereshefsky, 2010a, p. 261). Therefore, even if the **Biological Species Concept** allows for reasonable species designations with respect to sexually reproducing organisms, it fails to do so with respect to asexually reproducing life.

Since one of the criteria for a good species concept includes the view that it should range over all life forms (criterion #1), the **Biological Species Concept** falls short. Also, because the **Biological Species Concept** gives primary focus to reproductive isolation and gene flow dynamics, it captures, to a degree, the real (reproductive) features of organisms (criterion #3); however, its near-blind allegiance to reproductive isolation reveals that it does not give enough attention to organisms as evolutionary entities (criterion #2).

Lastly, it appears that the **Biological Species Concept** is compatible with nominalism (criterion #4). Specifically, it is the reproductive repertoire of individual organisms that determine how they are grouped. There is nothing beyond this sorting than the (sexual) activities of individuals; all of this is in keeping with the nominalist insistence that moving beyond particular individuals is to move in the direction of an opaque ontology. There is a worry, however, that Stamos brings to light related to Mayr's understanding of the nature of species. Since Mayr thinks that species should be understood like sibling relationships, he thinks that 'species' is a relational term. In the same way that 'brother' can only refer to a male that has at least one other sibling, population X can only refer to species Y because it is related to the set of populations Z while maintaining reproductive isolation from Z. Yet, Mayr does not hesitate to claim that it makes no sense to employ the species label to X if it is the sole surviving population relative to Z (Stamos, 2003, p. 196). The implication is that the nature of species, for Mayr, is that of a relational property. In response, Stamos argues that there is a logical problem for Mayr:

> The problem here is that if species are truly *individuals*, then it is extremely odd that their ontology is relational, that the logic of their existence requires other individuals of the same category. Logically there is no reason why at some time in the future I may not be the only living organism left in the universe. If species are individuals, then this logic ought to extend to them as well. (Stamos, 2003, p. 197)

First, notice that Stamos' criticism is a clear reminder that the realism/nominalism debate is pertinent to the species debate and that this debate should not be marginalized even if specific species concepts have neglected this philosophical concern. Second, one way to understand Stamos' worry is that Mayr appears to be equivocating on how he uses 'individual.' Sometimes Mayr is focused on the reproductive patterns of particular individuals; other times he seems to shift away from particular individual organisms in order to give attention to the relational property present as a result of the relationship between individual groups. It is this shift from individual organisms to individual groups that allows Stamos to inject his worry that it should be possible to allow for the existence of a single individual organism and a single individual population; yet, Mayr's strong endorsement of a need for a relational property to allow for the existence of distinct populations does not make that possible.

If Stamos' criticism is reasonable, then it is correspondingly evenhanded to claim that Mayr's **Biological Species Concept** is not really in keeping with exclusive nominalism (criterion #4). Such an ontological addition of a relational property implied by the **Biological Species Concept**, given the problems it already must confront, would only seem to make matters worse.[23] Table 5.5 summarizes all of this.

	CRITERION #1	CRITERION #2	CRITERION #3	CRITERION #4
Biological Species Concept	No	No	Yes	No

Table 5.5 *Biological Species Concept and The Four-Fold Species Criteria*

Species as Individual Lineages

With the nominalism objection in mind, I suspect that it may not seem odd to think of species as individuals. When we observe a bunch of ants on a mound of dirt, we acknowledge that there is a colony of ants, but the colony just is a bunch of individual ants. Yet, we do distinguish "*this* single/one colony over here, as opposed to that single/one colony over *there*." We also speak of individual companies or sports teams.[24] Maybe we need to be more sensitive to our own ways of speaking. As it turns out, 'individual' is a technical term that has different meanings. According to a number of scholars, one of these meanings does allow for treating species as individuals. For instance, David Hull defines 'individuals' as "spatiotemporally localized cohesive and continuous entities (historical entities)" (Hull, 2006, p. 364). Drawing on an evolutionary sense of time and a rejection of **Species as Essentialist-Natural Kind**, Hull's **Argument from Individuality** and defense of **Species as Individual Lineages** can be put forth as follows:

Hull's **Argument from Individuality**

P1. Either species are natural kind entities or species are individuals.

P2. Species are not natural kind entities (rejection of **Species as Essentialist-Natural Kind**).

P3. Species are individuals.

---

23 Defenders of the **Biological Species Concept** could endorse a kind of "trope nominalism." For more on the plausibility of trope nominalism, see Loux (2002, pp. 84–93) and the penultimate section of this chapter.
24 Ghiselin (1974, p. 538) also views species as individuals, drawing on economic analogies—for instance, he thinks that a species is like a firm.

P4. 'Individual' is defined as spatiotemporally localized cohesive and continuous entities (historical entities).

C1. It follows that species is defined as spatiotemporally localized cohesive and continuous entities (historical entities).

Let us grant that **P1** is true. Hull is well aware that there are more possibilities than natural kind entities and individuals. He is simply restricting the discussion to these two options to make the analysis manageable. So, for the sake of discussion, species are either natural kind entities or individuals. (Note: some of the other species concepts will reject **P1** on the grounds that it falsely assumes that the only alternative to **Species as Essentialist-Natural Kind** is this account of individual.)

Given the earlier discussion of **Species as Essentialist-Natural Kind**, let us also grant the truth of **P2**. This is a reasonable request. Recall that an integral part of **Species as Essentialist-Natural Kind** (the universal condition) is that natural kind entities are *not* spatiotemporally restricted. So long as the set of features X are present, then any being in any possible world that possesses set X is a member of that kind. From a Darwinian perspective, however, since organisms are historical entities that potentially manifest considerable change/variation, it is the case that biological entities on Earth are spatiotemporally restricted. Again, the truth of **P2** is pretty solid.

P3 logically follows from the truth of **P2**. Since the truth of **P2** is established, P3 follows accordingly. It is **P4** that is the crucial premise in this argument. In defense of the truth of **P4**, Hull offers two separate analogies, (1) genetic-individual analogy and (2) part-whole analogy, to unpack his claim that species are individuals and individuals are "spatio-temporally localized cohesive and continuous entities."

First, as part of his three-part genetic analogy, Hull explains the copying transitions from genes to individual organisms to species. His argument is that evolution is about copying. Those entities that are successful in their local environments will pass on (to the next generation) the relevant information that aided in their success. This copy process begins with genes. Successful genes get passed on to future generations. Although specific genes do not survive over long periods of time, the successful information does get passed on in the form of genetic replication—this successful genetic information, which will itself vary over time, is what Hull means by 'gene lineage' (Hull, 2006, p. 367).

Second, Hull claims that individual organisms are like genes. In the same way that gene lineages exist with a degree of variability, individual organisms form variable lineages by way of reproduction. Crucial to Hull's argument is that, because of both genetic and individual variations over long stretches of time, individuals can belong to the same species regardless of how different they may appear to the naked eye. Here, we see Hull clearly tipping his hat to criterion #2; that is, he acknowledges that organisms are the product of evolutionary history and forces and that, despite considerable variation in physical features (i.e., phenotypes), it is reasonable to think of some individuals belonging to a somewhat inclusive group, while others can reasonably be thought to belong to a somewhat distinct inclusive group (Hull, 2006, p. 368).

Third, drawing on both genetic lineages and individual lineages, Hull concludes that species must also be lineages. The thrust of his argument revolves around *conceptual constraint*. He thinks that species evolve because of selection at the level of genes and individuals. The result is that we are conceptually constrained to accept species as individuals because species lineage is made manifest as a result of gene lineages and individual lineages (Hull, 2006, p. 369).[25] So, when Hull tells us that species are "spatio-temporally localized cohesive and continuous entities (historical entities)," he means that they represent a unit that is a product of genetic and individual lineages, which are themselves a product of adaptations. It is the adaptations to local environments that account for both cohesiveness and continuity. This point reveals that Hull is sensitive to criterion #3; that is, he thinks that real features of organisms (i.e., their adaptations) help to make sense of group cohesiveness and continuity with respect to grouping individuals into species. Thus, according to this reading of Hull's **Argument from Individuality**, **P4** is true by way of gene and individual lineages.

PART-WHOLE ANALOGY

In what can be viewed as further support of **P4** of his **Argument from Individuality**, Hull proceeds to defend the view that species are individuals by relying on part-whole relationships. Recall that **P4** claims that individuals are cohesive and continuous entities. This is true, thinks Hull, because species maintain their cohesiveness and continuity in much the same way that parts of organisms work together to maintain the cohesiveness and continuity of the individual organism (Hull, 2006, p. 377 and p. 382). The justification is that, in the same way that an organism's development requires its interacting parts to function properly in order for the organism to be considered a single whole entity, species are also a single whole entity as a result of the proper functioning of their individual organisms. Thus, Hull thinks he can reasonably conclude that 'individual' is best understood as localized cohesive and continuous entities within the context of biology; that is, *species are individual lineages*. A formal definition is as follows:

> **Species as Individual Lineages** → Population X constitutes a species Y, if and only if, X is a spatiotemporally localized cohesive and continuous historically-based entity.

---

25   Hull's reference to "sets" is, in part, a rejection of the view that the species concept is best understood within the concept of set. For more on this debate over species as sets, see Kitcher (1984 and 1987), Sober (1984), and Stamos (2003, pp. 252–56).

Bringing all this together, Hull reminds the reader that appearance can be deceiving, and it is the overall continuity of the dynamics of populations that solidifies their status as species. So as to give clarity to his emphasis on both cohesiveness and continuity and how they connect with both his genetic/individual and part-whole analogies, he stresses that appearance should be ignored in determining what is and is not part of the same species. Rather, if a group has suffered internal disruption and a significant loss of membership, then one can claim that the existing species has changed. Specifically, if two or more large groups form from the original species and the two groups show significant differences in membership and internal cohesion from the original species, then Hull thinks that these two groups can be viewed as separate species (Hull, 2006, p. 374).[26]

REPLY TO SPECIES AS INDIVIDUAL LINEAGES

Hull's version of **Species as Individual Lineages** is compelling at first glance. Notice that it is designed to: range over all life forms (criterion #1), acknowledge the evolutionary history of populations (criterion #2), and incorporate real organism characteristics (criterion #3) by way of cohesiveness and continuity. Additionally, **Species as Individual Lineages** insightfully can navigate the challenge of nominalism without chucking the reality of species (criterion #4). Shouldn't we, then, proceed to give full endorsement to **Species as Individual Lineages**?

Not so fast! A number of players in this debate have preferred that we not be too quick to give full allegiance to **Species as Individual Lineages**. Their worries range over denying that it satisfies criterion #1 and that the cost of being able to accommodate criterion #3 is too high.[27] Let us explore each of these worries.

First, much like the **Biological Species Concept**, **Species as Individual Lineages** appears to be unable to accommodate asexually reproducing organisms into the species discussion. Primarily, asexual organisms (see the **Biological Species Concept** discussion for examples) cannot produce the requisite cohesiveness and continuity due to the fact that they are reproductively isolated from other individuals. The result is that **Species as Individual Lineages** is forced to accept that asexually reproducing organisms are not individuals and thus are not species (Ghiselin, 1987, p. 138). Furthermore, it is clear that Hull is of two minds on this issue of asexual groups. In the spirit of defense, Hull simply insists that, contrary to the adamant stance of taxonomists, not all organisms belong to a species (Hull, 1987, p. 179). Although not explicitly stated, Hull could be pointing out that the twists and turns of hundreds of millions of years of evolution could very well produce organisms that do

---

26 Note that Hull discusses (2006, pp. 370–76) various ways that membership and cohesion can be disrupted to form new species.

27 This does not exhaust the sorts of criticisms levied against **Species as Individual Lineages**. See Stamos (2003) for some of the other worries.

not have the degree of cohesiveness and continuity to be integrated fully into any particular species group.

Yet, in a somewhat tone of resignation, within the context of the species debate and given his own allegiance to evolutionary theory, Hull understands that "an adequate theory of biological evolution must apply to all organisms—plants as well as animals ..." (1992, p. 185; found in Stamos, 2003, p. 247). Hull is well aware that his concept of species is not immune to this requirement. Indeed, as he notes above, the concept of species is constrained by selection processes working at lower levels (i.e., genes and individuals). It seems, then, that Hull is trying to stave off the asexual-groups-criticism of **Species as Individual Lineages**, but is forced to back-pedal in the light of his own allegiance to evolutionary theory. Reasonably, it is this same hundreds of millions of years of evolution that account for the integrity present in a great many asexual groups, suggesting that they should be well-ensconced in any species account. Thus, all of this might suggest that **Species as Individual Lineages** cannot evade this criticism and should be abandoned.

In the same cautious approach, however, that we are being asked to take regarding our acceptance of **Species as Individual Lineages**, we should be equally circumspect with respect to our rejection of it. In particular, Hull's willingness to acknowledge that **Species as Individual Lineages** does not accommodate asexual groups is unwarranted in the light of both cohesiveness and continuity. Organisms can have methods or properties other than those related to sexual encounters to ensure cohesiveness and continuity. For instance, it should be sufficient that, so long as a group of organisms express a consistent set of reproductive strategies, hunting methods, or foraging techniques, each such expression might be sufficient to constitute being cohesive and continuous so long as such an expression is the product of selection mechanisms. If this is so, then Hull can assert that **Species as Individual Lineages** does range over all populations of cohesive and integrated organisms. Criterion #1 is thus salvageable and so is **Species as Individual Lineages**.[28]

Are we now ready to cast our ballot in favor of **Species as Individual Lineages**? Regrettably, the answer must remain in the negative. The reason is that, because of **Species as Individual Lineages**'s commitment to the historical nature of organisms, it is apparently forced to endorse erroneous views regarding the extinction of species. This implication suggests that Hull's account cannot readily handle criterion #3 because of his strong endorsement of criterion #1. Using an artefact analogy strategy, Hull argues that, in the same way that a building can only be Baroque based on its history and location, a species can only exist as a specific species based on its history and location. As Hull puts it, "*Dodo ineptus* is conceptually the sort of thing as the Baroque period. Both are gone and can never return. Extinction is necessarily forever." Hull's argument can be represented as follows:

---

28 Stamos has also suggested that both Hull and Ghiselin need not cling to sexual reproduction as the only means to attain cohesion and integration. Cautiously, he offers the lives of bacteria as an example (Stamos, 2003, pp. 248–51).

P1. The extinction of any historical entity is necessarily forever.

P2. Species are historical entities.

C1. The extinction of species is necessarily forever.

Hull thinks that the truth of **P1** is made clear by the analogy of the Baroque-style building example. An artefact counts as Baroque so long as it has a particular set of aesthetic features and was made during the seventeenth century. Both conditions need to be satisfied, argues Hull, in order for an item to be considered genuinely Baroque. By implication, if every Baroque item were destroyed, then it follows that not only is it the case that no Baroque item exists, but also that no Baroque item could *ever* exist again. In the same way, a species is an entity that has a particular history. Much like the Baroque-style entities, a species ceases to exist when all its members cease to exist and any species (like the Dodo bird species) that loses all of its members *ipso facto* remains out of existence forever. Thus, according to Hull, species-as-historical-entities implies that extinction is forever.

In reply to Hull's species-extinction argument, Stamos offers a language-argument-from-analogy. In short, Stamos argues that, since even dead languages can be resurrected and subsequently modified, so might species be (Stamos, 2003, p. 245). This is why Stamos thinks that "a genetically engineered dodo would not be like a replica of a Tiffany lamp; it would be like a Tiffany lamp, the real thing (Stamos, 2003, pp. 245–46). Stamos is here rejecting Hull's use of the history of an artefact to make sense of the possible resurrection of extinct biological entities. Rather, he thinks that language is a better way of making sense of species. Ultimately, Stamos' criticism is designed to show that **P1** (above) is false; namely that there are some historical entities, like languages, that are not necessarily the kinds of entities that can be genuinely extinct. What this means is that Hull's use of both "any" and "necessarily" in **P1** is simply too strong. What follows is that the extinction implication of **Species as Individual Lineages** is not nearly the boon that Hull considers it to be. If correct, Stamos has offered a plausible reason for rejecting **Species as Individual Lineages**. Again, because Hull gives so much emphasis to the historical nature of species (criterion #1), he does not give enough regard to the actual features of organisms (criterion #3). Stamos attempts to show this point by way of this exinction discussion.

It is difficult not to take Stamos' side on this argument. If Dr. Dodostein creates a dodo from existing dodo DNA, it seems hard to understand why this token dodo is not part of the dodo line. The same could be said for plants or other organisms. Now, it cannot be thought that the difference in location/environment is a difference that makes a species difference. What if Dr. Dodostein went back in time and simply brought back a dodo with him to our present time. Is this time-transported dodo still part of its species? If Hull responds in the negative, then he would be committed to location/environment and time as part of defining a species. Yet this seems problematic. What if a seventeenth century European explorer simply took a dodo back to his home country? Is this transported dodo still part of its species or does it lose its species status the moment it encounters any novelty that is not part of its species-specific environment? The worry is that Hull needs to provide a range of novelty that is legitimate and illegitimate with respect to a population's environment. This is going to be extremely difficult to do because the very history upon which Hull relies is littered with surprising twists and turns. Indeed, embracing such twists and turns is the hallmark of a Darwinian state of mind. At first glance, at any rate, Hull's rendition of **Species as Individual Lineages** must either bite the bullet and insist that loss of species designation with respect to environmental novelty is acceptable or abandon the claim that extinction is relevant to biological organisms. Ultimately, though, the former route appears unwarranted from an evolutionary perspective and being resigned to the latter alternative might very well require relinquishing **Species as Individual Lineages** in its present form.[29]

To complicate matters a bit further, what are we to make of **Species as Individual Lineages** in the light of the nominalist challenge? If species are individuals, then does Hull embrace the exclusive nominalist position? The answer to this question is not obvious. In fact, it appears that Hull wants to have it both ways; that is, he wants species to be both individuals and a distinct sort of unit out in the world. Primarily, Hull is clear that species can be distinguished by being cohesive and continuous entities. So, the properties of cohesiveness and continuity allow one to infer what organisms belong to what species. From this perspective, species are out in the world; this clearly tells against exclusive nominalism—specifically, the properties of cohesiveness and continuity are properties out in the world that assist in distinguishing one population from another as distinct species.

At the same time, Hull thinks that species are individuals in a way that would move him in the direction of exclusive nominalism. Specifically, he argues that, regardless of overall similarity, each organism is a unique individual—even identical twins—as a result of occupying a distinct location in space/time and ontogenetic continuity. In the same way that individual organisms are unique individuals, Hull argues that species are also unique individuals based on their spatiotemporal location and continuity (Hull, 1976, p. 176). So, Hull thinks he is able to conclude that species are individuals like individual organisms. What follows from this is that Hull can argue that his species concept not only allows for the existence of species out-in-the-world, but also that his account reveals the falsity of exclusive nominalism. The implication is that species are, for Hull, continuous and cohesive individual historical entities that are composed of unique individual organisms.

It is not clear that the exclusive nominalist defenders would be too quick to congratulate Hull on his ability to reconcile his rendition of **Species as Individual Lineages** with their brand of nominalism. Primarily, Hull has not shown that the presence of any additional feature associated with location and continuity cannot be reasonably explained by the activities of individual organisms. This is not to deny that causal

---

29 For more on this particular debate, see Zachos (2016, ch. 3).

relations are an integral part of organism activity and interactions. Rather, it is to deny that such activities and interactions are sufficient to warrant the addition of species into our ontological space. For example, the mutualism exhibited by the security guard look-out Goby fish and the house-building Pistol shrimp reveals a rather impressive level of continuity within a restricted space. This sort of symbiosis, however, would hardly suggest that these unlikely partners are part of the same species.[30] The same goes for the reciprocal benefits gained by the parasite-eating Remora fish and the resulting clean and healthy shark to which it attaches. The point is that as much as we should be on-guard with respect to appearances in terms of species ascription, we should also keep a watchful eye on the continuity of interaction and location of organisms as being species locators. What follows is that **Species as Individual Lineages** is not really able to show that species should be included in our ontology in the light of continuity of interaction between individuals. Thus, defenders of exclusive nominalism can insist that Hull's version of **Species as Individual Lineages** can readily align itself with nominalism (criterion #4), but cannot maintain species as part of the fabric of the natural world. It seems, then, that the challenge put forth by the exclusive nominalist remains intact.

Briefly, in terms of our criteria, **Species as Individual Lineages** welcomes the Darwinian turn of viewing organisms as evolutionary entities (criterion #2) and, with some tinkering, could possibly range over all life forms (criterion #1). Notably, however, **Species as Individual Lineages** gives considerable weight to population cohesiveness and continuity, but it does not focus on particular features of organisms in order to determine species inclusion, exclusion, and extinction (criterion #3). Finally, **Species as Individual Lineages** appears to struggle in its attempt to embrace nominalism (criterion #4) without fully jettisoning the metaphysical reality of species. Table 5.6 brings all of this together.

	CRITERION #1	CRITERION #2	CRITERION #3	CRITERION #4
Species as Individual Lineages	Yes	Yes	No (or not easily)	No (or not easily)

Table 5.6 *Species as Individual Lineages and The Four-Fold Species Criteria*

Phylogenetic-Cladistic Species Concept

Given the difficulties with the non-historical species concepts and **Species as Individual Lineages**, the search for a persuasive alternative continues. Specifically, the exploration

---

30 Briefly, the Pistol shrimp dig holes in which both it and the Goby fish live. As part of the deal, the Goby fish acts as a look-out (by way of tactile communication) for the poor-sighted shrimp while it is digging. The result is a shared domicile and a secure area in which to eat. For more on this interesting sort of mutualism, see Judson (2008) and Duerbaum (2013).

has shifted in the direction of an account that is respectful of organisms as evolved entities, but not merely as extended lineages. Also, given the roundabout paths that evolution has taken, the additional requirement is that a robust species concept will take seriously shared features of organisms, but not fall into the arbitrariness-trap of relying solely on observable phenotypic features as the linchpin of analysis. (We saw earlier why this proved to be a problem for pheneticism.) It is the **Phylogenetic-Cladistic Species Concept,** also known as cladism (from the Greek word *klados* [κλάδος], meaning branching or subdividing), that purportedly meets these requirements. Basically, the **Phylogenetic-Cladistic Species Concept** is the view that a group of organisms constitutes a species so long as there is a clear line of ancestral descent.[31] This line of descent, picturesquely, takes on a branching pattern. Consider the following generic branching pattern:

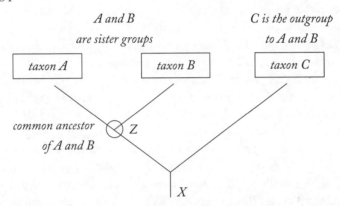

The above branching image is what is known as a *cladogram*. It is designed to show the various points at which populations branch-off from other populations and how closely related some populations are to other populations. At the bottom of the cladogram, there is an initial branching, **X**, that generated populations **A, B,** and **C**. This means that there was a distant common ancestor shared by these populations. For example, we can imagine that **A, B,** and **C** are rabbit populations. The cladogram suggests that there was a prior rabbit (or rabbit-like) population **X** off of which **A, B,** and **C** branched. We further observe that there is an additional split between **A** and **B** who shared a more recent common rabbit or rabbit-like ancestor **Z** not shared by **C**. **A** and **B** are considered sister groups because of the shared recent common ancestor. **C** is considered an outgroup because it lacks the recent common ancestor, **Z**, had by **A** and **B**.

So, how do defenders of the **Phylogenetic-Cladistic Species Concept** conceive of species in the light of cladogram thinking? There are two conditions that have to be satisfied for species inclusion according to the **Phylogenetic-Cladistic Species Concept**. (*At this point, some technical terms, unique to phylogenetics, need to be introduced. Please keep track of them; they are crucial to understanding the ensuing exchange of ideas.*) The

---

31 The cladistic approach has its origin in the work of Hennig (1979). See Panchen (1992, chapters 7–10) regarding Hennig's work and those who advanced it.

first condition is that the only groups that are included in a single species are those that belong to the same common ancestor. Such groups are known as *monophyletic* groups. This means that a species is the collection of groups that includes all of the descendants of a common ancestor. So, rabbit groups **A** and **B** in the above cladogram could be considered part of the same species because they share a common ancestor not shared by **C**. On the **Phylogenetic-Cladistic Species Concept** account, no matter how similar in appearance **C** is to **A** and **B**, **C** is not part of the same species as **A** and **B** (Garvey, 2007, p. 150).

We must plumb a bit further to grasp the insight of the second condition. There is a technical reason why defenders of the **Phylogenetic-Cladistic Species Concept** proceed as they do. Primarily, their goal is to capture the true unfolding of life's history and to do this, they think, we need to latch onto only the relevant ensemble of characteristics that forces us to distinguish some groups from others. And yes, at times, this compels us to accept counter-intuitive species groupings. For example, because some reptiles branch off at the same time as the Aves group (i.e., birds), there are no features present in the Aves group that are not present in the fossil record related to reptiles or dinosaurs. So, counter-intuitively (based on appearance), the Aves "class" has lost its class status and is now grouped within Reptilia class based on the phylogenetic-cladistic approach. Similarly, a crocodile looks more like a lizard than a bird, but counter-intuitively is more closely related to the latter than the former through the lens of the phylogenetic-cladistic approach. The primary justification given for this is the idea of derived shared character traits (*synapomorphies*) as distinct from character traits that are ancestrally similar (*symplesiomorphies*). To illustrate, let us assume that rabbit populations **A** and **B** both have dark hair due to the selective advantage of this feature, while rabbit population **C** has gray hair from an ancestral population. Upon close genetic examination, it is determined that a certain ancestor **Z** of **A** and **B** had dark hair. Two possibilities present themselves: (1) either **A** and **B** have this feature as a result of selection working on a single common ancestor **Z** or (2) two separate selection scenarios from distinct ancestral populations independently produced the presence of dark hair in the two populations. In our scenario, **A** and **B** are more closely related to each other than either is related to **C** (holding all other features constant), if the dark hair trait is the product of their common ancestor **Z**; that is, if dark hair is a *synapomorphy* (that is, a shared derived feature), then it is a difference that makes a difference with respect to species inclusion for the **Phylogenetic-Cladistic Species Concept**. If the dark hair trait in **A** and **B** is derived from distinct selection events from separate ancestral populations, then the feature is a *symplesiomorphy* and is not a difference maker with respect to determining **A** and **B**'s relationship to **C**. As our example assumes, rabbit populations **A** and **B** are part of the same species (and **C** is ruled out of this species) because dark hair is determined to be a *synapomorphy*. Of course, all of this gets complicated quickly when numerous ensembles of characteristics are used to determine species inclusion and exclusion (more on this below). Regardless, the value given to *synapomorphies* over *symplesiomorphies* does not change for the **Phylogenetic-Cladistic Species Concept** (Ridley, 1996, p. 383). Thus, the second condition of requiring the presence of some collection of *synapomorphies* is crucial for the **Phylogenetic-Cladistic Species Concept**'s determination of species inclusion or exclusion. A formal definition looks like this:

> **Phylogenetic-Cladistic Species Concept** → X and Y belongs to the same species Z, if and only if, X and Y have a common ancestor A based on a variety of derived-shared synapomorphic characteristics D.

Now, one might be tempted to balk at such a radical implication regarding the indifference with respect to appearance. You might be thinking to yourself: If an organism looks, acts, and sounds like a duck, then (damn it!) it is a duck! Yet, the natural world is replete with examples that would vindicate the **Phylogenetic-Cladistic Species Concept** advocates. For instance, some hedgehogs look strikingly similar to porcupines, but they are not part of the same species; the tasty non-toxic Viceroy butterfly looks remarkably like the toxic-tasting Monarch butterfly. Again, these butterflies are not part of the same species. This strategic form of mimicry (in the butterfly case) is not uncommon in nature and should stand as a reminder that one should not impulsively be enamored by mere appearances. Rather, claim the **Phylogenetic-Cladistic Species Concept** supporters, if we stick to organizing groups in a monophyletic way (i.e., in a way that understands a collection of organisms as a group of organisms that includes the most recent ancestor of the group and all of its descendants), then we can be confident that we have captured the actual branching pattern of life in the natural world as it is. This is the metaphysical claim put forth by the **Phylogenetic-Cladistic Species Concept** (Sterelny and Griffiths, 1999, p. 197). Undeniably, champions of the **Phylogenetic-Cladistic Species Concept**, fully willing to embrace the vagaries of evolution, insist that such supposed counter-intuitive results should not be of much surprise to the reflective evolutionist. This is pretty much what should be expected, think upholders of the **Phylogenetic-Cladistic Species Concept**, once we flip-on our Darwinian lenses—the world looks much more different (and potentially more transparent) than whatever other lenses through which we happen to be gazing!

REPLY TO THE PHYLOGENETIC-CLADISTIC SPECIES CONCEPT

It is easy to see why the **Phylogenetic-Cladistic Species Concept** is one of the dominant conceptual frameworks in the species debate today.[32] At first glance, it is able to draw on the importance of taking physical characteristics seriously (criterion #3), while avoiding the pitfalls of *pheneticism*; it is able to accommodate evolutionary history without being encumbered by the **Species as Individual Lineages**'s unconstrained endorsement of lineage-as-individual (criterion #2); and it is able to range over all organisms in a way that the **Biological Species Concept** is unable (criterion #1). Critics, however,

---

32 See Brandon and Mishler (1996) for this sort of defense.

have been unmoved by these supposed advances made by the **Phylogenetic-Cladistic Species Concept**.

One criticism put forth by Coyne and Orr is that an implication of the **Phylogenetic-Cladistic Species Concept** is that "it would tremendously increase the number of named species" (2010, p. 283). Their reasoning is that the **Phylogenetic-Cladistic Species Concept** is committed to the view that "any trait can serve to diagnose a new species, even one as trivial as a small difference in color or a single nucleotide difference in DNA sequence" (2010, p. 283). So, for instance, the rabbit example used in this discussion would fit their worry. If brown hair is the only substantive difference between populations **A**, **B**, and **C**, Coyne and Orr can see no reason to carve-out two distinct species. In a sense, Coyne and Orr are levying the same criticism of the **Phylogenetic-Cladistic Species Concept** that was levied against pheneticism, namely that the choice of features that constitute species inclusion is arbitrary. The upshot is that the **Phylogenetic-Cladistic Species Concept** is not really capturing significant species identities.

No doubt, the **Phylogenetic-Cladistic Species Concept** guardians will reply that its focus on only monophyletic groups renders the arbitrariness criticism dead in the water. There is nothing arbitrary about locating the gene sequences that reveal the presence of synapomorphies. If DNA sequencing reveals brown hair to be a unique difference between **A** and **B** with respect to **C**, then a genuine synapomorphy is present. Thus, a unique monophyletic group can be codified. The result is that the **Phylogenetic-Cladistic Species Concept**'s analysis will be in keeping with evolutionary history and its overall branching patterns of life. If more species are revealed to exist, then so be it! Thus, the **Phylogenetic-Cladistic Species Concept** guardians will insist that the arbitrariness criticism is completely off the mark.

Coyne and Orr are aware of this sort of response, but demur that it is very difficult, *in practice*, to determine what gene combinations are present and/or absent in **A** and **B** such that one can definitively show that **A** and **B** are more closely related to each other than either is to **C** (Coyne and Orr, 2010, p. 284). Their reasoning for why this is so difficult in practice is that branching can occur (as a result of geographic or reproductive isolation) without the entire elimination of ancestral genes. It is only after considerable time has passed, with the assistance of natural selection and genetic drift, that a unique set of gene copies from a specific common ancestor solidified ("coalesce" in their argot). So, how are the **Phylogenetic-Cladistic Species Concept** practitioners to know (with respect to a specific set of groups, like our rabbit demes) whether or not a set of gene copies has coalesced in order to be recognized as a part of a unique species grouping? Their answer, think Coyne and Orr, is that these practitioners will have great difficulty revealing/locating coalescence in practice because different sets of gene combinations will very likely produce conflicting genealogical trajectories (Coyne and Orr, 2010, p. 285); that is, the use of one set of gene combinations will show no substantial coalescence between **A**, **B**, and **C**, while another set might actually reveal a closer unity between **A** and **C** relative to **B**, etc. Any snapshot of gene sequencing will likely not tell you whether or not coalescence has been achieved (even if it has). The implication here is that **Phylogenetic-Cladistic Species Concept** practitioners cannot know or convincingly justify their species groupings. Thus, Coyne and Orr caution that, although the **Phylogenetic-Cladistic Species Concept** delivers the theoretical impression of offering the best species concept to date, it is not nearly as copacetic in practice as its adherents might lead us to believe.

It is not clear that this is a knockdown argument against the **Phylogenetic-Cladistic Species Concept**. Primarily, its defenders could simply concede that it is difficult to show the presence of sets of coalescent synapomorphies, which constitute unique monophyletic groups, but it is not impossible to do so; and, on those occasions when it is possible, unique species can be identified. Yet, Coyne and Orr counter that on the occasions that the **Phylogenetic-Cladistic Species Concept** does get it right, it will be on those very same occasions when a number of other species concepts would also get it right (Coyne and Orr, 2010, pp. 281–88). Thus, the **Phylogenetic-Cladistic Species Concept** does not appear to be a superior species concept than any other. For instance, imagine that in the presence of geographic isolation, the **Phylogenetic-Cladistic Species Concept** is able to establish a set of populations that constitute a unique species. Defenders of the **Biological Species Concept** and **Species as Individual Lineages**, for instance, will also arrive at the same conclusion because either the lack of gene flow in such populations or population continuity is in keeping with their respective accounts. Thus, the **Phylogenetic-Cladistic Species Concept** appears to be a viable species concept candidate, but it does not seem to single itself out as well as its proponents profess.

To make matters a bit worse, Garvey argues that the **Phylogenetic-Cladistic Species Concept** may have difficulty making sense of how to designate some plant hybrids. The issue is that depending upon how hybridization unfolds, a **Phylogenetic-Cladistic Species Concept** practitioner has the unenviable task of figuring out if a new hybrid belongs to its mother's or father's monophyletic group (Garvey, 2007, p. 154). Imagine a new hybrid of Phalaenopsis orchid (usually these orchids have been crossed with other species in order to produce the eye-catching large, white, purple and/or striped flat-petaled flowers). Does this new hybrid species belong to the monophyletic group on the mother's side or the monophyletic group on the father's side? There is no easy answer here, revealing that monophyly can lead to some conceptual problems (Garvey, 2007, p. 154) with respect to the proper designation of hybrids.

With respect to our criteria, the above concern regarding hybrids reveals that the **Phylogenetic-Cladistic Species Concept** is not able to range over all life forms as was initially thought (criterion #1). In the case of hybrids, it simply cannot make appropriate grouping sense of competing monophyletic groups. Since it can be assumed that there will be overlapping sets of synapomorphies (from mother monophyletic group and father monophyletic group) present in the hybrid, the **Phylogenetic-Cladistic Species Concept**'s strict reliance on monophyly—as a difference that makes a species-demarcation difference—proves not to be able to settle the matter in these hybrid scenarios. Thus, if worry about the proper designation of hybrids is on the mark, then this is yet another reason not to put all of our eggs in the **Phylogenetic-Cladistic Species Concept** basket.

Still, there is a reason why the **Phylogenetic-Cladistic Species Concept** has a strong allegiance (Wilkins, 2003). Not only is it sensitive to taking an evolutionary perspective with respect to biological life (criterion #2), but it does so in a way that focuses on actual features of organisms (criterion #3). Yet, in terms of the nominalism issue, it is unclear if the monophyletic methodology tells against the exclusive nominalist challenge; that is, drawing upon Garvey's worry, exclusive nominalists could insist that the **Phylogenetic-Cladistic Species Concept**'s species groupings do not settle the ontological concern about adding species to the world. For example, consider our previously discussed rabbit populations. Assume that there is a unique set of synapomorphies that distinguish **A** and **B** from **C**. Further assume that **A** and **B** are the product of hybrid populations. If a subset of the synapomorphies of **A** and **B** are from the maternal side of some hybrid population and another subset is from the paternal side of some other hybrid population, it is still arbitrary to treat **A** and **B** as distinctly related species even if they share very little with respect to **C**. Interestingly, such odd hybrid populations are not entirely uncommon with respect to plants and insects. Thus, the nominalist/individuality challenge could withstand attempts by the **Phylogenetic-Cladistic Species Concept** to insist that its species demarcations are genuinely capturing the way the world is. They can insist that the complications and arbitrariness concerns that stem from hybrid scenarios actually help to magnify why it might be epistemically soothing to carve-up organisms in this branching fashion, but it does not make any serious ontological dent in the exclusive nominalist individuality blockade. Thus, given the limitations of the **Phylogenetic-Cladistic Species Concept**, it might be better to keep searching for the illusive best-bet species concept. You can remind yourself of this discussion by briefly reflecting on Table 5.7.

	CRITERION #1	CRITERION #2	CRITERION #3	CRITERION #4
Phylogenetic-Cladistic Species Concept	No	Yes	Yes	No

Table 5.7 *Phylogenetic-Cladistic Species Concept and The Four-Fold Species Criteria*

Relational Biosimilarity Species Concept

It appears, then, that even the **Phylogenetic-Cladistic Species Concept** has numerous shortcomings, leaving us without a viable species concept. Recall that the non-historical accounts and the historical species concepts, including the **Phylogenetic-Cladistic Species Concept**, appear to fall short with respect to our discussed species criteria. Where, then, are we to find an account that is sensitive to evolutionary history, can incorporate both sexually and non-sexually reproducing organisms, and is respectful of taking actual physical features seriously? One location would be in the work of

Stamos, who provides what he thinks is a concept of species that may be the answer to our query. Primarily, he reasonably reminds the reader that biological organisms are constituted by a host of relations. These relations include interbreeding, mating, nurturing, developmental, social, competitive, phenotypic similarity, etc. (Stamos, 2003, p. 297). Stamos is arguing that the different ways in which organisms are similar (e.g., phenotypic features or ontogenetic development) are the product of the causal forces associated with the aforementioned host of relations—a bundle of relations! It is because of these causal relations that one can confidently claim that X is a different species from Y.

What is being described is Stamos' **Relational Biosimilarity Species Concept**. It is comprised of four elements. First, for Stamos, 'horizontal' (as opposed to vertical) refers to "species that exist at any one time" (Stamos, 2003, p. 67).[33] This captures what is meant by a non-historical perspective. Note, however, he does not want to give the reader the impression that organisms are lifeless; so, he adds the second point concerning species being dynamic. "Dynamic," in this context, appears to refer to the fact that biological systems evolve, but species delimitation requires focusing, not on protracted time scales, but on more short-term changes. Third, Stamos thinks that phenotypic similarities cannot be ignored with respect to locating species out in the world. Clearly, he thinks that organisms that look alike are most likely to be part of the same species. His justification for this is his fourth point—causal relations. That is, 'species' refers to those living beings—both sexual and asexual reproducing beings—who have phenotypic similarities produced by a shared set of causal relations. With respect to sexually reproducing organisms, this includes interbreeding relations, ontogenetic relations, ecological relations, and social relations; and for asexually reproducing organisms (e.g., plants), the causal relations would include ecological, gene transfer, and social relations (Stamos, 2003, p. 297). Stamos' **Relational Biosimilarity Species Concept** can be summarized as follows:

> **Relational Biosimilarity Species Concept** → An entity E necessarily belongs to species X, if and only if, E exhibits a set of phenotypic characteristics **P** that is circumscribed by a set of causal relations **R**, which will likely vary, depending upon whether or not E is a sexually or asexually reproducing organism.

Returning to our rabbit populations, Stamos can claim that his **Relational Biosimilarity Species Concept** best captures why **A** and **B** should be grouped together as a species and C should be ruled out. Primarily, the bundle of causal relations that exist in **A** and in **B** reveal a kind of isomorphism not present with respect to the bundle

---

33 Note that Stamos does not mean a few minutes or seconds when he uses the concept of horizontal. Rather, what he means is that the snapshot of the tree of life that is relevant to the species debate is the minimum time period required for the various causal relations to establish a species (Stamos, 2003, pp. 67–74).

of causal relations that **C** has in comparison. For instance, **A** and **B** may engage in a set of predator avoidance behaviors not present in **C**. These behaviors, the product of causal relations in the local environment, along with the mix of other causal relations, help to reveal species relatedness. The Relational Biosimilarity Species Concept can offer the same kind of analysis with respect to plants and other asexual entities. Thus, at first glance, it appears that Stamos has provided an insightful resolution to the species concern.

In terms of our criteria, since **Relational Biosimilarity Species Concept** is rather sensitive to organism features (in terms of the various causal relations that produce them), it clearly supports criterion #3. Plus, this species concept can be employed to demarcate both sexual and asexual reproducing organisms and, ultimately lay claim to ranging over all life forms (criterion #1). And, as a further boon, the **Relational Biosimilarity Species Concept** is respectful of organisms as evolutionary entities in terms of understanding that biological systems change and manifest an overall dynamism over long stretches of time (criterion #2).

REPLY TO THE RELATIONAL BIOSIMILARITY SPECIES CONCEPT

As a preliminary response, those who defend criterion #2 might be suspicious of the **Relational Biosimilarity Species Concept**'s supposed endorsement of organisms as evolutionary beings.[34] Rather, they could insist that its support is half-hearted at best. Most pressingly, the **Relational Biosimilarity Species Concept**'s ability to align itself with criterion #3 is precisely why it cannot fully incorporate criterion #2; that is, since causal relations requires including existing organism features and their various interactions, evolutionary time scale considerations appear to be rather secondary for the **Relational Biosimilarity Species Concept**. Stamos appears to concede as much when he claims that the horizontal perspective (i.e., the more immediate time scales) is the only legitimate perspective regarding the questions concerning the reality and number of species (Stamos, 2003, p. 317).

Yet, Stamos is aware that the above claim gives the impression that his **Relational Biosimilarity Species Concept** ignores entirely the view that biological organisms are the product of a long (i.e., vertical) evolutionary history. He responds that this sort of worry is the "deepest of all" misunderstandings (Stamos, 2003, p. 317). Stamos uses the example of three varieties of the Welsh ragwort (a flowering plant [*S. cambrensis*] found in the UK), to illustrate how his **Relational Biosimilarity Species Concept** is compatible with thinking of organisms in a vertical way. Briefly, he explains that phenotypic appearances, genetic, geographic, and enzyme analyses reveal that there were "two or three separate origins of *S. cambrensis* in Britain during this century"

---

34 In fairness to Stamos, he offers a remarkably sustained defense of his **Relational Biosimilarity Species Concept**. I cannot engage in all of the intricacies in this brief offering. I can only direct the reader to his impressive efforts (Stamos, 2003, ch. 5).

(Stamos, 2003, p. 320). He then goes on to conclude that "the moral of the story is that history reconstruction in species biology countenances a wide variety of methods and that the biosimilarity species concept is fully consistent with this fact" (Stamos, 2003, p. 320).

At first glance, it is hard to see the connection that Stamos is making. If he means that his **Relational Biosimilarity Species Concept** can accommodate all of these various methods in order to make sense of the historical trajectory of organisms, then he misses the point of the "vertical critic." The reason is that Stamos' **Relational Biosimilarity Species Concept** can, as he claims, acknowledge the efforts of the historical camp, but does not in any clear way employ their efforts in his causal relations account of species designation. Stamos might have more to offer in reply, but the preliminary conclusion is that his **Relational Biosimilarity Species Concept** does not really do much to take seriously criterion #2. So, although the **Relational Biosimilarity Species Concept** does well to incorporate criteria #1 and #3, it struggles to handle criterion #2.

Given all the species concepts thus far, Stamos has the most to say about the nominalist challenge (criterion #4). Before directly addressing Stamos' analysis, it is important to remind ourselves of what is at stake. Recall, exclusive nominalists insist that there are only individual organisms out in the world, while universals in the form of sets, properties, features, etc. are merely useful labels constructed by the mind. For example, there are individual brown horses out in the world, but the-set-of-brown-horses is a convenient labeling device. Realists, however, insist that there are sets, properties, features etc. out in the world; for instance, the-set-of-brown-horses is out in the world. The concern at this point is whether or not Stamos' **Relational Biosimilarity Species Concept** can tell against the exclusive nominalist challenge. More specifically, it remains to be seen if Stamos' causal relations are universals out in the world. In order for his account to be successful, he must show that his causal relation properties are repeatable; that is, in the same way that whiteness can be repeated across space and time without changing (this is known as the requirement of repeatability/numerical identity discussed earlier in the chapter [p. 179]), he has to show, for example, that chimpanzee social and sexual relations generated by hunting patterns are also repeatable. To illustrate, female chimpanzees who beg to eat captured prey are only allowed a portion of meat after they have acquiesced to copulate with the male(s) in possession of the meat.[35] In this example, there is a complex set of relations working together. For instance, diet and sexuality relations come together such that the bringing to fruition of the meat-diet relation further causes the existence of the sexuality relation. In what sense, then, can Stamos' **Relational Biosimilarity Species Concept** deal with this criterion of repeatability/numerical identity? His answer is as follows:

---

35   For further details about chimpanzee hunting behavior, see Stanford (1995, p. 256).

> [A]lthough an individual organism, for example, cannot recur while retaining its numerical identity, perhaps each of its properties can. And if each of its properties can, then perhaps the something more in each of the similarity relations of which they are a part can also recur while retaining their numerical identity. (Stamos, 2003, p. 351)

Using the chimpanzee example, Stamos is arguing that, although the same chimps cannot exist again and still be the same chimps, it is possible that their properties can. For example, being sexually subservient in the presence of a male chimp holding monkey meat is a relational property of female chimps to sexually stimulated meat-carrying male chimps. This relational property can recur, according to Stamos' account, in the presence of a sex-demanding male chimp in possession of monkey meat and a monkey-meat begging female chimp in close proximity. Thus, if this account is correct, then Stamos can claim that he has located universals out in the world. As Stamos puts it, "All species ... share the same ontology: they are intricate complexes of phenotypic similarity relations *supervening* on a partially disjunctive base of causal relations" (Stamos, 2003, p. 314; my italics). More will be said about this claim below, but the upshot is that exclusive nominalism could very well be vanquished by Stamos' **Relational Biosimilarity Species Concept**.

There are two separate responses to Stamos' account that may tell against his attempt to fight off exclusive nominalism. The first is related to the specifics of his reliance on causal relations. Let us grant that these various causal relations are irreducible and are required to appreciate fully the nature of organisms. The concern, as noted above, is whether or not there is genuine replication of such relations. It is replication of features, properties, relations, etc., that is a necessary condition to make the realist claim that such properties are mind-independent parts of the world. So, if we grant Stamos the existence of relational properties, has he vanquished exclusive nominalism? Well, the *inclusive nominalist* (Loux uses the phrase *trope nominalism* for inclusive nominalism) would be quick to respond in the negative. Loux explains that inclusive nominalists insist that: there are only token instances of attributes such that colors, shapes, properties, relations, etc., are simply a bunch of unique colors, shapes, properties, relations, etc. (Loux, 2002, pp. 84–85). For instance, there are individual instances of the color red and these instances are unique to the extent that they are exemplified by unique individual things. Two instances of the color red on Ball A and Ball B respectively are not replications of redness, but are token instances of Red Ball A and Red Ball B; however similar Ball A may appear to Ball B, the inclusive nominalist insists that there must be some difference with respect to any property—*otherwise any so-called identical properties would really only be one property. For instance, if both Ball A and Ball B really have the identical color red, then there cannot really be two instances of red*. Similarly, the relational property (related to sexuality and diet) present in the female monkey and sex-seeking and meat-clutching male monkey is unique to that specific female and male interaction and can never be replicated. Thus, contrary to Stamos, the inclusive nominalist can grant him his causal relations, but still claim that nominalism remains true. The upshot is that Stamos' **Relational Biosimilarity Species Concept** will have a tough time trying to circumvent the nominalist challenge because the exact property or causal relation can never be repeated. If the inclusive nominalist critique is on the mark, then talk of species and their supervenient properties being "out in the world" remains that—just talk.

I will not try to resolve this metaphysical quandary further here.[36] The inclusive nominalist may not have the last word against the **Relational Biosimilarity Species Concept**, but suffice it to say that this sort of challenge from the nominalist will not go away easily. Still, Stamos may be able to claim victory over the exclusive nominalist if he is willing to accept inclusive nominalism.

Now, the second issue is related to Stamos' claim above that relational properties *supervene* over their base constituents. 'Supervenience' is a term that is peculiar to philosophy and requires explication. Principally, it is the idea that Y-properties naturally supervene on X-properties if any two naturally possible scenarios with the same X-properties have the same Y-properties (Chalmers, 1996, p. 36). For example, the property of being charitable can supervene over both the giving of money to those less fortunate and the giving of one's time to those who need assistance of some sort. In both instances, the base property of giving something to those in need (X) produces the same property of being charitable. Similarly, the causal relation (Y) of the sex-and-diet chimp scenario supervenes over the base property (X) related to male-female chimp sexual dynamics with respect to food. So long as the same base property (X) is present, the same causal relation property (Y) will be present. In a sense, Stamos is arguing that species-level causal relations are supervenient on base properties. Since those base properties can be replicated by future organisms, it is ontologically reasonable, Stamos could argue, to claim that the species-level causal relations are natural properties out in the world.

The problem with supervenience is that it assumes what requires independent justification; namely, it needs to be shown that base properties can really be multiply realized. Restated, it is reasonable to take supervenient properties seriously (i.e., including them in our ontology), if multiple realizability can actually occur. If multiple realizability is dismissed as metaphysically extravagant, then supervenience is rendered moribund at least with respect to the species debate. If supervenience is dead in the water, then so is the **Relational Biosimilarity Species Concept**'s attempt to fight off some forms of nominalism. As we have already seen, exclusive nominalists deny the possibility of multiple realizability because of their insistence on the nominalist nature of all properties and relations (note that it is not being assumed that multiple realizability is a necessary condition for supervenience, but only that it is relevant to Stamos's analysis). No doubt, this by itself is not a clear vindication of exclusive nominalism. Rather, it should be seen as a challenge that Stamos needs to overcome. Thus, at first glance, unless Stamos can offer a persuasive reply against the exclusive nominalist, his **Relational Biosimilarity Species Concept** (and its reliance on supervenience) does not to tell against the exclusive nominalist challenge (criterion #4). Table 5.8 offers a reminder.

---

36   For more on inclusive (trope) nominalism, see Schaffer (2001).

	CRITERION #1	CRITERION #2	CRITERION #3	CRITERION #4
Relational Biosimilarity Species Concept	Yes	No	Yes	No (But a really good try!)

Table 5.8 *Relational Biosimilarity Species Concept and The Four-Fold Species Criteria*

VII. A PLEA FOR PLURALISM AND CONCLUDING REMARKS

It appears that we have arrived at an impasse. There does not seem to be a persuasive concept of species to be found in this bunch. Yet, I must remind the reader of two points. First, there are numerous species accounts that have not been explored at all in this chapter.[37] One of these may prove to be quite robust. Second, however forceful some of the arguments may seem to be with respect to the species accounts discussed, this analysis is certainly open to further objections and replies. One such response is that the very criteria used in this chapter are over-demanding. For instance, *pluralists* claim that, in the light of the depth of variation produced by evolutionary forces, different species concepts are needed to capture nature's diversity.[38] Some pluralists think that, for example, criterion #1 sets the bar too high because the various subsets of complex causal relations involved in bringing organisms together are not ontologically any better than any other possible subset of causal relations (Dupré, 1993 and Kitcher, 1984). From this perspective, biologists can produce different groupings of the same population of organisms based upon the subset of causal relations that suits their interests. Other pluralists (e.g., Mishler and Brandon, 1996 and Ereshefsky, 2001) reject criterion #4 and insist that species are lineages that are distinguished by "interbreeding, ecological, or phylogenetic relations" (discussed in Richards, 2008, p. 182) that cannot be unified by any ranking system. Yet another pluralist camp combines epistemic species concepts with a particular ontological thrust. For instance, Mayden (1997) thinks that species are one long series of ancestors that vary based on particular evolutionary trajectories (this is the ontological claim). He stresses, however, that the subtleties of evolution have made it very difficult to carve apart these trajectories. The result is that epistemically useful species concepts must be conjoined with his evolutionary account to distinguish different species (this is the epistemic claim). From this perspective, different species concepts can be employed to organize different sorts of populations.

---

37 For example, just for starters, see the species concepts offered by Mallet (1995), Templeton (1989), and Van Valen (1976).

38 Pluralism in the species debate has many faces that cannot be put on display here. Ereshefsky (1992) does a fine job of examining many of these different versions.

How palatable you find these versions of pluralism may depend upon how seriously you take the four-fold criteria. If you think the criteria must be satisfied, you will not only reject all of these versions of pluralism, but you will also be suspicious of the various species concepts discussed in this chapter. Alternatively, if you reject some or all of the criteria, you may find one of the species concepts much more appealing than the others. Maybe, contrary to Darwin's cautious pessimism, we are not merely trying to define the undefinable. Hopefully, at the very least, this chapter has sparked an interest in you to continue to reflect on the species debate.

CHAPTER REVIEW: DISCUSSION AND QUESTIONS

1. Talk amongst a few of your classmates. Do any of your animal encounters produce the same kind of response like my zoo experience described at the beginning of this chapter? [pp. 161-62]
2. Describe why the species debate is so complicated. More specifically, drawing upon Brandon and Mishler, *why* is the species debate so complex? [pp. 164–65]
3. As discussed early in the chapter, provide at least two reasons why one should care about the species debate? Can you think of any other ones than those provided in the text? [pp. 165–66]
4. What is the nominalist challenge with respect to the species debate? Where, at first glance, does Darwin fit into this discussion? Try to answer these questions by way of the exclusive nominalism challenge [pp. 166–70]
5. In addition to the nominalist challenge, what are the three criteria that a good species concept should be able to incorporate or adequately address? Be able to explain each criterion as it relates to the species debate. [pp. 171–72]
6. In a non-technical way, explain **Species as Essentialist-Natural Kind** to another class member. Now, with greater precision, you should be able to explain **Species as Essentialist-Natural Kind** by way of the following: **Indispensability Condition, Causal Efficacy Condition, Static Condition, Universal Condition**. [pp. 174–75]
7. How does Darwinism tell against **Species as Essentialist-Natural Kind** (the four conditions noted in question 6 above)? Restated, given an evolutionary view of organisms, how is/is not **Species as Essentialist-Natural Kind** able to accommodate the four-fold criteria? [pp. 176–79]
8. Briefly explain the **Phenetic Species Concept**. How does the **Phenetic Species Concept** determine species inclusion and exclusion? In terms of the criteria, which one poses the most serious problem for the **Phenetic Species Concept**? [pp. 179–81]
9. How does the **Biological Species Concept** make sense of species? [p. 182] The **Biological Species Concept** seems like a promising approach, but how does it fall short (in terms of our criteria; don't forget the challenge posed by nominalism)? [pp. 183–85]
10. How would you distinguish **Species as Individual Lineages** from the previous species concepts? [pp. 185–86] In terms of our criteria, how does it contend with the nominalist challenge? Although not rigorously discussed in the text, what do you think of Hull's genetic-individual and part-whole analogies? [pp. 186–88]

11. What is the extinction objection to **Species as Individual Lineages**? Are you persuaded by Hull's attempt to deal with this objection? Use Dr. Dodostein to help clarify the concern. [pp. 189–91]
12. Explain the **Phylogenetic-Cladistic Species Concept**. Try to see if, with the help of an example, your explanation can include the difference between *synapomorphies* and *symplesiomorphies*. In what way is the **Phylogenetic-Cladistic Species Concept** thought to be superior to **Species as Individual Lineages**? [pp. 192–95]
13. The two main criticisms of the **Phylogenetic-Cladistic Species Concept** are related to its reliance on branching patterns. What is the first criticism raised by Coyne and Orr related to the **Phylogenetic-Cladistic Species Concept** not being operational? What is the second criticism by Garvey related to plant hybridization? How are these two criticisms related to how the **Phylogenetic-Cladistic Species Concept** addresses the four-fold species criteria? [pp. 196–97]
14. How could you use the rabbit example in this chapter to help make sense of the **Relational Biosimilarity Species Concept**? [pp. 199–200]
15. Although not specifically discussed in the chapter, in what way is the **Relational Biosimilarity Species Concept** similar to the **Phenetic Species Concept** discussed earlier in the chapter? [Think]
16. Why does the **Relational Biosimilarity Species Concept** struggle to accommodate criterion #2? [pp. 199–200]
17. What is *inclusive (trope) nominalism* and why does it pose a challenge to the **Relational Biosimilarity Species Concept**? [pp. 202–03]
18. Can you use your own example to explain supervenience? (You should include the idea of multiple realizability in your answer.) How is this concept of supervenience being used with respect to the **Relational Biosimilarity Species Concept**? [p. 202–03]
19. In what way does the notion of multiple realizability/exemplification potentially work against the **Relational Biosimilarity Species Concept**? [p. 203]
20. Summarize the different versions of pluralism briefly discussed at the end of the chapter. [p. 204] What are your thoughts about the four-fold criteria? [pp. 204–05]

REFERENCES

Ananth, Mahesh. *In Defense of an Evolutionary Concept of Health: Nature, Norms, and Human Biology*. Aldershot, UK: Ashgate Publishing, 2008.

Barnes, Jonathan, trans. *Aristotle: Posterior Analytics*. In *The Complete Works of Aristotle*, vol. 1, ed. Jonathan Barnes. Princeton: Princeton University Press, 1984.

Baum, David A., and Michael J. Donoghue. "Choosing among Alternative 'Phylogenetic' Species Concepts." *Systematic Botany* 20 (1995): 560–73.

Brand, Allan M. "Racism and Research: The Case of the Tuskegee Syphilis Study." *The Hastings Center Report* 8, 6 (1978): 21–29.

Brandon, Robert N., and Brent D. Mishler. "Individuality, Pluralism, and the Phylogenetic Species Concept." In *Concepts and Methods in Evolutionary Biology*, ed. Robert N. Brandon, 106–23. Cambridge: Cambridge University Press, 1996.

Burkhardt, Frederick, and Sydney Smith, eds. *The Correspondence of Charles Darwin*, vol. 6, 1856–57. Cambridge: Cambridge University Press, 1991.
Burma, Benjamin. "The Species Concept: A Semantic View." *Evolution* 3, 4 (1949): 369–70.
Chalmers, David J. *The Conscious Mind*. Oxford: Oxford University Press, 1996.
Copi, Irving. "Essence and Accident." *The Journal of Philosophy* 51 (1954): 706–19.
Coyne, Jerry A., and H. Allen Orr. "Speciation: A Catalogue and Critique of Species Concepts." In *Philosophy of Biology: An Anthology*, ed. Alexander Rosenberg and Robert Arp, 272–92. West Sussex, UK: Wiley-Blackwell, 2010.
Cracraft, Joel. "Speciation and Its Ontology: The Empirical Consequences of Alternative Species Concepts for Understanding Patterns and Processes of Differentiation." In *Speciation and Its Consequences*, ed. D. Otte and J.A. Endler, 28–59. Sunderland, MA: Sinauer Associates, 1989.
Darwin, Charles. *The Origin of Species by Means of Natural Selection or the Preservation of Favoured Races in the Struggle for Life*. 6th ed. New York: Mentor/Penguin, 1958.
De Queiroz, Kevin, and Michael J. Donoghue. "Phylogenetic Systematics and the Species Problem." *Cladistics* 4 (1988): 317–38.
Dennett, Daniel C. *Darwin's Dangerous Idea: Evolution and the Meanings of Life*. New York: Simon & Schuster, 1995.
Dover, Gabby. "A Species Definition: A Functionalist Approach." *Trends in Ecology and Evolution* 10, 12 (1995): 489–90.
Duerbaum, Johannes. "The Symbiotic Relationship between Gobies and Pistol Shrimp." 2013. http://www.fishchannel.com/saltwater-aquariums/species-info/goby/gobies-and-pistol-shrimp.aspx.
Dupré, John. *The Disorder of Things: Metaphysical Foundations of the Disunity of Science*. Cambridge, MA: Harvard University Press, 1993.
Elder, Crawford L. "Biological Species Are Natural Kinds." *Southern Journal of Philosophy* 46, 3 (2008): 339–62.
Eldredge, Niles. *Time Frames*. New York: Simon & Schuster, 1985.
Ereshefsky, Marc. "Species, Taxonomy, and Systematics." In *Philosophy of Biology: An Anthology*, ed. Alexander Rosenberg and Robert Arp, 255–71. West Sussex, UK: Wiley-Blackwell, 2010a.
———. "Darwin's Solution to the Species Problem." *Synthese* 71, 3 (2010b): 405–25.
———. *The Poverty of the Linnaean Hierarchy: A Philosophical Study of Biological Taxonomy*. Cambridge: Cambridge University Press, 2001.
———. "Eliminative Pluralism." *Philosophy of Science* 59, 4 (1992): 671–90.
Garvey, Brian. *Philosophy of Biology*. Montreal and Kingston: McGill-Queen's University Press, 2007.
Ghiselin, Michael "Species Concepts, Individuality, and Objectivity." *Biology & Philosophy* 2, 2 (1987): 127–43.
———. "A Radical Solution to the Species Problem." *Systematic Biology* 23, 4 (1974): 536–44.
Grene, Marjorie, and David Depew. *The Philosophy of Biology: An Episodic History*. Cambridge: Cambridge University Press, 2004.

Griffths, Paul. "Squaring the Circle: Natural Kinds with Historical Essences." In *Species: New Interdisciplinary Essays*, ed. Robert A. Wilson, 187–208. Cambridge, MA: MIT Press, 1999.

Hardie, R.P. and R.K. Gayle, trans. *Aristotle: Physics*. In *The Complete Works of Aristotle*, vol. 1, ed. Jonathan Barnes. Princeton: Princeton University Press, 1984.

Hennig, Willi. *Phylogenetic Systematics*. Urbana: University of Illinois Press, 1979.

Hodge, Felicia Schanche. "No Meaningful Apology for American Indian Unethical Research Abuses." *Ethics & Behavior* 22, 6 (2012): 431–44.

Hull, David "A Matter of Individuality." In *Conceptual Issues in Evolutionary Biology*, 3rd ed., ed. Elliott Sober, 363–86. Cambridge, MA: MIT Press, 2006. Originally published in *Philosophy of Science* 45, 3 (1978): 335–60.

———. "Contemporary Systematic Philosophies." In *Conceptual Issues in Evolutionary Biology*, 2nd ed., ed. Elliott Sober, 308–33. Cambridge, MA: MIT Press, 1994. Originally published in *Annual Review of Ecology and Systematics* 1, 1 (1970): 19–54.

———. "Individual." In *Keywords in Evolutionary Biology*, ed. Evelyn Fox Keller and Elisabeth A. Lloyd, 180–87. Cambridge, MA: Harvard University Press, 1992.

———. "Are Species Really Individuals?" *Systematic Zoology* 25, 2 (1976): 174–91.

———. "The Effect of Essentialism on Taxonomy: Two Thousand Years of Stasis." *British Journal for the Philosophy of Science* 15, 2 (1965): 1–18.

Judson, Oilivia. "A Mutual Affair." *New York Times*, April 8, 2008.

Kitcher, Philip. "Ghostly Whispers: Mayr, Ghiselin and the 'Philosophers' on the Ontology of Species." *Biology & Philosophy* 2, 2 (1987): 184–92.

———. "Species." *Philosophy of Science* 51 (1984): 308–33.

Kitts, David B., and David J. Kitts. "Biological Species as Natural Kinds." *Philosophy of Science* 46, 4 (1979): 613–22.

Locke, John. *An Essay Concerning Human Understanding*. Ed. P.H. Nidditch. Oxford: Clarendon Press, 1975.

Loux, Michael J. *Metaphysics*. 2nd ed. London: Routledge, 2002.

Luckow, Melissa. "Species Concepts: Assumptions, Methods and Applications." *Systematic Botany* 20, 4 (1995): 589–605.

Mallet, J.L.B. "Isolating Mechanisms." In *Encyclopaedia of Ecology and Environmental Management*, ed. Peter P. Calow, 379–80. London: Wiley-Blackwell, 1998.

———. "A Species Definition for the Modern Synthesis." *Trends in Ecology and Evolution* 10, 7 (1995): 294–99.

Mayden, Richard. "A Hierarchy of Species Concepts: The Denouement in the Saga of the Species Problem." In *Species: The Units of Biodiversity*, ed. M.F. Claridge, H.A. Dawah, and M.R. Wilson, 381–424. London: Chapman and Hall, 1997.

Mayr, Ernst. *What Makes Biology Unique?* Cambridge: Cambridge University Press, 2004.

———. *Toward a New Philosophy of Biology*. Cambridge, MA: Harvard University Press, 1988.

———. *The Growth of Biological Thought*. Cambridge, MA: Belknap/Harvard University Press, 1982.

Michener, C.D. "Diverse Approaches to Systematics." *Evolutionary Biology* 4 (1970): 1–38.

Mishler, Brent D. "Species Are Not Uniquely Real Biological Entities." In *Contemporary Debates in Philosophy of Biology*, ed. Francisco J. Ayala and Robert Arp, 110–22. West Sussex, UK: Wiley-Blackwell, 2010.

Okasha, Samir. "Darwinian Metaphysics: Species and the Question of Essentialism." *Synthese* 131, 2 (2002): 191–213.

Panchen, Alec L. *Classification, Evolution and the Nature of Biology*. Cambridge: Cambridge University Press, 1992.

Paterlini, Marta. "There Shall Be Order. The Legacy of Linnaeus in the Age of Molecular Biology." *EMBO Reports* 8, 9 (2007): 814–16.

Paterson, H.E.H. "Perspective on Speciation by Reinforcement." *South African Journal of Science* 78 (1982): 53–57.

Paterson, H.E.H., and M. Macnamara. "The Recognition Concept of Species: Macnamara Interviews Paterson." *South African Journal of Science* 80 (1984): 312–18.

Quine, W.V.O. *Elementary Logic*. Cambridge, MA: Harvard University Press, 1965.

Richards, Richard A. "Species and Taxonomy." In *The Oxford Handbook of Philosophy of Biology*, ed. Michael Ruse, 161–88. New York: Oxford University Press, 2008.

Ridley, Mark. *Evolution*. 2nd ed. Cambridge, MA: Blackwell Science, 1996.

———. "The Cladistic Solution to the Species Problem." *Biology and Philosophy* 4, 1 (1989): 1–16.

Rojas, Martha. "The Species Problem and Conservation: What Are We Protecting?" *Conservation Biology* 6, 2 (1992): 170–78.

Rosenberg, Alexander, and Robert Arp, eds. *Philosophy of Biology: An Anthology*. West Sussex, UK: Wiley-Blackwell, 2010.

Schaffer, Jonathan. "The Individuation of Tropes." *Australasian Journal of Philosophy* 79, 2 (2001): 247–57.

Shapiro, Lawrence. "Multiple Realizations." *Journal of Philosophy* 97, 12 (2000): 635–54.

———. "How to Test for Multiple Realization." *Philosophy of Science* 75, 5 (2008): 514–25.

Shaw, K.L. "The Geneological View of Speciation." *Journal of Evolutionary Biology* 14, 6 (2001): 880–82.

———. "Species and the Diversity of Natural Groups." In *Endless Forms: Species and Speciation*, ed. D.J. Howard and S.J. Berlocher, 44–56. Oxford: Oxford University Press, 1998.

Simpson, G.G. *Principles of Animal Anatomy*. New York: Columbia University Press, 1961.

Sneath, Peter H.A., and Robert R. Sokal. *Numerical Taxonomy*. San Francisco: W.H. Freeman and Company, 1973.

Sober, Elliott "Sets, Species, and Evolution: Comments on Philip Kitcher's 'Species.'" *Philosophy of Science* 51, 2 (1984): 334–41.

———. "Evolution, Population Thinking, and Essentialism." *Philosophy of Science* 47, 3 (1980): 350–83.

Sokal, Robert R., and Theodore J. Crovello. "The Biological Species Concept: A Critical Evaluation." *American Naturalist* 104, 936 (1970): 127–53.

Stamos, David. *The Species Problem*. Lanham: Lexington Books, 2003.

———. "Was Darwin Really a Species Nominalist?" *Journal of the History of Biology* 29, 1 (1996): 127–44.

Stanford, Craig. "Chimpanzee Hunting Behavior and Human Evolution." *American Scientist* 83, 3 (May–June 1995): 256.

Sterelny, Kim, and Paul E. Griffiths. *Sex and Death*. Chicago: University of Chicago Press, 1999.

Templeton, Alan A. "Species and Speciation: Geography, Population Structure, Ecology, and Gene Trees." In *Endless Forms: Species and Speciation*, ed. Daniel J. Howard and Stewart H. Berlocher, 32–43. New York: Oxford University Press, 1998.

———. "The Meaning of Species and Speciation: A Genetic Perspective." In *Speciation and Its Consequences*, ed. Danielle Otte and John A. Endler, 3–27. Sunderland: Sinauer Associates, 1989.

Van Valen, Leigh. "Ecological Species, Multispecies, and Oaks." *Taxon* 25, 2/3 (1976): 233–39.

Wake, David. "What Salamanders Have Taught Us about Evolution." *Annual Review of Ecology, Evolution and Systematics* 40 (2009): 333–52.

———. "Problems with Species: Patterns and Processes of Species Formation in Salamanders." *Annals of the Missouri Botanical Garden* 93, 1 (2006): 8–23.

Weiner, Jonathan. *The Beak of the Finch*. New York: Knopf, 1994.

Wheeler, Quentin, and Norman Platnick. "The Phylogenetic Species Concept, *sensu* Wheeler and Platnick." In *Species Concepts and the Phylogenetic Theory: A Debate*, ed. Quentin Wheeler and Rudolf Meier, 55–69. New York: Columbia University Press, 2000.

Wiley, E.O. "The Evolutionary Species Concept Reconsidered." *Systematic Zoology* 27, 1 (1978): 17–26.

Wilkins, John S. "How to Be a Chaste Species Pluralist-Realist: The Origins of Species Modes and the Synapomorphic Species Concept." *Biology and Philosophy* 18 (2003): 621–38.

Zachos, Frank E. *Species Concepts in Biology: Historical Development, Theoretical Foundations and Practical Relevance*. Switzerland: Springer, 2016.

Chapter 6

NATURALIZING THE MORAL SENSE

Evolution and Ethics

A moral being is one who is capable of comparing his past and future actions or motives, and of approving or disapproving of them. We have no reason to suppose that any of the lower animals have this capacity.

—Charles Darwin (1981 [1871], pp. 88–89)

I. INTRODUCTION

The idea of naturalizing human morality by connecting it to nonhuman animal behavior is not a new endeavor. Aristotle, in his attempt to distinguish the human animal from other animals, offers the following account:

> It is clear why humans are political animals to a greater degree than are any of the bees or gregarious animals.... For it is a distinguishing feature of humans that only they are able to perceive good and bad, justice and injustice, and other such things, and it is the common awareness of these things that produces a household and a city-state. (*Politics*, Book I.2, 1253^{a}8–15)

Notice that Aristotle thinks that humans are animals and that they differ *by degree* with respect to non-human animals primarily in terms of their ability to perceive normative (i.e., value-laden) aspects of living. This human moral sensibility, which for Aristotle is a natural feature, includes both self-regarding and social-regarding elements. These "selfish" and "altruistic" features of human nature are, for Aristotle, an integral part of capturing humans as political and social animals. It is reasonable, then, to wonder what is the source of this "moral sense" given that animals, like humans, display a variety of other-regarding and self-regarding behaviors.

One answer to the above query emerged as a result of Darwin's work. Much like Aristotle, Darwin recognized that humans displayed both self-regarding and social-regarding behaviors. In terms of the source of these behaviors, Darwin argued that natural selection retained self-regarding and social-regarding instincts that assisted in the production of more offspring in early human evolution. As he puts it, "The following proposition seems to me highly probable—namely, that any animal whatever, endowed with well-marked social instincts, would inevitably acquire a moral sense or conscience, as soon as its intellectual powers become ... nearly as well developed, as in man" (1871, pp. 71–72). Interestingly, Darwin also argued that the set of moral sentiments (i.e., social instincts) that comprised the human moral sense evolved for the good of the community (1871, p. 85; Stamos, 2008, p. 163) more so than the individual organism. Darwin summarizes all of this as follows:

> It must not be forgotten that although a high standard of morality gives but a slight or no advantage to each individual man and his children over the other men of the same tribe, yet an increase in the number of well-endowed men will certainly give an immense advantage to one tribe over another. There can be no doubt that a tribe including many members who, from possessing in a high degree the spirit of patriotism, fidelity, obedience, courage, and sympathy, were always ready to give aid to each other and to sacrifice themselves for the common good, would be victorious over most other tribes; and this would be natural selection. At all times throughout the world tribes have supplanted other tribes; and as morality is one element in their success, the standard of morality and the number of well-endowed men will thus everywhere tend to rise and increase. (1871, p. 166)

Yet, prior to and after Darwin's ruminations on the relationship between evolution and human morality, many naturalists defended a staunch rugged-individualism account of evolutionary ethics. Specifically, these scholars explained much animal and human behavior in terms of individual competition. This fixation on competition gave special attention to how individuals could garner scarce resources and secure reproductive advantages at the expense of others. Even seemingly "altruistic" behavior was interpreted in a self-regarding way.[1] As the saying goes, "scratch an altruist and watch a hypocrite bleed." The implication is that other-regarding behavior (i.e., altruism) was explained by being explained away; the idea of humans expressing a cooperative morality was cast aside in favor of viewing humans as possessing a self-regarding strategic moral sense that appeared to be in keeping with a Darwinian-struggle-for-life worldview. Put another way, once we flip-on our Darwinian lenses, the life and death struggles of non-human animals presents a rather competitive and dire picture of life

---

[1] For details on this history, see Bradie (1994), Ruse and Maienschein (1999), and Boniolo and De Anna (2006).

that should count as the foundation of human morality—not quite the picture that Aristotle or Darwin envisioned!

Given the above rendering of evolution and ethics, it is not surprising that some scholars consider evolution a "universal acid" to the extent that it can penetrate (i.e., influence) just about every aspect of life—even human life.[2] From this perspective, evolution not only makes sense of human origins, but it presumably plays an integral role in understanding most human activities. In particular, the activities associated with humans as moral beings have been thought to be best understood from a Darwinian angle. Of recent vintage, a prominent defender of evolutionary ethics, Ruse, offers the following truculent pronouncement:

> The time has come to take seriously the fact that we humans are modified monkeys, not the favored Creation of a Benevolent God on the Sixth Day.... The question is not whether biology—specifically, our evolution—is connected with ethics, but how. Thanks to recent developments in biological science, we can now throw considerable light on this problem. (Ruse, 1995, p. 93)[3]

Clearly, Ruse thinks that the "evolutionary acid" finds its way into the moral dimension of human activity. He does, however, point out that the interesting question is not whether or not evolution helps to make sense of humans as moral beings, but *how* evolution can make sense of humans as moral beings. Drawing upon the insights of Kitcher (1994, p. 440), this how-aspect of Ruse's query can be located. To this end, Kitcher explicates four distinct projects that can be grouped under the evolutionary ethics umbrella.[4] Specifically, defenders of evolutionary ethics [hereafter EE] engage in explaining how

1. people come to acquire ethical concepts, make ethical judgments, and formulate moral systems **[EE1]**

2. to derive new moral principles in the light of a proper understanding of human nature and existing moral norms **[EE2]**

3. to make sense of the nature of (moral) goodness (i.e., metaethics) **[EE3]**

4. humans can revise their existing scheme of ethical concepts, judgments, principles, or moral systems by revealing new moral principles *not* derived from an existing scheme of ethical principles **[EE4]**

---

2 See Dennett (1995) for his use of "universal acid."
3 For a recent attempt to argue for a close moral connection between humans and their primate kin, see De Waal (2006).
4 Throughout this chapter I will use 'morality' and 'ethics' interchangeably. For more on these terms, see Hinman (2007).

In order to appreciate these four **EE** projects, a brief reminder of the subject matter of morality will prove useful. Let us, then, turn to this directly and then connect it back to the four **EE** projects (**EE1–EE4**) noted on the previous page.

II. THE THREE FACES OF MORALITY

With respect to the subject matter of morality, I think a movie reference will help. Many years after it came out, I watched the film *Butch Cassidy and the Sundance Kid*. Broadly, the movie is about the travails of two likeable criminals, who engage in a long series of train and bank robberies in the early 1900s. Both criminals are well aware of the potential consequences of their actions. Regardless, they do not seem to mind that not only are they risking their lives, but also that they are living *unethical* lives. Oddly, the film does an excellent job of moving the audience to cheer for these charming villains, despite the fact that they are both *morally bad*. Their *badness* primarily resides in their desire to steal property that does not belong to them. Put another way, we think that Butch and Sundance are *bad* because they *should* not take property that does not belong to them, unless they have permission to do so from the relevant property owner.

Our concern about Butch and Sundance's motivations and our corresponding moral condemnation can be understood by way of Nagel's general account of ethics. In summary, Nagel's account reveals three ways in which ethics is divided. First, he points out that ethics is about how we should live in the world with others; that is, ethics is about how our *everyday actions* and interactions with others *should* proceed from a moral point of view. Second, Nagel concedes that developing a more accurate picture of the world is part of ethics. One way of trying to understand this claim is that there may be "moral facts" out in the world that require an explanation. The question of whether or not there are "moral facts" out in the world falls under the purview of *meta-ethics*. Third, the methods of bringing to fruition what we ought to be and what we should desire fall within the scope of *ethical theory* (Nagel, 1980, pp. 198–99). The three can be outlined as follows:

> (1) Everyday Moral Issues → This sense of ethics refers to the everyday moral challenges with which people might contend and the relevant *moral principles* associated with such challenges.[5] The following are examples of the kinds of questions that capture the day-to-day moral issues and possible moral principles with which people grapple:

---

5   In terms of everyday morality, I am ignoring the distinction between *situational ethics* and *norm-driven ethics*. For more on these concepts and how they are related to evolutionary biology, see Bradie (1994, p. 12) and Pope (2007, pp. 133–34).

DAY-TO-DAY CIRCUMSTANCE	AND	MORAL PRINCIPLE
"Should I have an abortion?"		"Is it moral to kill a person?"
"Should I lie to my boss?"		"Is lying ever just?"
"Should I steal bread to feed my child?"		"Is stealing ever morally acceptable?"
"Should Sundance cheat on his wife?"		"Is it moral to break a promise?"

(2) Ethical Theory → This aspect of ethics deals with the *moral theories* that govern human action. For instance, **Utilitarianism** is the view that one should perform those actions that promote the greatest good for the greatest number of people. In sharp contrast, **Kantianism** prescribes that we should only perform those actions that are in accordance with the principle known as the "Categorical Imperative." Alternatively, **Virtue Ethics** gives pride of place to the development of a set of characteristics (e.g., honesty, bravery, gratitude, etc.) which reflect the morally good person. At first glance, the actions of Butch and Sundance would not be endorsed by any of these moral theories.

(3) Meta-ethics → This way of thinking about ethics revolves around understanding the nature or essence of ethical properties and ethical statements. "Is 'good' a natural property or is it a non-natural property?" "Are ethical statements capable of being true or false or are they really just disguised commands?" "Is there a moral fact out in the world that tells against the purloining ways of Butch and Sundance?" are questions that are addressed within Meta-ethics.

III. ETHICS PROJECTS AND EVOLUTIONARY ETHICS PROJECTS

With this brief sketch of how to think about different senses of ethics, we can connect the ethical projects to the **EE** projects. Recall that the **EE1** project is focused on explaining how people *have come to have* ethical concepts, moral judgments, and moral systems. This is a reminder that human moral principles and moral theories do not pop up out of nowhere. Rather, they are the product of the evolutionary history of *Homo sapiens*. So, if it were possible to go back far enough in human evolution, then it should be possible to determine the genetic and corresponding proto-social influences that account for nascent biological and moral principles and moral theories. It is an account of this deep historical development of human morality that is of primary interest to the **EE1** project. In terms of the above three senses of ethics, **EE1** is interested in Everyday Ethics and Ethical Theory (but not so much meta-ethics). What this means is that **EE1** attempts to explain, for example, how the evolution of the taboos against lying or incest gained legitimacy in early human populations or how, for instance, the evolution of how virtue theory may have come to gain ascendency (if it did) in many early human populations. Given that **EE1** is primarily concerned with determining the evolutionary "facts" related to evolutionary ethics, it appears to be the most benign version of evolutionary ethics. So long as speculative claims about human biological and social evolution are put forth *as speculative*, **EE1** is an eminently reasonable research program.

EE2 is somewhat similar to **EE1** with the caveat that our existing moral principles can be expounded or applied in ways that had not been anticipated in the light of facts about humans and their environments. Kitcher tells us that fields like medical ethics and environmental ethics are pursuing projects related to **EE2**. He explains that **EE2**:

> can reveal to us our deepest and most entrenched desires. By recognizing those desires, we can obtain a fuller understanding of human happiness and thus apply our fundamental principles in a more enlightened way. (Kitcher, 1994, p. 442)

He goes on to use the utilitarian moral theory as an example:

> Card-carrying Utilitarians who defend the view that morally correct actions are those that promote the greatest happiness of the greatest number, who suppose that those to be counted are presently existing human beings, and who identify happiness with states of physical and psychological well-being will derive concrete ethical precepts by learning how the maximization principle of happiness can actually be achieved. (Kitcher, 1994, p. 442)

For example, an evolutionary analysis could be offered for how principled exceptions to the general prohibition against killing or promise keeping came to fruition and the corresponding punishments that emerged. This account could include, for example, an evolutionary understanding of human nature that includes an entrenched desire for retribution as well as the known anthropological studies regarding human systems of punishment. Or a deeper understanding of our evolved desires related to suffering might inform either our support or rejection of certain forms of euthanasia. Regarding the three senses of ethics, **EE2** is primarily focused on making sense of the emergence of new moral norms related to Everyday Ethics and Ethical Theory (meta-ethics to a lesser extent). The attempt to make claims about human nature reveals that **EE2** does take a radical departure from **EE1** and can fall prey to certain objections if not handled carefully.

Unlike, **EE1** and **EE2**, **EE3** is exclusively concerned with the nature of moral goodness and how to make evolutionary sense of it—that is, meta-ethics. As stated above on p. 215, defenders of **EE3** either attempt to validate an existing meta-ethical theory by supplementing it with relevant evolutionary biology concepts and facts or they attempt to offer a unique meta-ethical perspective via evolutionary biological concepts and facts. To illustrate, evolutionary considerations might reveal moral goodness to be a natural property out in the world or that moral statements refer only to a set of emotional states of the person expressing the moral statement. As we will come to see, this project runs into numerous difficulties. Indeed, depending upon the **EE3** put forth, Kitcher thinks that such an enterprise should be dismissed as "deeply confused" (Kitcher, 1994, p. 445).

EE4 is the most ambitious **EE** project of the bunch. It attempts to make sense of new moral principles, judgments, systems, and/or meta-ethics claims that have shown stability in human populations, but have not been influenced by an existing moral framework. For example, one can imagine an evolutionary account that makes sense of a new moral principle against human cloning or the meta-ethical principle that moral goodness is a second-order quality like color—principles that came about via insights from evolutionary biology *independently* of existing human moral characteristics. Much like **EE3**, **EE4** must contend with some rather stiff objections that render it less than palatable.

Table 6.1 helps to illustrate the relationship between **EE1–EE4** and the three senses of ethics. The 'X' used in each box refers to the aspect(s) of ethics that falls within the domain of a particular **EE** project.

	EE1	EE2	EE3	EE4
Everyday Ethics	X	X		X
Ethical Theory	X	X		X
Meta-Ethics			X	X

Table 6.1 *Ethical Theories and Evolutionary Ethics Projects*

IV. FACES OF EVOLUTION: EVOLUTIONARY ALTRUISM AND HUMAN MORALITY

Having a clearer account of the connection between ethical projects and **EE** projects, it is now possible to turn to some of the relevant evolutionary biology concepts used to justify particular **EE** projects. In general, those who espouse an evolutionary ethic draw on animal social behavior as part of their defense (Ruse and Wilson, 1998). Much like many other animal species, the argument goes, humans not only have concern for themselves, but they also express other-regarding behavior toward kin and non-kin. This expression of other-regarding behavior—be it from kindness, cooperation, or moral duty—is generally labeled as *altruism* (some of these concepts should be familiar from the units of selection discussion in ch. 3). We help family, friends, and even strangers; in sum, humans are altruistic to a certain degree. The argument can be formalized as follows:

P1. Animals engage in social behavior and such behavior is both genetically controlled and promotes survival and reproductive success.

P2. Humans are animals.

C1. Humans engage in social behavior and such behavior is both genetically controlled and promotes survival and reproductive success.

Is **P1** true? It may appear false because, from a Darwinian perspective, engaging in other-regarding behavior is costly to the individual donating his resources and energy. Sacrificing one's time and energy reduces one's survival and reproductive success to some extent. Why would an individual risk reproductive success to assist others? To deal with this sort of conceptual difficulty, Darwinians have offered the following mechanisms that are supposed to reveal the truth of **P1**: (1) kin selection, (2) reciprocal altruism, and (3) group selection.

Kin selection is the idea that an individual can have more of his genes passed on to future generations by helping or sacrificing for others who are genetically related (*inclusive fitness*; see Hamilton, 1964). For example, an individual who sacrifices for two siblings or eight cousins would help ensure that his genes are represented in the next generation (under the assumption that those relatives who benefit from such sacrifices go on to reproduce).[6] With respect to humans, the inference made is that kin selection explains why we are willing to help other family members. So, the reason why parents would be willing to invest so much energy and time into their children and other kin is because the kin carry many of their same genes, which these kin can pass on during their reproductive years. In this way, even if one makes considerable sacrifices to the detriment of one's life, some subset of one's genes moves on to future generations, partly as a result of one's willingness to donate one's resources to family members. Thus, non-human and human other-regarding behavior toward family can be beneficial to individual fitness and, therefore, understood from an evolutionary perspective.

This takes care of family, but what about the fact that non-human animals and humans cooperate and help non-family members? This is explained in terms of reciprocal altruism or what is commonly known as "you-scratch-my-back-and-I-will-scratch-your-back."[7] Again, the inference is that the human animal helps other people with the expectation that they will get some assistance in return; thus, the label reciprocal altruism (Trivers, 1971). Notice that one is still advancing one's survival and reproductive success while helping other non-family members. Ultimately, one is able to pass on one's genes as a result of the benefits reaped from others having reciprocated as

---

6   In different ways, hymenoptera (sawflies, bees, ants, and wasps) and squirrels (Sherman, 1977) reveal the truth of the kin selection mechanism.
7   Depending upon the criteria one uses (Stephens, 1996), cleaner-fish (Feder, 1996), warning calls of birds (Trivers, 1971), and the feeding practices of bats (Wilkinson, 1984 and 1988) are examples of reciprocal altruism.

a result of one's assistance. Thus, it is possible to make evolutionary sense of human other-regarding behavior toward non-kin, further suggesting the truth of **P1**.

In contrast, group selection is natural selection working at the level of the group (see ch. 3). There are at least two ways to understand group selection. One is the idea that there exists an altruistic trait that no one member of the group possesses, but is present in the group as a whole. This unique group-level trait gives it an advantage over other groups. This is an odd way to think of group selection because (even if we granted such a trait) it is hard to make sense of how such a trait could be reproductively passed on if it was not possessed by any of the members of the group. So, alternatively, the second way to think of group selection is the idea that there are groups in which members are largely composed of altruists; that is, individuals that will sacrifice for the promulgation of the group. Put another way, groups with altruistic individuals can out-produce groups with primarily selfish individuals. From this perspective, there is selection pressure in favor of a set of traits that produce other-regarding behaviors in individuals that ensure the survival of the group to which they belong. These altruistic groups can avoid extinction and survive and reproduce better than selfish groups. The implication is that, much like kin selection and reciprocal altruism, this second version of group selection, according to some (Sober and Wilson, 1998), shows why it can pay off to engage exclusively in other-regarding cooperative behavior.[8] So, if there is any substance to group selection, then this would give additional evidence to the truth of **P1**.

In terms of the three different faces of morality noted above, it is reasonable to think that these various forms of altruism can be employed at the level of everyday morality, moral theory, and meta-ethics. For example, if we think that humans are generally concerned about the consequences or utility of their actions and practices, then it might be the case that group selection makes best sense of humans as utilitarians. Humans evolved in fairly small demes (i.e., local populations of the same sort of organisms), requiring a kind of group dynamic that might have required considerable sacrifice by some in order to keep the group alive. Such sacrifices or other-regarding actions would likely be the result of numerous interactions with another individual. This would reasonably aid in creating cooperation and recognizing defection. Thus, one could argue that group selection makes best sense of the kind of altruistic sacrifices and support required to sustain our early ancestors. Alternatively, reciprocal altruism might best make sense of how different hominid groups cooperated with one another during heavy environmental stress; for example, sharing of food in exchange for assistance with shelter. At this point, these suggestions are speculative, but give a flavor for how these different evolutionary mechanisms could be used to understand the different senses of morality. Keeping in mind the connection between ethical projects and **EE** projects, Table 6.2 helps to make sense of the inclusion of distinct selection mechanisms into the discussion as follows:

---

8   Whether this is a genuine case of group selection is debatable. For further discussion, see Okasha (2001).

	FACES OF MORALITY & EE PROJECTS	EVERYDAY MORALITY EE1, EE2, AND EE4	MORAL THEORY EE1, EE2, AND EE4	META-ETHICS EE3 AND EE4
MECHANISMS OF EVOLUTION	Kin Selection			
	Reciprocal Altruism			
	Group Selection			
	Combination Selection			

Table 6.2 *Moral Projects, Evolutionary Ethics Projects, and Selection Mechanisms*

Notice that a given ethical project can be the focus of distinct **EE** projects, which, in turn, can employ different combinations of evolutionary mechanisms to produce distinguishing **EE** projects. For instance, one could insist that everyday morality is best understood in terms of **EE1** and reciprocal altruism, kin selection makes best sense of **EE2** at the level of moral theory, and group selection offers a reasonable foundation for **EE3** and a particular meta-ethical account. Of course, care must be taken not to pigeon-hole any particular account discussed, but the discussion to follow will show that some of the scholars fit quite neatly into this matrix (and their names would go in the corresponding empty slots in Table 6.2). Also, care must be taken not to misrepresent what a particular author is claiming. For example, an author might claim that group selection makes best sense of humans as duty-based beings, but insist that the focus is only on the biological pre-conditions for moral theory (i.e., **EE1**). The point here is that an author may not be saying anything substantive about the fully normative dimension to human morality. This point will be made more precise in the discussions below. As we will see, this is a position taken by many authors. Lastly, mixed or combination accounts are also possible. For instance, reciprocal altruism and kin selection could have been working together to make sense of why, at the level of moral theory and **EE2**, humans evolved to endorse a utilitarian outlook (or some other moral theory).

V. OBJECTIONS TO NATURALIZING ETHICS

Before turning to particular **EE** projects, it is crucial to point out three major objections to biologizing ethics. What this means is that those who think that evolution plays a *direct role* in any of the three aforementioned ethical projects (i.e., E2–E4) must address (1a) the is-ought objection (naturalistic fallacy), (1b) the open-question objection (2) the objection from agency, and (3) the objection to evolutionary progress; *for part of doing philosophy well includes the ability to be aware of major (and minor) objections to one's* *position and to be able to argue persuasively against them.* A glimpse into each objection will reveal that it is no trivial task to try to defend the claim that evolution, in any of its guises, plays a *direct role* in human morality.

1a. The Is-Ought Objection: Hume's Hatchet

As we have seen, the language of morality reveals that morality is normative—it is about how we *ought* to act or *should* act in order to be considered good. Importantly, thinking ethically is about thinking in terms of prescriptions. In the same way a medical doctor prescribes what medicine his patient ought to take in order to secure or regain physical health, a "moral doctor" prescribes what actions we ought or should perform in order to secure or regain moral health (Hinman, 2007, ch. 1). These prescribed actions may be governed by concern for consequences, duties, rights, character, justice, or some other framework. For now, the ethical point about normativity and prescriptivity is that they are at the heart of ethics and stand as a unified challenge to those who would try to make sense of morality from a non-normative perspective. Yet, the naturalistic perspective put forth by many **EE** projects is just the kind of non-normative account that runs into this normativity challenge. The challenge, known as *The Naturalistic Fallacy*, has been famously articulated by the philosopher David Hume. Briefly, Hume expresses his astonishment that most ethical arguments that he has scrutinized have employed factual claims in defense of normative claims; that is, authors have erroneously moved from using "is-claims" to defend "ought-claims" (Hume, 1740/2007, Book III). Let us consider an example to illustrate Hume's insightful point.

P1. Bank robbery **is** punishable by life imprisonment.

P2. If another crime is committed while robbing a bank, then the death penalty **is** a possible consequence.

C1. Thus, people **ought** not rob banks.

At first glance, the above argument looks unproblematic. Assuming **P1** and **P2** are true, it seems eminently reasonable that **C1** should follow. Indeed, people do not want to spend their life in prison and they do not want the death penalty as a result of ancillary activities related to a bank heist. So, it is rather innocuous to claim that people should not rob banks.

Upon closer inspection, however, the above argument is invalid because, as Hume warned above, the transition from "is-statements" to an "ought-statement" simply does not follow. Note that the concern here is not about the truth or falsity of **P1** and **P2**. Even if we grant that **P1** and **P2** are true, they are true as factual claims; that is, they are "is-statements" in terms of the punishments that accompany the crime(s) of bank robbery and its ancillary criminal activities. These "is-statements" say absolutely nothing about what one *ought* to do. So, concluding with an "ought-statement" (**C1**) reveals that the argument is not set up correctly—it is invalid and thus unsound.

Hume's point is that an "is-relationship" is different from an "ought-relationship." The "is-relationship" is about facts, while the "ought-relationship" is about what should be the case—that is, normativity. Hume correctly goes on to demand that additional justification (i.e., premises) is needed for such an argument to be valid. What this means is that ought-statements are required in the premises in order to generate an ought-conclusion. To see this, consider the following modified version of the above argument:

P1. Bank robbery **is** punishable by life imprisonment.

P2. If another crime **is** committed while robbing a bank, then the death penalty **is** a possible consequence.

P3. If people **ought** to be motivated not to rob banks in the light of **P1** and **P2**, then people **ought** not rob banks.

P4. People **ought** to be motivated not to rob banks in the light of **P1** and **P2**.

C1. Thus, people **ought** not rob banks.

Some set of premises, like **P3** and **P4**, is required in order to generate **C1**. To the point, in order to defend conclusions that have normative claims, it is necessary to have value/normative premises. The upshot of all this is that it is incumbent upon the naturalist—in our case the Darwinian—to show how biological evolution can make sense of normative claims without reducing them to biological claims. This appears to be a most daunting or completely impossible task. "Hume's Hatchet," argue many philosophers, is waiting to cut off any attempt by Darwinians to make sense of normativity in terms of biological evolution.[9]

1b. The Open-Question Objection: Moore's Machete

The philosopher, G.E. Moore, offers what he considers a knock-down argument against any attempt to naturalize (moral) goodness. Before moving to Moore's specific objection, it is worth noting that he makes pretty much the same claim as Hume when he argues against the leading evolutionary ethicist of his day, Herbert Spencer. He critically notes that Spencer and others like him argue fallaciously because evolution, "while it shews us the direction in which we *are* developing thereby and for that reason shews us the direction in which we ought to develop" (Moore, 1903, p. 46). Here we see Moore offering-up the same "is-ought" concern as Hume. Moving beyond Hume's

---

[9] See Farber (1994) and Woolcock (1999).

Hatchet, Moore brandishes his own ingenious "Machete" to strike down those who would dare to entertain seriously the idea that goodness can be reduced to a natural property.[10] His version of the naturalistic fallacy is known as "the open-question argument." The argument can be understood as follows:

> **P1.** If for any definition of 'goodness' it is possible to produce a corresponding sensible concern/question about such a definition, then any such definition of 'goodness' cannot fully capture the nature of goodness.
>
> **P2.** It is possible to produce a sensible concern/question about any definition of 'goodness.'
>
> ---
>
> **C1.** Thus, any definition of 'goodness' cannot fully capture the nature of goodness.

If we allow for the truth of **P1**, it must be determined what Moore has in mind with respect to **P2**. Restated, what sort of concern or question could be posed to tell against any definition of 'goodness'? *Moore's answer is simply: "But is it good?"* And if such a reply is coherent, thinks Moore, then it reveals that the definition up for consideration is off the mark. Let's consider two examples to illustrate Moore's concern.

> Example #1: Is the *spider* in the corner an *arachnid*?
>
> Example #2: It is morally required of Tom to behave altruistically, because evolution has produced a genetic tendency to altruism in humans.

Example #1 is a question that does not really make much sense. By definition, since any air-breathing invertebrate arthropod with eight legs is an arachnid, a spider is an arachnid. Now, one may not know that a spider is an invertebrate arthropod with eight legs, but once armed with this knowledge, it makes no sense to ask whether or not a spider is an arachnid. Similarly, to pose the question: "Is Tom the bachelor an unmarried male?" is incoherent because being a bachelor means that such a person is an unmarried male. From this perspective, it is not open to a rational agent to ask such questions about spiders and bachelors without further welcoming a charge of unintelligibility.

In contrast, the second example reveals no similar sort of conceptual snafu because it seems eminently reasonable/intelligible to query as to whether or not the biological source of a person's moral make-up actually constitutes a moral requirement for a

---

10 In some of the recent literature, there appears to be some controversy regarding how to interpret Moore's claim that the moral status or worth of X is not understood in terms of its natural properties. See G.E. Moore (1903). With respect to some of the recent literature, see Sturgeon (2003), Casebeer (2003), Curry (2006), and Kraut (2007).

person to behave altruistically or in some other moral manner. So, one could imagine Moore responding as follows to example #2: "I understand that the following gene sequence is designed to make sense of Tom's moral requirement to behave altruisically, but is being genetically disposed in this way really good?" Moore thinks that this kind of response is always available to any sort of attempt to substitute moral goodness with any natural property. Take, as another example, the claim that "That which one desires is good." One can imagine Moore replying to this desire-based account of goodness by claiming: "Yes, but is it good that goodness be understood in terms of one's desires?" Again, this question does not appear to suffer from any sort of conceptual confusion. *Rather, Moore thinks that the "but-is-it-good?" rejoinder actually reveals that the nature of goodness remains open and cannot be closed entirely by a naturalistic account of goodness* (Darwall, Gibbard, and Railton, 1992, p. 116).

Whether or not one thinks that Moore has hit on an important insight, we can demand of ethical naturalists that they be willing to address Moore's open-question worry.[11] This can be done by directly replying to Moore's argument or indirectly *by offering arguments* that a particular naturalistic account of the good can be defended as opposed to being endorsed as unquestionably right. For the sake of our discussion, although it would be grand if scholars would respond to both Hume's and Moore's distinct challenges, it will be enough that either Hume's or Moore's rendering of the naturalistic fallacy be taken seriously by those who wish to espouse an evolutionary ethic.

2. The Objection from Agency: Korsgaard's Katana

Putting to one side both Hume's and Moore's naturalistic fallacy objections, there is an additional roadblock awaiting those who want to naturalize ethics. This obstruction is related to the ontological status of reasons as causes and the issue of moral responsibility. The concern is that any attempt to Darwinize ethics will result in treating reasons as causes and such a move will ensure that human actions cannot be free. If the reasons for an action are actually causes of it, we can assume that these causes themselves are caused; then it is not possible to have done otherwise because every choice—that is, a set of reasons, beliefs, desires, etc., governing an action—came about from preceding causal influences.

Yet, if none of our choices are up to us—that is, they are all determined—then it is not possible to be held responsible for our actions (van Inwagen, 1983, p. 56). Imagine if a child rapist were to claim that his brutal sexual assaults on children were "written in the stars" and that it was determined from the beginning of the universe that he would behave in this way. I am not responsible, he could claim, for these actions are entirely out of my control—"it was not up to me," he could add. "My reasons are the product of a long causal chain of events." What if Butch and Sundance offered a similar assessment of their criminal actions? My suspicion is that we would look at such claims as a complete sham and insist that such people are responsible for their

---

11 For further citations and a forceful attempt to offer a Moorean challenge to naturalists, see Gampel (1996). For standard objections to Moore's Machete, see Miller (2003).

actions and the corresponding punishments should be brought to fruition as quickly as possible. Why do we think this way? As Woolcock (1999, pp. 286–87) makes clear, humans have the capacity to resist their beliefs and desires and change them. It is these features that make us moral agents (as opposed to, for example, the programmed attacking behavior of spiders when an insect is ensnarled in their webs), and responsible for our actions. What we can glean from this is that there must be something about the nature of reasoning that cannot be fully captured by talk of causation.

Korsgaard can be understood to offer just this "something." She explains the normative dimension to human morality as distinct from other non-human animals. She emphasizes that, in terms of practical reasoning, it is the ability to create separation from both possible means and ends that allows humans to reflect on whether or not to pursue any particular set of means and ends. *This is the control aspect of rational reflection that does not seem to be present in a purely causal account* (Korsgaard, 2006, p. 112 and p. 113).[12] So, we will be unimpressed by the rapist's claim that he could not control his actions because it is far from obvious that such behaviors can be justified in terms of being unable to resist the desires that aid in such behaviors. Indeed, with respect to Butch and Sundance, we will say the same thing to them about reflecting on means and ends if they claim that they could not resist their desires to steal.[13]

Now, granting the above account from Korsgaard about the human agency and control, the following argument related to reason and causes should be clear:

P1. If evolutionary ethics is committed to the view that reasons are causes with respect to human behavior, then evolutionary ethics either misunderstands the implications of evolutionary theory with respect to human behavior or evolutionary ethics is a mistake.

P2. Evolutionary ethics is committed to the view that reasons are causes with respect to human behavior.

C1. So, evolutionary ethics either misunderstands the implications of evolutionary theory with respect to human behavior or evolutionary ethics is a mistake.

---

12   It is important to note that there is a long tradition in philosophy of defending free-will while granting a thoroughgoing determinism. This position, known as compatibilism, locates moral responsibility on *how* actions are performed rather than the causal antecedents of such actions. So, as long as one's choices and reasons are the product of rational reflection, argues the compatibilist, this is all that is required for choosing freely. See ch. 8 of Double (1999) and Frankfurt (1971) for more on the compatibilist alternative.
13   If it could be shown by professional psychologists, psychiatrists, or neurobiologists that the rapist is a "sex addict" and he could not control his actions and that Butch and Sundance are kleptomaniacs and could not have done otherwise, then we may have to concede that their actions could not be governed by reasons. The assumption with these examples, however, is that these people are not inflicted by such compulsive manias.

Notice that if we take seriously the claim that reasons are not entirely understood in terms of causes, then anytime an evolutionary ethicist insists or assumes that reasons are causes, it follows that the evolutionary ethicist has erroneously ignored the objection from agency (Woolcock, 1999). Looking back at the animal-social-behavior argument at the start of Section IV, **P2** of that argument would be considered false or vague by human-agency defenders because even if it is true that humans are animals, we are unique in terms of the rational life we exhibit. Importantly, the point here is that reasons are distinct from first-order desires such that Butch and Sundance can desire to rob a bank, but it is also the case that they can employ various reasons or higher-order desires that can thwart first-order desires from producing bank-robbing behaviors (this is pretty much the point that Korsgaard is making). For instance, they could reason that it is better to act on their desires to not desire robbing banks because of the financial ruin of innocent people that would result. Thus, even if it is the case that first-order desires are causal forces, which are the product of evolution, it is not the case, according to the objection from agency, that reasons and second-order desires are merely causes; rather, they also have justificatory significance. What this implies is that **P2** directly in the above argument is false. This means that reasons (and second-order desires) are not completely understood by way of causes that are the product of evolution. Presumably, then, this also means that a plausible evolutionary ethic had best stay clear of treating reasons as causes.

3. The Objection to Progressive Evolution: Gould's Guillotine

Darwin appears to suggest that the process of evolution—that is the unfolding of life over millions of years—reveals a kind of progress (Darwin, 1859, p. 345). Although Darwin's considered view on progress is a bit inconsistent, this quotation gives the impression that the forces of evolution produce more advanced organisms than their predecessors.[14] At first glance, this is not an absurd claim. Moving from prokaryotic cells to eukaryotic cells and then to complex organisms appears to reveal a kind of progressive movement from simple to complex entities. Picking up on Darwin's suggestion, Wilson boldly pronounces "Progress, then, is a property of evolution of life as a whole by almost any conceivable intuitive standard, including the acquisition of goals and intentions in the behavior of animals" (Wilson, 1992, p. 187).

Before moving any further in the discussion, a definition of evolutionary progress is in order.[15] Drawing on Wilson's account above, evolutionary progress can be defined as follows:

Evolutionary progress refers to the inevitable movement from less complexity to greater complexity of organisms on the whole in Earth's life history.

---

14   For more on Darwin's position on progress, see Richards (1988).
15   For a detailed exchange of ideas on the notion of progress and evolution beyond this explication, see Nitecki's (1988) collection of essays and corresponding citations, Gould's (2002, ch. 6) final book, and McShea and Brandon's (2010) defense of diversity and complexity.

In the above definition, there are two crucial points to keep in mind. First, in order to claim progress as a property of evolution, there must be clear movement from less complex to more complex organisms. Although 'complex' is a difficult term to pin down, let us follow Wilson's lead for now and accept that 'complex' refers to those changes in organism features he notes in the above quotation.[16] Second, if the first condition is met, defenders of the progressive view insist that progress is an inherent feature of evolution. What this means is that no matter how often you re-run the tape of life, the same or very similar set of complex features of organisms—like those noted by Wilson above—would always emerge. The details surrounding these two characteristics of progress will become clearer as the discussion unfolds.[17] To assist us in getting to these details, the following argument reconstruction should help make sense of Wilson's claims and how all of this is related to human morality:

Argument from Progress

P1. If progress is an inherent property of evolution, then the effects of evolution are a product of the inherent property of progress.

P2. Progress is an inherent property of evolution.

P3. The effects of evolution are a product of the inherent property of progress.

P4. Human morality is an effect of evolution.[18]

C1. Human morality is a product of the inherent property of progress.

Let us grant the truth of **P1**, although it is debatable. It is not obviously true that all effects of evolution need be the product of its "progressive force." Rather, some features of organisms could be side effects of other adaptive features. The point here is that the truth of **P1** may be plausible with respect to adaptive features, but not side effects. So, the above argument may need to change 'effects' to 'adaptive effects' in order to gain credibility. Yet, one could imagine that a defender of the progressive view would claim that the property of progress is all-encompassing and ranges over all evolutionary forces. It is this all-encompassing feature of progress that ensures

---

16 For more on this progress debate, see Ruse's discussion (1993) regarding the distinction between *absolute progress* from *comparative progress*.

17 It is worth noting here that Wilson's account does not seem to follow, even if progressive complexity is granted. The reason is that different sorts of complexity could emerge. This suggests that the necessity of complexity does not guarantee the inevitability of particular complex features.

18 Morality is included here in the light of Wilson's phrase, "the acquisition of goals and intentions in the behavior of animals." In the case of humans, such goals and behaviors can and do take on a moral dimension.

that certain features of life—including human morality—are inevitable. Thus, they may insist on 'effects' rather than the modified 'adaptive effects.' So, for the moment, let us grant the truth of **P1**.

What if **P1** were true? **P2**, then, becomes the crucial premise of the argument. Wilson and Conway Morris would claim that **P2** is obviously true looking at the unfolding of life on Earth.[19] Specifically, Wilson claims that body size, feeding and defensive techniques, brain and behavioral complexity, social organization, and precision of environmental control reveal that progress is fundamental to evolution. Similarly, Conway Morris makes the point in terms of the human species, "We may be unique, but paradoxically those properties that define our uniqueness can still be inherent in the evolutionary process. In other words, if we humans had not evolved then something more-or-less identical would have emerged" (Conway Morris, 2003, 196). Since morality is thought to be one of those properties that defines human uniqueness, it would emerge inevitably (in some shape or form) through evolution. Thus, Wilson and Conway Morris would contend that **P2** of the **Argument from Progress** is true.

GOULD'S GUILLOTINE: THE SERENDIPITY RESPONSE

Over much of his career, Gould has attempted to dispel the notion of progress as an integral part of evolution. Indeed, he makes clear that the concept of natural selection provides little help in locating progress in evolutionary theory (Gould, 1996, pp. 136–40). There are two parts to Gould's account that can be understood as arguments against the truth of **P2** of the **Argument from Progress**. First, he points out that the basic tenets of evolution by natural selection give no scope for inherent progress. Why is this so? Well, consider these three elements of natural selection:

(1) Organisms show trait variations.

(2) Some trait variations are more successful in contending with local environmental tensions than other trait variations.

(3) Some of the more successful trait variations, which are heritable, are passed on to future generations by way of reproduction.

Gould is challenging his readers to locate *inherent progress* in any of the above statements.[20] First, with respect to (1), it is true that organisms show variation in

---

19 Note that I will assume, for the sake of this argument, that **P4** is true. What this means is that evolution is the direct cause of human morality and that human morality is an adaptive feature. It is not a side effect, but a trait that assists in human survival and reproductive success. What this means in detail will emerge as the chapter unfolds.

20 These three features of natural selection are not exactly the same as those used by Gould. Surprisingly, he resists using 'successful' in his account. This omission may be a mere oversight, but it needs to be present to give an accurate depiction of natural selection.

similar features. For example, different sub-species of Finches show remarkable variation with respect to the shape and length of their beaks. Yet, even if one granted that variation is an inherent feature of biological systems due to the exchange of genetic material during reproduction, it does not follow that specific advantageous and more complex variations will necessarily emerge from the shuffling of genetic material. For example, Williston's Law tells us that "in any lineage where there are serially homologous parts, the number of those parts *tends* to decrease, while the diversity and specialization of different variations *tends* to increase" (Garvey, 2007, p. 69). Notice that this so-called law uses 'tends to.' The reason why is that biological "laws" are not exceptionless laws. For instance, it may be true that many species that have homologous parts (e.g., crayfish) show less number of these parts in favor of more variation and specialization in the evolution of the species. As Garvey points out, however, many species of millipedes and centipedes (having long rows of near-identical legs) do not show more diversification and specialization than their predecessors (Garvey, 2007, p. 70). Since 'progress' refers to more complex and more advantageous characteristics than predecessors, Gould's rejection of inherent progress being part of (1) is on the mark.

In terms of (2), one should be willing to agree that some variations prove to be more useful than other variations in the light of various environmental stressors. For instance, having a thick beak to crack open strong-shelled nuts has proven to be an advantage for some finches over other finches. In this case, it is true that greater thickness proves to be an advantage, but it is not at all clear why thickness counts as more complex. In fact, Gould (1996, p. 139) provides an example to illustrate this point:

> As an adult, the famous parasite, *Sacculina*, a barnacle by ancestry, looks like a formless bag of reproductive tissue attached to the underbelly of its crab host ... but surely less anatomically complex than the barnacle on the bottom of your boat, waving its legs through the water in search of food.

So the *Sacculina* has become "simpler" in its features as it has adapted to its local environment—to the point that it no longer possesses the legs of its barnacle predecessors. Not having legs is advantageous, but the absence of such appendages would suggest a less complex organism. Gould's point is that the random beneficial variations of life reveal no obvious progress from simple to complex. Of course, Gould acknowledges that there are directional trends as a result of adaptations to local environments, but he points out that these examples "in the fossil record do not establish a pattern" (Gould, 1994a, p. 304). What such examples do reveal is that evolution by natural selection can produce advantageous characteristics that are more complex than similar characteristics possessed by predecessors, but it does not follow (as the *Sacculina* example illustrates) that the complexity condition is a necessary condition. Thus, without the complexity condition, progress (as defined above) can no longer be seen as an ineliminable feature of evolution.

Finally, with regard to (3), it is true that if a trait is heritable and beneficial (and let us also grant more complex), it can be passed on to future generations. But is this a necessary effect produced by progress being an inherent feature of natural selection? In what can be viewed as further support of the truth of **P2** of the **Argument from Progress** and a reply to Gould, Conway Morris argues that there is enough evidence to suggest that certain traits do, in fact, reveal an inherent progressive aspect to evolution (Conway Morris, 2003, pp. 281–82). Specifically, Conway Morris is arguing that there has been independent evolution—separate convergences—of the same feature in different species in different geological epochs. For example, compound eyes have evolved several times in different species; so has silk/web production in arthropods and various insects; similar dagger-like canines are present in placental cats and South American marsupials; comparable plumage can be found in tropical seabirds (see Conway Morris, 2003). There are many other examples, but Conway Morris' point is that all this convergence of similar characteristics (including behavioral repertoires and intelligence) suggests that evolution produces what can be called constrained variation—similar features will almost always emerge to contend with a range of environmental hurdles. In terms of (3) above, Conway Morris can be understood to claim that convergence is the crucial evidence that suggests that certain features will be passed on inevitably. Armed with these examples of convergence, we can think of Conway Morris affirming the truth of **P2** of the **Argument from Progress** a few pages back.

Drawing on his understanding of mass extinctions, Gould responds to this type of inevitability claim by pressing the notion of sheer luck (Gould, 1994b, p. 316). Some advantageous traits acquired during normal evolutionary processes may be beneficial during mass extinction events and some may not. For example, organisms that were able to lay dormant during the darkness that resulted in the extinction of dinosaurs were able to survive the resulting harsh climate changes. There is very little, if anything, inherent in evolution, argues Gould, moving in this direction; just lucky that some organisms had this (and other) characteristics—features secured during normal evolutionary transitions. As Gould (1989) has said elsewhere, and contrary to Conway Morris, if we rewound the tape of life, its evolution would likely be radically different than what has come to pass. For example, if the mass extinction event that took out the dinosaurs and most other life forms had not occurred, it is very unlikely that mammals would have evolved as they have or at all. This means that humans and their intelligent bipedal ancestors would have most likely not had the chance to grace this planet. The absence of the dinosaurs (especially the predatory ones) allowed for mammalian evolution to expand and this is simply a matter of serendipity. This is why Gould thinks that the human species is merely a "tiny twig on an improbable branch of a contingent limb on a fortunate tree" (Gould, 1989, p. 291). Thus, contrary to Conway Morris and others, according to Gould's account, all of this suggests that talk of inherent progress with respect to evolution should be substituted with talk of dumb luck.

CRITERIA	EXPLANATION
1a. Is/Ought Objection (**Hume's Hatchet**)	It is an error in reasoning to move from "is" premises to an "ought" conclusion. This suggests that a normative conclusion requires a corresponding specific set of normative premises.
1b. Open-Question Argument (**Moore's Machete**)	"But-is-it-good?" reveals that any attempt to reduce goodness to a natural property is doomed to failure. Whatever goodness may be, it is not identical to any natural property or set of natural properties.
2. Objection from Agency (**Korsgaard's Katana**)	Since reasons are not causes and moral agency is tied to reasons, no naturalistic causal account can fully make sense of human morality.
3. Objection to Progressive Evolution (**Gould's Guillotine**)	It is a mistake to think of evolution as a progression from simpler forms to more complex forms. Correspondingly, it would be a mistake to couple human morality to a progressive view interpretation of evolution.

Table 6.3 *Objections to Naturalizing Morality/Normativity*

How does all this connect to morality? Recall that human morality is an inevitable evolutionary occurrence, if and only if, progress is an integral aspect of evolution. Gould has shown, however, that it is misguided to believe that progress is ineliminably connected to evolution. What this reveals for our discussion is that **P2** of the **Argument from Progress** could very likely be false, rendering Wilson's argument (and to a large extent Conway Morris' argument) unsound. I say likely because Gould's reply, although compelling, may not be a knock-down argument against the progressivists like Wilson and Conway Morris because it does rely on speculating on how life would unfold under a different possible world. The result is that, for example, Conway Morris' convergence argument is not directly addressed. Yet, Gould's argument from mass extinction, which would change local environments dramatically and, correspondingly, the adaptive features related to those environments, is rather compelling. For instance, imagine that there were no trees as a result of a particular mass extinction. It is possible that a human-like organism might not have evolved

(Ruse, 2006b, p. 202). So, what plausibly follows from all this, then, is that those Darwinians who think that an evolutionary ethic can be defended on the grounds that evolution is inherently progressive are, at the very least, relying on a potentially unstable foundation. Thus, we should be suspicious of (and be prepared to reject) any evolutionary ethic that merely draws on an inherent progressive concept of evolution.[21]

To summarize this section, we have looked at three standard objections to evolutionary ethics. The first is Hume's is-ought objection. Here we have an objection based on a logical error. A variation on this first objection is Moore's open-question argument. This criticism dares anyone to provide a natural property as a legitimate substitute for 'goodness' by invoking "But-is-it-good?" The second is the objection from agency. If we think reasons are not causes, then we must grant a unique status—agency—to those beings whose lives are governed by reasons. And the third is the argument against progress. As we just saw, it is a mistake to think that progress is an inherent feature of evolution. What follows from this is that it is also a mistake to think that humans and their moral ways are an inevitable effect of evolution. Thus, it is *prima facie* the case that an evolutionary ethic that violates any of these objections is thought to be less than persuasive. Table 6.3 on the previous page summarizes all of this.

VI. THE EVOLUTION OF THE MORAL SENSE: CONTEMPORARY EVOLUTIONARY ETHICS

Let us take stock of what we have before us. The faces of morality, the distinct evolutionary ethics projects, the relevant evolutionary mechanisms and how the three interconnect, and the main objections to evolutionary ethics have been explicated in the previous sections. The table is now set to move to specific evolutionary ethics projects. To this end, this section will provide an example of **EE1** and everyday ethics, **EE2** and moral theory, and **EE3** and meta-ethics. This should provide an adequate flavor for how distinct **EE** projects are defended and to what extent each is able to avoid the major objections noted above. The hope is that students will have enough of the evolutionary ethics landscape so that they can traverse other regions of its terrain.

Ruse on Innate Mental Dispositions, the Moral Sense Faculty, and Justice

Recall that defenders of **EE1** pursue the project of finding *the preconditions* that aid in how humans come to acquire ethical concepts, make ethical judgments, and formulate moral systems. Although in the wrong hands an **EE1** project can wield rather unsubstantiated speculations about human biological and moral evolution, its empirical focus keeps it fairly immune to the objections to more ambitious evolutionary ethics projects. Ruse (1995) pursues just this conservative version of **EE1** and he begins by

---

21 Admittedly, the debate on progressive evolution is hotly contested. I have not offered a knockdown argument against the progressivists here. For a good glimpse into this debate, focused on Gould and Dawkins, see Sterelny (2009).

trying to find the middle ground circumstances that account for a human *moral sense* faculty. At one extreme, given changing social and environmental circumstances, Ruse argues that humans likely would not have fared well in the struggle for survival if they were genetically pre-programmed to behave "altruistically." Alternatively, he submits that if human altruism were the product of purely case-by-case and point-by-point rational decision-making processes, then such a super brain would require a tremendous expenditure of energy. Such high energy production and consumption would have likely proved detrimental to our survival and reproductive success. In response to these extreme possibilities, Ruse hypothesizes the presence of an adaptive *moral sense faculty* in the aid of survival and reproductive success. As Ruse suggests, this moral sense faculty "is a cost-effective way of getting us to cooperate, which avoids both the pitfalls of blind action and the expense of a super-brain of pure rationality" (Ruse, 1995, p. 97).

Ruse makes it clear that it is *the moral sense* that has evolved. According to Ruse, 'moral sense' refers to a set of innate mental dispositions (hereafter **IMDs**) to behave morally. Specifically, the set of **IMDs** moves us toward producing value judgments that correspondingly guide us in the direction of cooperating with others.[22] "Conscience" is another term he uses for the moral sense faculty that is the product of **IMDs**. Although we do not have any choice about the presence of these **IMDs**, we are able to exercise choice over whether or not to follow the dictates of our moral sense.[23] So, according to Ruse, we really do make choices about our moral lives, but those choices are in response to a set of **IMDs** that are moving us in one direction or another. It is the **IMDs** that are the product of evolution by natural selection. The general form of this kind of reasoning has the following steps:

Innate disposition **D** in organism **O** tends to produce → behavior **B**, which tends to produce → result **R**.[24]

Additionally, Ruse here relies on the moral and political ruminations of John Rawls, regarding evolution and justice. Briefly, Rawls (1971) argues that justice is best understood in terms of the following principles that *free and rationally self-interested persons* would accept: (1) regulation of agreements, (2) certain kinds of social cooperation, and (3) certain forms of government. The details of these principles would be determined, in part, by also accepting the idea that one would embrace only those principles that make better the worst positions of society. Rawls argues in this fashion so as to ensure that our thinking about justice is disinterested rather than biased via class, education, race, gender, career, etc. So, according to Rawls, we would make sure

---

22 Some recent work on the neurobiology of emotions and how these brain circuits and chemistries are relevant to human morality would be relevant here as well. See Panksepp (1999), Nesse 2001, and Barber (2004, ch. 6) for further details.
23 Whether or not Ruse has offered enough to establish the presence of an evolved moral sense is debatable. See Joyce (2006, p. 51) for a critical assessment.
24 These steps are a variation of those given by Rottschaefer (2007, p. 292).

that the details related to (1)–(3) include details about assisting the least advantaged of society and guarding against their exploitation.

Given this Rawlsian framework, Ruse goes on to claim that the evolved human moral sense provides the foundation for Rawls' principles of justice (1–3 above) and makes sense of human cooperative and social ventures (Ruse, 2006a, p. 20). Upon reflection, humans have come to realize that it is in their own best interest to help others. Put in Darwinian terms, survival and reproductive success is increased when one can secure one's own interests by securing the interests of others. As Ruse goes on to explain, "What excites the evolutionist is the fact that we have feelings of moral obligation laid over our brute biological nature, inclining us to be decent for altruistic reasons" (Ruse, 1995, p. 97). From this perspective, these moral feelings, which include the fact that people have a natural proclivity to care for themselves, motivate us to endorse contractual regulations, social cooperation, and governmental entities so long as these help to secure (for the most part) everyone's interests.

Importantly, Ruse has made it clear that our moral sense—that is, our set of **IMDs**—is an adaptive feature. It is present because it increases human survival and reproductive success. Does Ruse think the same thing applies to the principles of justice noted above? Well, Ruse is committed to the view that the evolutionist moves beyond the Rawlsian by "linking" the principles of justice to our biological past (Ruse, 1995, pp. 98–99). What does Ruse mean here by "linking"? Does our evolved moral sense cause us to accept the principles of justice? Are the principles of justice themselves adaptive moral features? Ruse's answer is simply to endorse Rawls' speculation. Rawls hypothesizes about human nature as follows:

> The theory of evolution would suggest that it [human nature] is the outcome of natural selection; the capacity for *a sense of justice* and the moral feelings is an adaptation of mankind to its place in nature. (Rawls, 1971, p. 503; quoted from Ruse, 1995, p. 99; my italics)

Ruse responds, "This is precisely the evolutionist's approach" (Ruse, 1995, p. 99). This reply makes clear that Ruse is not claiming that humans have principles-of-justice adaptations. Rather, he is claiming, much like Rawls, that humans have an evolved moral sense of justice, which motivates us to follow this set of principles; however— and this crucial—it is our reasoning and reflection which guide our decision to accept these principles of justice; principles that support our enlightened self-interested nature.[25] This moral sense of justice is the product of variations on *reciprocal altruism*. As Ruse puts it, "You scratch my back and I will scratch yours" (2010, p. 299). Table 6.4 shows Ruse's overall version of **EE1**.

---

25 Ruse relies, in part, on the work of Lumsden and Wilson (1981), who argue that certain codes (they use the term 'culturgen') are learned as a result of innate biases in the teaching and acquisition processes. See both Lumsden and Wilson (1980) and Robert (2006) for further details.

NATURALIZING THE MORAL SENSE

		FACES OF MORALITY & EE PROJECTS	EVERYDAY MORALITY (MORAL SENSE OF JUSTICE) EE1
MECHANISMS OF EVOLUTION		Kin Selection	
		Reciprocal Altruism	X
		Group Selection	
		Combination Selection	

Table 6.4 *Ruse's EE1 Project*

RUSE'S EE1 PROJECT AND THE STANDARD OBJECTIONS TO EE

How does Ruse's **EE1** project fare with respect to the standard objections noted earlier? At first glance, Ruse has neither committed the naturalistic fallacy nor violated the argument from agency. He has not moved from "is" to "ought." He is only claiming that our other-regarding tendencies have their origin in the set of **IMDs**—a set that is itself the product of evolution. Even when Ruse applies his moral-sense-faculty theory to Rawls' principles of justice, he makes it clear that it is *the moral sense of justice* that is an adaptive feature that motivates us to embrace something like Rawls's principles. Cautiously, he acknowledges that reason and reflection ultimately determine whether or not Rawls' account will be endorsed. So, even if our **IMDs** are moving us to act in a certain way, we have the free choice to embrace or disregard such urgings. Thus, how we ought to act, even keeping in mind the influence of our **IMDs**, is not being substituted for what is the case. Thus, Hume's Hatchet draws no blood with respect to Ruse's **EE1** project.

Much the same can be said of the version of the naturalistic fallacy represented by Moore's Machete. Ruse is emphasizing the preconditions that account for the benefits of reciprocal altruism and how reciprocal altruism moves us in the direction of Rawls' principles that govern self-regarding strategic advantages of social justice practices. Thus, even if we have an evolved "desire" for something like Rawls's account, it is reason and reflection that ultimately do the heavily lifting in terms of endorsement. From this perspective, there is no need to be concerned about the "but is this desire for Rawls' account really good," because this is being determined by reason and reflection—not merely our evolved desire set. Thus, Ruse's **EE1** project is left unscathed by Moore's Machete.

Additionally, Ruse's claim that we exercise free choices reveals that he is sensitive to the argument from agency. Recall that the argument from agency requires that reasons not be reduced to causes. This objection is elaborated, in part, by noting our ability to resist desires that might assail us. Given Ruse's willingness to acknowledge free choices and his belief that we can reject the influence of our **IMDs**, he should be willing to acknowledge that reasons are not causes (Ruse, 1995, p. 97). The result, then, is that Ruse has offered a general evolutionary account of the preconditions of human morality that evades the swing of Korsgaard's Katana.

Finally, Ruse is not claiming that the adaptive moral sense is an inevitable effect of some kind of progressivist theory of evolution. To the contrary, he rejects an evolutionary ethic that relies on progress. As he puts it, "It is far from obvious either that natural selection promotes progress or that progress actually occurs, at least in any clear definable or quantifiable way" (Ruse, 2006a, p. 19). Ruse defends this claim by drawing on Gould's argument from mass extinction. He concludes that "there is enough truth to make one very wary about biological progress as a basis for one's moral code" (Ruse, 2006a, p. 19). Thus, along with steering clear of the naturalistic fallacy and the argument from agency, Ruse's **EE1** also escapes Gould's Guillotine by openly not embracing a progressivist theory of evolutionary ethics.[26]

Richards and Moral Theory

A core aspect of Ruse's Rawlsian account is a consequentialist element; namely, according to Ruse, the moral sense faculty aids in the production of a full-blown moral theory of justice that gives prominence to benefits given and received. "You scratch my back and I'll scratch yours" is a pithy aphorism unique to those forged alliances that are based on mutually beneficial *consequences* between self-interested individuals in social contexts.

In what can be viewed as a rejection of this sort of consequentialist thinking in terms of evolutionary ethics, Richards (1987) advocates a non-consequentialist account. Since the consequentialist perspective gives little or no emphasis to one's motives, Richards shuns this aspect of utilitarianism, insisting that human altruism produces actions that are designed to assist others in need. Thus, according to Richards, we come to the aid of others, not because we are concerned about the consequences, but because we have either an evolved predilection and/or have been culturally molded (in a non-consequentialist way) to care for others in our community. He states his position, which closely resembles **EE2**, by emphasizing that "the evolutionary ethics that I am advocating regards an action as good only if it is intentionally performed from a certain kind of motive and can be justified by that motive. I will assume as an empirical postulate that the motive has been established by community or kin selection" (Richards, 1987, p. 609).

---

26 Note, however, that it is not obvious why Rawls' principles need be the relevant ones to which we are disposed, as Ruse suggests; for on Ruse's and Rawls' accounts, there is a consequentialist self-interested component to the actions of humans that is crucial to their willingness to provide assistance to others in the name of justice.

It is unclear why Richards is committed to a solely non-consequentialist moral theory. From a Darwinian perspective, features are retained due to their usefulness— that is, the benefits they provide to the individual or group. It would not be surprising, then, if the altruistic motive included a consequentialist component. Put another way, the evolved motive to help others could include a concern for duty in terms of benefiting kin, but could include a consequentialist-based component to help non-kin. This latter consequentialist-type consideration for non-kin could be the product of cultural forces. Of course, we should not rule out the possibility that a duty-based moral theory could be the product of cultural forces as well. Thus, although possibly unstable in particular circumstances, a mixed account of moral theory (which includes both Darwinian and social components) is as, if not more, plausible than Richards' strict non-consequentialist account.

In fact, Richards' own account appears to endorse a consequentialist outlook at the level of moral theory.[27] He defends his revised account (RV) as follows:

> [I]t might have been deemed in the community interest to sacrifice virgins, and this ritual might in fact have contributed to community cohesiveness and thus have been of continuing evolutionary advantage. But RV does not therefore sanction the sacrifice of virgins, but only acts that, on balance, appear to be conducive to the community good. (Richards, 1986, p. 614)

From an EE2 perspective, this qualification by Richards is not surprising. Recall that EE2 attempts to derive additional moral insights by drawing upon existing aspects of human evolution. So, Richards appears to be claiming that intending to act for community benefit is an evolved feature (buttressed by cultural forces) of an earlier self-interested aspect of human morality. This is one way of reading Richards' understanding and condemnation of virgin sacrifice—like applying evolutionary thinking to some environmental ethics or medical ethics scenario. In this way, Richards can introduce a non-consequentialist evolutionary ethic (as the additional moral insight about recent human evolved morality) without ignoring a possible role of consequentialist thinking (this would be the older human/hominid evolved morality) as part of the history of human morality. This is clearly in keeping with an EE2 project. So, even if early human moral psychology was predominantly consequentialist due to reciprocal altruism, this psychology, argues Richards, continued to evolve to produce a chiefly other-regarding community-based intentional psychology by way of kin selection or group/community selection. The upshot is that both kin selection and/or group selection were likely forces at work in Richards' Kantian/intention based evolutionary ethic. All of this is captured in Table 6.5.

---

27 I have merely scratched the surface of what is a rather turbulent debate between consequentialists and Kantians. I simply cannot do justice to this aspect of this debate in the pages here. For a general glimpse into both camps, see Hinman (2007).

	FACES OF MORALITY & EE PROJECTS	MORAL THEORY (KANTIAN/INTENTION-BASED) EE2
MECHANISMS OF EVOLUTION	Kin Selection	X
	Reciprocal Altruism	
	Group Selection	X
	Combination Selection	X

Table 6.5 *Richards' EE2 Project*

RICHARDS' EE2 PROJECT AND THE STANDARD OBJECTIONS

Richards' **EE2** project does not hold-up so well to certain standard objections. This becomes clear when we examine his overall conclusion: "Since, therefore, human beings are moral beings—an unavoidable condition produced by evolution—each ought to act for the community good" (Richards, 1986, p. 289).

Before turning to the overall argument, notice that Richards understands human morality as an "unavoidable condition" of evolution. This clearly violates Gould's objection from progress. Recall that the presence of humans and the set of their adaptive features are the result of numerous fortuitous events. It follows, then, that the set of adaptive features associated with human morality is also subject to the very same or related set of accidental happenstances. This means humans could have evolved an entirely different moral outlook than the one suggested by Richards and could have not evolved a moral sense at all. Richards does not expand on why he thinks the unavoidability condition is so crucial to his account. In order for it to be taken seriously, he needs to address Gould's Guillotine. Instead, his **EE2** is greatly weakened as a result of not addressing why the vagaries associated with evolving systems are not relevant to the evolution of humans as moral beings. Thus, at the very least, to the extent that Richards' **EE2** relies on the unavoidability condition, it is correspondingly less than persuasive.

Still, the general argument he offers above can be reconstructed as follows:

P1. If humans have an evolved moral sense that functions to enhance the good of the community, then humans ought to act for the sake of the good of the community.

P2. Humans have an evolved moral sense that functions to enhance the good of the community.

C1. Humans ought to act for the sake of the good of the community.

Richards asks us to grant him the truth of **P2**. (Recall that he concedes that something like **P2** is true based on a kin and/or group selection account of human evolution.) He argues that if **P2** is accepted as true, then his argument is sound. Now, at this point in the chapter, I hope that you can see that **P1** appears to commit the naturalistic fallacy; for even if we grant the truth of **P2** (and the antecedent of **P1**), it simply does not follow that humans *ought* to act for the sake of the community in the light of the fact that humans have an evolved moral sense that functions to enhance the community good. In other words, we can grant that the human moral sense is an evolved feature for the sake of benefiting the community, but we can still deny that we *ought* to benefit the community. Richards has clearly moved from "is" to "ought" in **P1**, rendering **P1** false and the argument unsound.

Richards is well aware of the above sort of objection. He responds in defense of **P1** with the following argument from analogy explained to us by Lemos (200, p. 50):

> Just as we can reasonably infer that thunder *ought* to occur given that lightening has occurred, so too can we reasonably infer that we *ought* to act altruistically given that we have evolved to do so. (Richards, 1986, pp. 287–91; my italics)

Scholars have been quick to pounce on the above argument (Voorzanger, 1987, Williams, 1990, Joyce, 2006, and Lemos, 2008). First, however, let us understand it. Richards is suggesting that if lightening occurs, we can reasonably infer that thunder also ought to occur. In the same way, for example, if we see that someone in our community needs help, then we can reasonably infer that we ought to help. So, at first glance, Richards can claim that **P1** is true in the light of the soundness of his argument from analogy.

There is something amiss, however, with this argument. Should we accept Richards' (1) 'lightening ought' and (2) 'moral ought' as having the same meaning? The critics correctly reply with a uniform "No!" The reason is that Richards is equivocating in his use of 'ought.' In the lightening-thunder case, 'ought' translates into "will likely occur." There is no moral value/normative force to this "ought" at all. This "ought" is best described as a predictive or mechanistic ought (Williams, 1990). What this means is that the physics behind the lightening-produced thunder ensures that a powerful enough lightening strike will deterministically produce thunder—thunder will necessarily follow lightening of a certain magnitude as long as no other intervening forces are present. In the moral case, however, the "ought" is a normative "ought." It is not describing how caring for our neighbor-in-need occurred or even the likelihood that such assistance will occur. Rather, it is prescribing what should be done (and intended) for our neighbor. It is this prescriptive sense of ought that makes Richards' "necessary-mechanistic-ought" very different from his "necessary-practical-ought."

Now, with respect to **P1**, "our evolved moral sense to care for our community" is like the lightening case—it is describing the moral mechanisms that are currently present as a result of human evolution. However, that we ought to act in the way our moral sense dictates is (again) not a predictive or mechanistic ought, but a prescriptive/normative ought. Thus, all of this takes us to a stage where it should be clear that Richards is moving from "is" to "ought"; maneuvering directly under the path of Hume's Hatchet. He has committed the naturalistic fallacy, rendering his argument from analogy rather emaciated and **P1** false. Thus, his evolutionary ethic is less than persuasive; indeed, the argument he offers is unsound.[28]

What about the objection from agency? Richards does not directly address it, but he seems to violate it; since he views "oughts" through the lenses of his lightening analogy, this seems to reinforce the view that reasons-to-do-X would be predictable in the same sort of lightening fashion. If this is correct, then reasons are causally potent in a way that violates the moral agency objection. Charitably, this assessment can only be offered in the form of a conditional since Richards does not put forth a positive reply to the objection. At any rate, Korsgaard's Katana looms large over the plausibleness of Richards' **EE2** project to the extent that the agency objection is as damaging to naturalizing ethics as some philosophers believe.

Richards' account is mentioned not so much as a more reasonable alternative to Ruse's **EE1** offering, but more so as a reminder of how difficult it is to defend an **EE** proposal that moves beyond **EE1**. No doubt, others have tried. For example, Arnhart (1998) favors coupling Darwinism and a neo-Aristotelian conception of rights and virtues; Rottschaefer (2007, 1991a, and 1991b), who is sympathetic with part of Richards' account, gives considerable weight to Darwinian-type teleology (recall Chapter 4) in arguing that the human moral sense evolved to bring about the goal of community survival; Sober and Wilson (1998), in contrast, have tried to defend a non-egoistic picture of human morality that is coupled with their own rendition of group selection; in different ways, more recently, Casebeer (2003), Joyce (2006), and Lemos (2008) have tried to advance distinct evolutionary ethics wedded to Aristotle's virtue theory.

What has emerged from this literature is that there is little consensus regarding the relationship between evolution and moral theory. Some scholars favor a utilitarian perspective, while others have endorsed a more duty-based account. And still others have drawn on moral rights, contractarianism, or virtue theory. Additionally, some endorse kin selection, while others embrace either reciprocal altruism, group selection, or some mixed-selectionist account. Furthermore, all of these accounts have had to navigate the objections to naturalizing ethics with only a modicum of success. At any rate, although the arguments put forth by Ruse and Richards should be taken seriously, they are not—by a long shot—the only **EE** games in town![29]

---

28   Richards (1987, pp. 612–20) further attempts to fight off the naturalistic fallacy by arguing that it is not a fallacy at all. Like Arnhart, Richards claims that, ultimately, normative arguments rest on factual foundations. See Gampel (1996) for a counter-reply to this sort of argument by Richards.

29   Due to space limitations, I have said nothing about the field of evolutionary game-theory. For those interested in this literature, see the works of Axelrod (1994), Alexander (2007), D'Arms (2000), Harms and Skyrms (2008), and Skyrms (1996 and 2004).

Arnhart and Meta-ethics

As part of his justification of an evolved moral sense related to natural rights, Arnhart (1998) offers an **EE3** project; that is, he pursues a meta-ethical inquiry when he draws upon the sciences to defend a naturalistic account of the nature of moral goodness.[30] He thinks that the moral nature of humans is the product of natural selection and that this constitution is foundationally understood in terms of brain function (this is similar to Ruse's **IMDs**). As Arnhart puts it, the twenty virtues[31] that comprise the human moral sense "have evolved by natural selection over four million years of human evolutionary history to become components of the species-specific nature of human beings, [and] that they are based on the physiological mechanisms of the brain, and that they direct and limit the social variability of human beings as adapted to diverse ecological circumstances" (Arnhart, 1998, p. 36). For example, as part of their practical reasoning skill set, most (if not all) human populations appear to have evolved mental "cheating detectors" (some set of brain circuitry) that allows them to recognize and control various forms of communal defection, like those of Butch and Sundance (see ch. 7 for more on this point). The EE3 aspect of Arnhart's account is evident when he claims that "I appeal to this natural moral sense in defending a naturalistic ethics based on my claim that *the good is the desirable* ... what is 'desirable' for human beings is whatever promotes their human flourishing" (Arnhart, 1998, pp. 81–82; my italics). For Arnhart, human flourishing is associated with a complete life; that is, a life in which a person expresses (in speech, thought, and action) his natural twenty virtues. His argument, let's call it **The Argument from Naturally Good Virtues**, is as follows:

P1. The human sense of justice arises from emotional and rational brain processes.

P2. The human sense of justice justifies the claim that the good is the desirable.

P3. The desirable is whatever promotes human flourishing.

P4. Human flourishing is associated with a complete life.

P5. A complete life is the actualization of our natural virtues.

C1. Thus, by substitution, the good is the actualization of our natural virtues.

---

30 Arnhart is referring to the work of Damasio (1994) and Masters and Gruter (1992).
31 This moral sense, he explains, is made up of twenty natural desires (Arnhart, 1998, pp. 31–36). To varying degrees, Arnhart thinks that all twenty natural desires are found in all human societies and are the product of natural selection. It is reasonable, then, to infer that the moral elements (whatever they may be), present in each of these twenty desires, encompass Arnhart's version of the human moral sense faculty.

It is worth noting that **P1** is justified by Arnhart in terms of natural selection. Recall that he also claims that variation in human morality is explained by four million years of human adaptations to "diverse ecological circumstances." Although the inference is tentative, it is reasonable to suggest that every selection mechanism is part of Arnhart's account. The primary reason for drawing this inference is that if humans evolved a variety of moral sense faculties as a result of adaptations to distinct and diverse local environments, then it is likely that different selection mechanisms may very well have been at play in these diverse environments. For example, kin selection could have been the driving force behind molding the moral sense of certain human populations, while group selection could have been molding the moral sense in other human populations in different environments. Alternatively, a combination of selection forces could have been at play in yet other environments. At any rate, since Arnhart does not wed himself to any particular selection mechanism, all of the discussed selection mechanisms could reasonably be included in his account. Table 6.6 captures Arnhart's **EE3** project.

	FACES OF MORALITY & EE PROJECTS	MORAL THEORY (KANTIAN/INTENTION-BASED) EE3
MECHANISMS OF EVOLUTION	Kin Selection	X
	Reciprocal Altruism	X
	Group Selection	X
	Combination Selection	X

Table 6.6 *Arnhart's EE3 Project*

ARNHART'S EE3 PROJECT AND THE STANDARD OBJECTIONS

In terms of the standard objections, Gould's Guillotine seems to not be an issue for Arnhart's **EE3**. Since his account can tolerate numerous selection processes and the likelihood of the formation of distinct moral senses due to local adaptations to different local environments, his analysis seems in line with the vagaries and accidents that contribute to the tapestry of life's evolutionary tale. Unless Arnhart were to endorse openly the view that the moral sense faculty, as a complex adaptation and in all of its guises, is an inevitable characteristic of human evolution, it is difficult to see how he is necessarily wedded to a progressive view of evolution. (But see the objection from agency below!) So, Arnhart's **EE3** does not appear to find its way to the chopping block of Gould's Guillotine.[32]

---

32 Not a lot of time has been spent on the objection to evolutionary progress because Arnhart does not directly address the evolutionary progress debate. In chapter 9 (Arnhart, 1998), however, he does distinguish his brand of Darwinian teleology from cosmic teleology. If his rejection of the

Yet, at the level of meta-ethics, has not Arnhart fallen victim to the naturalistic fallacy? Has he not avoided the blunt force of Hume's Hatchet or Moore's Machete? He clearly equates "the good" with "the desirable" and "the desirable" simply is the actualization of our natural virtues. Thus, the good appears to be the actualization of our natural virtues. If this is correct, then **P2** of **The Argument from Naturally Good Virtues** is false because this violates a version of Hume's is-ought objection and Moore's persistent "but is that which is desired good?" query.

Apparently, however, Arnhart (1998, pp. 82–83) rejects this entire line of reasoning. In what can be viewed as further defense of **P2** of **The Argument from Naturally Good Virtues**, Arnhart is rejecting the more-or-less modern view that facts are different kinds of "things" than values—a view that is part of the crux of the naturalistic fallacy. His argument, **The Argument from Facts and Values**, can be reconstructed as follows:

The Argument from Facts and Values

P1. If facts are distinct from values, then we would have no reason to obey normative judgments.

P2. If it is not the case that facts are distinct from values, then the naturalistic fallacy is not really a fallacy at all.

P3. It is not the case that we have no reason to obey normative judgments.

P4. It is not the case that facts are distinct from values.

C1. The naturalistic fallacy is not really a fallacy.

What are we to make of this argument? For starters, **P1** is false as stated. Even if it is true that facts are distinct from values, it does not follow that we would have no reason to obey normative judgments. For instance, there is no reason why we cannot be motivated to embrace additional normative judgments in support of a particular normative judgment. For example, "Parents ought to care for their children" is a normative claim that can be justified in the following way:

P1. One ought to act in such a way that would provide an overall benefit to one's community.

---

latter can be translated into a rejection of inherent progress, as I think it does, then he has not violated the objection to evolutionary progress.

P2. One way of providing a benefit to one's community is by being a caring parent for one's children.

C1. Therefore, parents ought to care for their children.

Assuming that **P2** is true, **P1** is a normative claim in support of why parents ought to care for their children. If one takes seriously the truth of **P1**, then one has ample justification for **C1**. The above argument seems sound. This line of reasoning further suggests that **P1** of **The Argument from Facts and Values** is false; that is, even if facts are distinct from values, we may have reasons to obey normative judgments. If this is so, then Arnhart's **Argument from Facts and Values** is rendered unsound.

Now, Arnhart would reply that **P1** in the above argument begs the question of why one ought to care about benefiting one's community. He responds that, ultimately, one will have to rely on some facts about the human condition (e.g., humans are naturally gregarious and their mental health necessitates securing community welfare). It is these facts that are doing the motivational work to get people to act morally—to give prescriptive force to the relevant normative judgment. Thus, Arnhart could insist that his **Argument from Facts and Values** remains unscathed.

Yet, the obvious response is to ask: "Why ought I care about human gregariousness?" It is not the fact of human gregariousness that drives people to be moral, but the normative imperative, "we ought to promote mental health and this can best be accomplished by securing community welfare," that drives us to act morally. Thus, Arnhart's **Argument from Facts and Values** seems to be in jeopardy all over again.

Of course, Arnhart might very well insist that even the "we ought to promote mental health and this can best be accomplished by securing community welfare" claim relies on some fundamental desire possessed by humans and it is this desire that secures the force of the normative claim. Yet again, however, the counter-reply could be that there is no reason to care about such a desire. The point here is that it is eminently reasonable to think that people could be resolutely motivated by normative claims. If this is so—and there is no reason to think that it is not—then **P1** of the previous argument is false.

Moreover, has not Arnhart committed the naturalistic fallacy by insisting that normative claims require factual claims to ensure appropriate moral action? The answer is "yes" based on his own analysis. To see this, it is first important to note that Arnhart is aware that his account could be vulnerable to the criticism that the good cannot be equated with human natural desires because humans frequently have bad desires. For example, Butch and Sundance desire to engage in high-risk and high-reward robbery. Since, for the most part, this type of stealing is considered a serious moral failure, this kind of desire is bad. Thus, a moral theory that is based on giving primacy to desires runs into these sorts of counter-examples.

In reply, Arnhart points out that the fulfillment of any old desire will not do (1998, pp. 82–83). Only those desires, he submits, that contribute to a complete life count as good desires. Since stealing does not promote a complete life, such desires are straightaway ruled out of the category of morally good desires. So, Arnhart thinks he gets around the bad-desire objection by stipulating that only those desires that contribute to a complete life are morally good desires.

Does this reply really get around the naturalistic fallacy? It is not clear that it does. For instance, I could desire all sorts of bad things in pursuit of a genuinely good thing. Put another way, a life could be viewed as "overall complete" even if some parts of it are neither perfect nor morally savory. For example, if Butch and Sundance's pilfering ways are designed to support financially a particular orphanage, then maybe their Robin Hood-like ideals and actions move them toward a complete life (see Lemos, 2008, p. 46 for a similar criticism). Thus, Arnhart's attempt to block the bad-desires argument by restricting desires to only those that promote a complete life does not seem to work.

Drawing upon both Hume and Moore, the additional inference to be made here is that the good cannot be reduced to or derived solely from facts. On the one hand, Hume's Hatchet reveals an is/ought fallacy present in Arnhart's **EE3** project. On the other hand, Moore's Machete divulges that desire-based accounts do not mitigate worries about the goodness of desires; that is, contrary to Arnhart's contention, the good is not necessarily the desirable. This further suggests that **P1** and **P4** of **The Argument from Facts and Values** are false. Thus, it is reasonable to claim that both Hume's and Moore's versions of the naturalistic fallacy have been violated by Arnhart, rendering his **EE3** less than persuasive.[33]

An additional nail in the coffin of Arnhart's account is that he violates the objection from agency—the idea that reasons are not causes. To see this, it is important to point out that Arnhart is implicitly endorsing a kind of *moral compatibilism* in which all of our actions are caused, but so long as they are caused in the right way, they are considered free. Specifically, in his own words, "Moral freedom should be identified not as the absence of determinism but as a certain kind of determinism" (Arnhart, 1998, pp. 84–85).[34] In this case, so long as our choices are the product of deliberation, this is all one can demand with respect to freedom and responsibility.

Yet, this clearly violates the objection from agency because thinking of reasons, desires, and reflective beliefs as determined causal forces, as Arnhart openly does, makes moral responsibility difficult to justify. For example, Butch and Sundance could claim that they are not responsible for their pilfering ways because, for instance, they were forced to rob Bank A, as opposed to doing charity work, as a result of their reflective desires and beliefs moving them in the direction of robbing Bank A. Now, if we assume, as Arnhart would like us to, that peoples' reflective beliefs and desires determine their action, then it is not at all clear that they have acted freely; that

---

33 For more on Hume and Moore, see Ball (1988) and Darwall (2003). No doubt, this discussion barely scratches the surface of the debates surrounding the naturalistic fallacy and meta-ethics. See Fisher (2011) and his citations for a good introduction to these and other meta-ethics related issues.

34 Dennett (2003) has developed an evolutionary rendition of free-will along these lines.

is, that they could have done otherwise. At the very least, Arnhart needs to offer a much more rigorous defense of his compatibilist position, if he hopes to persuade us that causal determinism does not tell against free will. Unfortunately, he does not offer such an account, rendering his **EE3** vulnerable to Korsgaard's Katana.[35] Thus, since causal determinism is part of his overall analysis, and this allegiance violates the objection from agency, we have a further reason to reject Arnhart's **EE3** project.

By way of Arnhart's **EE3** project, this section has offered a glimpse into another aspect of **EE**. Much like alternatives to Richards' **EE2**, there have been assorted attempts to defend some version of **EE3** while engaging some of the complex arguments embedded in the meta-ethics literature (e.g., Ruse, 2006a and 1995, Rottschaeffer and Martinsen, 1990 and 1991, and Joyce, 2006). With this analysis of Arnhart's account in place, students should be able to examine some of these different approaches to **EE3**.[36]

VII. HOW SHOULD WE PROCEED? KITCHER TO THE RESCUE?

The picture of modern human origins is not complete (see Sterelny, 2012). There is much about our early ancestors that we do not know. Still, if we accept that humans are a product of the tree/bush of life, then it is not unreasonable to claim that current human morality is, to some extent, a product of this history (Nesse, 2001). At the very least, exploring versions of **EE1** seems like a worthy endeavor. Indeed, Kitcher does just this when he conjectures as follows:

> Our ability to transcend the limited size and extreme fragility of early hominid social life rests on a capacity for articulating rules and using those rules to shape our wishes, plans, and intentions, so that the frequency with which the altruistic tendencies that underlie cooperation are overridden is diminished. (Kitcher, 2006, p. 172)

Kitcher is offering a dual-capacity account of the human moral sense or what he calls our capacity for "normative guidance." We have the capacity not only to formulate rules about our wishes, desires, plans, etc., but we also have altruistic capabilities. It is this combination of rule formulation and altruistic capacities that makes cooperation both possible and long-lasting. Kitcher sums up his normative guidance account (or his rendition of **EE1**) as follows: "If we suppose that a capacity for normative guidance served the evolutionary function of promoting social cohesion, then we

---

35 I do not object to Arnhart's allegiance to compatibilism. What is objectionable is his lack of a rigorous defense of it. Note that Arnhart's discussions of teleology and the hierarchical nature of biological systems could have been used to make sense of his compatibilist account (see Arnhart, 1998, ch. 9), but he makes no clear connection.

36 For the sake of space, no **EE4** project has been presented in this chapter. Still, for those curious to see a critical reply to an **EE4** account, see Kitcher's (1994) assessment of E.O. Wilson's **EE4** project.

might expect that the rules shaping individual attitudes would have specified the conditions under which one is to act with one's allies ..." (Kitcher, 2006, p. 172). Kitcher's self-described speculative account can be summarized in three parts. First, we should grant that humans have the capacity for being guided by values. Second, we should accept that this capacity for normative guidance has the function of promoting social cooperation. Third, we should also "expect" that the rules, which mold individual attitudes, would assist in determining how, when, and why we should cooperate with those individuals who are on our side.

We can assume that the capacities in the three parts above are the product of evolutionary biology. It is not unreasonable to think that some combination of kin selection and reciprocal altruism were at work (see Ananth, 2005 and Boniolo and Vezzoni, 2006). For instance, in order to be moved to accept assistance from non-kin allies and to offer assistance to them, a number of iterated "trust-forming" interactions would have had to occur. In this way, scratching one-another's back can be secured. Yet, as this trust-forming is occurring, kin would be negotiating with one another about the content of rules needed to trust both each other and non-kin allies. In this way, it is then reasonable to think that his account can be understood as a dual-kin-reciprocal altruism account that makes sense of the presence of the human capacity for normative guidance, the promotion of social cohesion, and the requisite rule-following details. True, this gives us merely a souped-up version of Ruse's **EE1**, but the hazards of moving beyond such an **EE** project may prove to be too perilous.

VIII. CONCLUSION

If humans are in possession of a moral sense, then it is worth exploring the origin of this feature. Restricting the domain of discourse to naturalistic accounts, it is worth determining whether or not evolution has anything interesting to contribute to this moral sense faculty. This chapter has endeavored to offer a critical glimpse into the evolutionary ethics debate. Upon explicating the various senses of morality, the different **EE** projects, the possible evolutionary selection forces, and the common objections that should be avoided, a number of contemporary evolutionary ethical theories were examined. Notably, almost all the authors accepted the view that humans are in possession of some sort of moral sense. Ultimately, all of these accounts beyond the **EE1** projects ran into serious difficulties as a result of falling victim to one or many of the common objections. The result of this analysis, then, is that many of the contemporary attempts at an evolutionary ethic are philosophically interesting but correspondingly difficult to defend philosophically.

CHAPTER REVIEW: DISCUSSION AND QUESTIONS

1. You should know the differences between the four evolutionary ethics projects (EE1–EE4). [p. 213]
2. Briefly explain "the three faces of morality." [pp. 214–215]

3. Table 6.1 summarizes the connection between evolutionary ethics projects and the faces of morality. Can you explain these connections? [p. 217]
4. In the altruism section, what reason is given for thinking that humans and non-human animals could be understood in terms of altruism? [pp. 217–18]
5. You should be able to distinguish the following in terms of other-regarding behavior: (1) kin selection, (2) reciprocal altruism, and (3) group selection. [pp. 218–19]
6. You should be able to explain the different faces of morality, the **EE** projects, and the role of the different selection mechanisms. Put another way, can you explain Table 6.2? [p. 220]
7. How is Hume's Hatchet distinct from Moore's Machete? [pp. 221–24]
8. Use Korsgaard's discussion to make sense of the objection from agency. [pp. 225–26]
9. What is Gould's Guillotine regarding evolutionary progress? Are you persuaded by it or do you think Conway Morris has the upper hand? By the way, what does all this talk about evolutionary progress have to do with ethics/morality? [pp. 226–32]
10. How does Ruse understand the "moral sense" faculty (think in terms of **IMDs**)? How does this faculty assist humans in being moral? Why does Ruse's account fit in the **EE1** category? [pp. 232–34]
11. Could you explain Table 6.4 to one of your classmates? [p. 235]
12. In terms of the standard objections to **EE** projects, why has Ruse not committed the naturalistic fallacy, the objection from agency, and the objection from progressive evolution? [pp. 235–36]
13. Why is Richards' account explained as a form of an **EE2** project? Use Table 6.5 to guide you. [pp. 236–38]
14. What objections appear to damage Richards' **EE2** project? [Try taking one of your classmates through the progressive evolution objection and the lightning example and its related objection.] [pp. 238–40]
15. What is the nature of goodness in Arnhart's **EE3** project? [pp. 241–42]
16. You should be able to explain Table 6.6 and why every selection mechanism might be at play in Arnhart's account. [p. 242]
17. Explain why the naturalistic fallacy and the objection from agency tell against Arnhart's **EE3** project. Make sure your explanation includes Arnhart's **Argument from Facts and Values**. [pp. 242–45]
18. If the objection from agency tells against Arnhart's account, why might his account have to contend with the objection from evolutionary progress? [pp. 245–46]
19. What is the Kitcher-type dual-kin reciprocal altruism suggestion at the end of this chapter? [p. 246]
20. How is Kitcher's **EE1** project both similar and distinct from Ruse's? [Please feel free to speculate a little here!] [pp. 246–47]
21. If the standard objections to **EE** projects are reasonable, what does this say about the possibility of an evolutionary ethic beyond an **EE1** project? [p. 247]

REFERENCES

Ananth, Mahesh. "Psychological Altruism vs. Biological Altruism: Narrowing the Gap with the Baldwin Effect." *Acta Biotheoretica* 53, 3 (2005): 217–39.
Arnhart, Larry. *Darwinian Natural Right: The Biological Ethics of Human Nature*. Albany: SUNY Press, 1998.
Ball, Stephen W. "Evolution, Explanation, and the Fact/Value Distinction." *Biology and Philosophy* 3 (1988): 317–48.
Barber, Nigel. *Kindness in a Cruel World*. Amherst: Prometheus Books, 2004.
Boniolo, Giovanni, and Gabrielle De Anna, eds. *Evolutionary Ethics and Contemporary Biology*. Cambridge: Cambridge University Press, 2006.
Boniolo, Giovanni, and Paolo Vezzoni. "Genetic Influences on Moral Capacity." In *Evolutionary Ethics and Contemporary Biology*, ed. Giovanni Boniolo and Gabriele De Anna, 77–96. Cambridge: Cambridge University Press, 2006.
Bradie, Michael. *The Secret Chain: Evolution and Ethics*. Albany: SUNY Press, 1994.
Casebeer, William C. *Natural Ethical Facts: Evolution, Connectionism, and Moral Cognition*. Cambridge, MA: MIT Press, 2003.
Conway Morris, Simon. *Life's Solution: Inevitable Humans in a Lonely Universe*. Cambridge: Cambridge University Press, 2003.
Curry, Oliver. "Who's Afraid of the Naturalistic Fallacy?" *Evolutionary Psychology* 4 (2006): 234–47.
Damasio, Antonio. *Descartes' Error: Emotion, Reason, and the Human Brain*. New York: G.E. Putnam's Sons, 1994.
Darwall, Stephen. "Moore, Normativity, and Intrinsic Value." *Ethics* 113, 3 (2003): 468–89.
Darwin, Charles. *The Descent of Man, and Selection in Relation to Sex*. Princeton: Princeton University Press, 1981 [1871].
———. *On the Origin of Species by Means of Natural Selection, or, the Preservation of Favoured Races in the Struggle for Life*. John Murray, 1859.
Dawkins, Richard. *The God Delusion*. Boston: Houghton Mifflin, 2006.
Dennett, Daniel. *Freedom Evolves*. New York: Viking Adult, 2003.
———. *Darwin's Dangerous Idea*. New York: Simon and Schuster, 1995.
De Waal, Frans. *Primates and Philosophers: How Morality Evolved*. Princeton: Princeton University Press, 2006.
Double, Richard. *Beginning Philosophy*. Oxford: Oxford University Press, 1999.
Farber, Paul Lawrence. *The Temptations of Evolutionary Ethics*. Berkeley: University of California Press, 1994.
Feder, H.M. "Cleaning Symbioses in the Marine Environment." In *Symbiosis*, vol. 1, ed. S.M. Henry, 327–80. New York: Academic Press, 1996.
Fisher, Andrew. *Metaethics: An Introduction*. Durham: Acumen, 2011.
Frankfurt, Harry. "Freedom of the Will and the Concept of a Person." *Journal of Philosophy* 68, 1 (1971): 5–20.
Gampel, Eric H. "A Defense of the Autonomy of Ethics: Why Value Is Not like Water." *Canadian Journal of Philosophy* 26, 2 (1996): 191–209.

Garvey, Brian. *Philosophy of Biology*. Montreal and Kingston: McGill-Queen's University Press, 2007.

Gould, Stephen Jay. *The Structure of Evolutionary Theory*. Cambridge, MA: Belknap Press of Harvard University Press, 2002.

———. *Life's Grandeur*. London: Jonathan Cape, 1996. First published in the United States in 1996 by Harmony Books as *Full House*.

———. "The Wheel of Fortune and the Wedge of Progress." In *Eight Little Piggies*, 300–12. New York: W.W. Norton, 1994a.

———. "Tires to Sandals." In *Eight Little Piggies*, 313–24. New York: W.W. Norton, 1994b.

———. *Wonderful Life*. New York: W.W. Norton, 1989.

Hamilton, W.D. "The Genetical Evolution of Social Behavior, I." *Journal of Theoretical Biology* 7, 1 (1964): 1–16.

Hinman, Lawrence. *Ethics: A Pluralistic Approach to Moral Theory*. 4th ed. Belmont: Wadsworth, 2007.

Hume, David. *A Treatise of Human Nature*. Book III. Ed. David Fate Norton and Mary J. Normton. Oxford: Oxford University Press, 2007 [1740].

Joyce, Richard. *The Evolution of Morality*. Cambridge, MA: MIT Press, 2006.

Kitcher, Philip. "Biology and Ethics." In *The Oxford Handbook of Ethical Theory*, ed. David Copp, 163–85. Oxford: Oxford University Press, 2006.

———. "Four Ways of 'Biologicizing' Ethics." In *Conceptual Issues in Evolutionary Biology*, ed. Elliott Sober, 439–50. Cambridge, MA: MIT Press, 1994.

Korsgaard, Christine. "Morality and the Distinctiveness of Human Action." In *Primates and Philosophers: How Morality Evolved*, ed. Stephen Macedo and Josial Ober, 98–119. Princeton: Princeton University Press, 2006.

Kraut, Richard. *What Is Good and Why*. Cambridge, MA: Harvard University Press, 2007.

Lemos, John. *Common Sense Darwinism: Evolution, Morality, and the Human Condition*. Chicago and La Salle: Open Court, 2008.

Lennox, James G. "Aristotle on the Biological Roots of Virtue: The Natural History of Natural Virtue." In *Biology and the Foundation of Ethics*, ed. Jane Maienschein and Michael Ruse, 10–31. Cambridge: Cambridge University Press, 1999.

Lumsden, Charles J., and E.O. Wilson. "Translation of Epigenetic Rules of Individual Behavior into Ethnographic Patterns." *Proceedings of the National Academy of Science, USA* 77, 1 (1980): 4382–86.

———. *Genes, Mind and Culture: The Coevolutionary Process*. Cambridge, MA: Harvard University Press, 1981.

Masters, Roger D., and Margaret Gruter, eds. *The Sense of Justice: Biological Foundations of Law*. Newbury Park: Sage, 1992.

McShea, Daniel W., and Robert N. Brandon. *Biology's First Law: The Tendency for Diversity and Complexity to Increase in Evolutionary Systems*. Chicago: University of Chicago Press, 2010.

Moore, G.E. *Principia Ethica*. Cambridge: Cambridge University Press, 1903.

Nagel, Thomas. "Ethics as an Autonomous Theoretical Subject." In *Morality as a Biological Phenomenon*, ed. Gunther S. Stent, 196–205. Berkeley: University of California Press, 1980.

Nesse, Randolph M. *Evolution and the Capacity for Commitment*. New York: Russell Sage Foundation, 2001.

Nitecki, Mathew H., ed. *Evolutionary Progress*. Chicago: University of Chicago Press, 1988.

Nozick, Robert. *Philosophical Explanations*. Cambridge, MA: Harvard University Press, 1981.

Okasha, Samir. "Why Won't the Group Selection Controversy Go Away?" *British Journal for the Philosophy of Science* 52, 1 (2001): 25–50.

Panksepp, Jaak. *Affective Neuroscience*. New York: Oxford University Press, 1998.

Pope, Stephen J. *Human Evolution and Christian Ethics*. Cambridge: Cambridge University Press, 2007.

Rawls, John. *A Theory of Justice*. Cambridge: Cambridge University Press, 1971.

Richards, Robert J. "Darwin, Spencer, and the Neo-Darwinians." In *Evolutionary Progress*, ed. Mathew H. Nitecki, 129–48. Chicago: University of Chicago Press, 1988.

———. *Darwin and the Emergence of Evolutionary Theories of Mind and Behavior*. Chicago: University of Chicago Press, 1987.

———. "A Defense of Evolutionary Ethics." *Biology and Philosophy* 1, 3 (1986): 265–93.

Robert, Jason Scott. *Embryology, Epigenesis, and Evolution*. Cambridge: Cambridge University Press, 2006.

Rottschaefer, William A. "Scientific Naturalistic Ethics." In *Science and Ethics*, ed. Paul Kurtz, 285–305. Amherst: Prometheus Books, 2007.

———. "Evolutionary Naturalistic Justifications of Morality: A Matter of Faith and Works." *Biology and Philosophy* 6, 3 (1991): 341–49.

Rottschaefer, William A., and Martinsen, D. "The Insufficiency of Supervenient Explanations of Moral Actions: Really Taking Darwin and the Naturalistic Fallacy Seriously." *Biology and Philosophy* 6, 4 (1991): 439–45.

Ruse, Michael. "Is Darwinian Metaethics Possible (and If It Is, Is It Well Taken)?" In *Evolutionary Ethics and Contemporary Biology*, ed. Giovanni Boniolo and Gabriele De Anna, 13–26. Cambridge: Cambridge University Press, 2006a.

———. *Darwinism and Its Discontents*. New York: Cambridge University Press, 2006b.

———. "Evolutionary Ethics." In *Biology, Ethics, and the Origins of Life*, ed. Holmes Rolston, III, 90–112. Boston: Jones and Bartlett Publishers, 1995.

———. "The New Evolutionary Ethics." In *Evolutionary Ethics*, ed. Mathew H. Nitecki and Doris V. Nitecki, 133–62. Albany: SUNY Press, 1993a.

———. "Evolution and Progress." *Trends in Ecology and Evolution* 8, 2 (1993b): 55–59.

Ruse, Michael, and Jane Maienschein, eds. *Biology and the Foundation of Ethics*. Cambridge: Cambridge University Press, 1999.

Ruse, Michael, and E.O. Wilson. "The Evolution of Ethics." In *Philosophy of Biology*, ed. Michael Ruse, 313–17. Amherst: Prometheus Books, 1998.

Savellos, Elias E., and Ümit D. Yalçin, eds. *Supervenience: New Essays*. Cambridge: Cambridge University Press, 1995.

Sherman, Paul. "Nepotism and the Evolution of Alarm Calls." *Science* 197 (1977): 1246–53.

Sober, Elliott. *The Nature of Selection*. Cambridge, MA: MIT Press, 1984.

Sober, Elliott, and D.S. Wilson. *Unto Others: The Evolution and Psychology of Unselfish Behavior*. Cambridge, MA: Harvard University Press, 1998.

Stephens, Christopher. "Modeling Reciprocal Altruism." *British Journal for the Philosophy of Science* 47, 4 (1996): 533–51.

Sterelny, Kim. *The Evolved Apprentice: How Evolution Made Humans Unique*. Cambridge, MA: MIT Press, 2012.

———, ed. *Dawkins vs Gould: Survival of the Fittest*. 2nd ed. Lanham: Icon/Totem Books, 2009.

Sturgeon, N.L. "Moore on Ethical Naturalism." *Ethics* 113, 3 (2003): 528–56.

Trivers, Robert, L. "The Evolution of Reciprocal Altruism." *Quarterly Review of Biology* 46, 1 (1971): 35–57.

van Inwagen, Peter. *An Essay on Freewill*. Oxford: Oxford University Press, 1983.

Voorzanger, Bart. "No Norms and No Nature—The Moral Relevance of Evolutionary Biology." *Biology and Philosophy* 2, 3 (1987): 253–70.

Wilkinson, G.S. "Reciprocal Altruism in Bats and Other Mammals." *Ethology and Sociobiology* 9, 2–4 (1988): 85–100.

———. "Reciprocal Food Sharing in the Vampire Bat." *Nature* 308 (1984): 181–84.

Williams, Patricia. "Evolved Ethics Re-Examined: The Theory of Robert J. Richards." *Biology and Philosophy* 5, 4 (1990): 451–57.

Wilson, E.O. *The Diversity of Life*. Cambridge, MA: Harvard University Press, 1992.

Woolcock, Peter G. "The Case against Evolutionary Ethics Today." In *Biology and the Foundation of Ethics*, ed. Jane Maienschein and Michael Ruse, 276–306. Cambridge: Cambridge University Press, 1999.

Chapter 7

EVOLUTIONARY PSYCHOLOGY

In the distant future I see open fields for far more important researches. Psychology will be based on a new foundation, that of the necessary requirement of each mental power and capacity by gradation. Light will be thrown on the origin of man and his history.

—Charles Darwin (1985 [1859], p. 458; my emphasis)

I. INTRODUCTION: JUST WHAT THE PROPHET ORDERED!

In our daily interactions with others, I suspect that many of us have either come across or heard of scenarios like the awkward and morally embarrassing ones noted below.

→ I remember my good friend disdainfully and painfully telling me that he "just knew" that his significant other was *cheating* on him after numerous scheduled dates were missed. He just could not understand how this could happen given the attention, time, and care he invested in her.

→ Along the lines of *cheating*, I distinctly recall my frustration and contempt for a number of students (during my undergraduate days) who were running an exam exchange scam in a number of courses.

→ While glancing at one of the scantily clad actresses in one of the tabloid papers in the grocery market check-out line, I can call to mind a rather jovial and an approximately 80-year-old (young?) gentlemen telling me that it was time for him to find a "young" paramour for himself. In the midst of this public revelation, I also noticed (with much amusement) the look of disgust on the young woman standing behind this grandfather figure.

→ I also recollect my friend's father telling me that the actress, Sophia Loren, was the most beautiful woman he had ever seen and that the young kids of today have no "real sense" of beauty!

→ Much to my surprise, I can recall some acquaintances telling me—and not simply in jest—that their "job" is to spread their seed around to as many women as possible. All of this bravado and posturing included an obvious lack of respect for women and also a rather callous disregard for the well-being of the potential childcare needs of the women they were so rapturous to impregnate.

→ Being an annoying and nosey 10-year old, I would sneak into my sister's sleepover parties and listen-in on the conversations that she and her girlfriends would have. Since most conversations revolved around boys, it was interesting to hear how all of these females agreed that height, athletic "v-shaped" physical appearance, a sense of humor, and wealth were the top characteristics that a potential boyfriend/husband should possess.

I mention these seemingly random and somewhat startling happenstances for a specific reason. Primarily, there is a school of thought within psychology, which draws on evolutionary considerations, that embraces the idea that there is a core set of principles that unifies these scenarios. This approach offers an evolutionary account of brain mechanisms that can help make sense of why, for example, humans are very good at spotting various sorts of "cheaters," parameters that make sense of human mating strategies, child-care practices, etc. If correct, such proclamations about the human psyche would be impressive to say the least. Actually, the opening quotation above makes Darwin look like something of a prophet! Having confidently thought to have established the importance of natural selection with respect to the production and modification of life, he boldly envisages research endeavors that link evolution to the underlying nature and origin of human psychology.

We can suspect that Darwin would not have been entirely surprised to find out that the late twentieth century would pursue the very area of research, *evolutionary psychology*, that he forecasted would burst onto the scientific landscape. For instance, the eminent psychologist, Steven Pinker, echoes to a contemporary audience Darwin's forecast about the long-reaching sway of evolution with respect to human psychology. Pinker argues from analogy that, in much the same way that natural selection was crucial to the production/design of eyes, hearts, limbs, etc., which have distinct sets of functional capacities, it was also crucial to the production/design of the human mind—which must have its corresponding set of distinct functional capacities (Pinker, 2002, p. 52). To the extent that this is true, Pinker somewhat glibly concludes that the field of psychology has always had evolutionary biology as a bedmate.

This presumption of unity between evolutionary biology and psychology is put on prominent display by two of the leading evolutionary psychologists, Cosmides and

Tooby. They summarize this field of evolutionary psychology (that they have, in part, spearheaded) as a discipline that continues to provide a unique account of the human mind as a bundle of adaptive problem-solving systems (Cosmides and Tooby, 2013, p. 202). To make sense of all this, we must first make a distinction between *Weak Evolutionary Psychology* (**WEP**) and *Strong Evolutionary Psychology* (**SEP**). **WEP** is an area of research that draws upon the insights of evolutionary biology without any specific commitment to the nature of human behavior and mind. Rather, whatever set of theories and principles are developed in making sense of the human mind, **WEP** cautiously reminds us *not* to ignore evolutionary biology as relevant to that set.[1] In contrast, **SEP** is the ambitious research crusade that not only employs evolutionary theory, but does so in a way that is committed to a unique set of principles designed to reveal the nature of human mind and behavior. As Buller summarizes, "The goal of evolutionary psychology [**SEP**] ... is to discover the mental organs that constitute our universal human nature and to articulate how those mental organs function to solve evolutionary problems" (Buller, 2005, p. 9; bracketed addition mine).[2] **SEP** is the research framework pursued by Cosmides and Tooby (and Pinker to a certain extent) in the light of their commitment to *specific* sets of evolved problem-solving mental systems.

It is **SEP** that will be the focus of this chapter. Special emphasis will be given to the reasonableness of **SEP**'s defense of the human mind as "a bundle of modules" and the nature of the environment in which these supposed modules evolved. Ultimately, upon displaying some of the core criticisms associated with each of these areas, this chapter will reveal that **SEP**, as a research program, is not nearly as powerful as its adherents boast.

II. PRINCIPLES OF SEP

In order to appreciate the scope of **SEP**, it is important to get a clear sense to which theoretical principles it is committed. There are four major tenets:

(1) The human mind is a sort of computer that is composed mostly of a bundle of evolutionary adaptive "mental modules."

(2) There is a time-frame and range of location—an evolutionary environment of adaptedness (**EEA**)—from which these evolutionary adaptive modules emerged and by which they were evolutionarily set and calibrated.

---

1 For more on the different faces of evolutionary psychology, see Buller (2007).
2 David Buller's *Adapting Minds* (2005) is one of the best efforts to make sense and rigorously critique both the theoretical and experimental efforts of **SEP** with respect to mating preferences, marriage, and parenting strategies. Just as informative and critically insightful, Robert Richardson's *Evolutionary Psychology as Maladapted Psychology* (2007) examines specific philosophy of biology concerns with respect to **SEP**. For those interested in the many philosophical and biological nuances of this topic, I highly recommend reading these texts back-to-back. For more textbook-type discussions, see Gaulin and McBurney (2004) and Workman and Reader (2008).

(3) The combination of the first two tenets secures the view that there is an "innate" universal nature specific to human beings.

(4) This single, universal, and adaptive module-constructed mind can assist in making sense of the "surface" cultural-psychological variation we observe in the world today.[3]

Immediately, the reader should see how heavily dependent (3) and (4) are on (1) and (2). If (1) and (2) can be shown to be misguided, then (3) and (4) will correspondingly be rendered problematic. Restated, if it is false that the mind is composed almost solely of self-contained mental modules and it is false that the **EEA** best captures the time frame in which the adaptations of the human mind were formed and solidified, then it is also unlikely that **SEP** is able both to offer a specific human universal nature and to make foundational sense of the psychological variation we observe in the world today. Let us get a closer look at (1) and (2) of **SEP** and the critical responses that have followed. Again, but in a more specific way, this chapter reveals that **SEP**'s defense of (1) and (2) is inadequate, further suggesting that its endorsement of (3) and (4) is premature.

1. Mental Modules: Silly SEP, Tricks Are for Kids!

Many, if not all defenders of **SEP**, insist that, along with some general-purpose processors, the mind is composed predominantly of a large bundle of domain-specific neural networks. These mental modules, they argue, are designed to solve specific environmental problems for the sake of survival and reproductive success—an evolved "bag of tricks." This position stands in contrast to the more traditional social scientific principle that the mind is best understood as a small collection of general processing modules (Gaulin and McBurney, 2004, ch. 1 and Richardson, 2007, p. 23). The following argument, let us call it **The Argument from Domain-specificity**, captures **SEP**'s resistance to the traditional social science perspective on the structure of the mind:

The Argument from Domain-specificity

P1. Either the mind is composed *predominantly* of a large bundle of domain-specific mental modules or it is composed *predominantly* of a small number of domain-general mental modules.

P2. If the mind is composed *predominantly* of domain-general mental modules, then it is because such a design is in keeping with humans as evolutionary entities.

---

3   This is a compressed version of the **SEP** principles as articulated by Tooby and Cosmides (2005).

P3. It is not the case that the presence of domain-general mental modules is in keeping with humans as evolutionary entities.

P4. It further follows that the mind is not composed *predominantly* of domain-general mental modules.

C1. Thus, the mind is composed *predominantly* of a large bundle of domain-specific mental modules.

As we have done with other arguments, let us assume (for the moment) that **P1** of **The Argument from Domain-specificity** is true; that is, we shall accept that these two options regarding modules are the only two available and they are exclusive of each other. As we will see (Fodor, 1985), this is a big assumption that may very well have to be rejected. **P2** will also be accepted as true, but (again) this may prove to be difficult to maintain as the analysis unfolds. Asking the reader to proceed in this manner should raise a few red flags and we will return to these two premises shortly.

For now, the focus will be on **P3**, the rejection of the mind as a composition of a small bundle of general purpose mental modules.[4] What evidence could a supporter of **SEP** give in defense of its truth? Symons (1992, p. 139) offers us an answer that is wrapped around precision and specificity. In terms of defending **P3** of **The Argument from Domain-specificity**, he is arguing that evolution must be understood as a process of maximizing those activities that are crucial to survival and reproductive success. Given that organisms must overcome specific environmental obstacles, it reasonably follows that specific mechanisms via specific ensembles of genes would be required to reconcile particular obstacles. Some specific variations on these mechanisms were more effective than others in achieving their specific goals and were retained due to their reproductive advantage. Yet, the maximizing element includes the idea that maximization and precision work hand-in-hand. Given this understanding of evolution, Symons argues that no general mechanism can produce precise maximizing effects with respect to unique and specific conflicts that require precise resolution. Rather, thinks Symons, it is specific gene-produced mechanisms that engage in specific and precise activities that are crucial to survival and reproduction. This line of argument, *mutatis mutandis*, applies to psychological mechanisms as well, thinks Symons. For example, the only way my friend could have "known" that his significant other was cheating on him or that I was able to "recognize" that a covert exam-cheating scam was being perpetrated is because my friend and I have specific cheating-detector mental modules that were able to pick-up environmental/social stimuli that revealed to us that unearned advantages and rewards were being accumulated by others.

---

4   A defense of **P3** would also be a defense of **P4**. The way that **The Argument from Domain-specificity** is structured, however, **P4** follows if **P3** is true. This is why the focus, for now, is being placed on **P3**.

So, in the same way that the human body's kidneys filter out toxins in fairly precise ways, the human mind's psychological mechanisms engage in their specific functions in fairly precise ways (Tooby and Cosmides, 1992, p. 80). From this perspective, the putative flexibility of the human mind is more accurately captured by the presence of "so many different modules, each with provisions to learn in its own way" (Pinker, 1995, p. 410). Thus, with respect to **P3** of **The Argument from Domain-specificity**, Symons and others can claim to have justified its truth, revealing the implausibility of the mind being comprised of a small bundle of domain-general psychological mechanisms.

REPLY TO P3 OF THE ARGUMENT FROM DOMAIN-SPECIFICITY

This "massive modularity of mind" thesis defended by Symons, Pinker, and Tooby and Cosmides is not entirely unreasonable. If we accept that environmental obstacles can be rather complex, then the idea that there are domain-specific and domain-sensitive dedicated processors that come "pre-packaged" with "innate" problem solving information does not appear to be a preposterous suggestion (Buss, 1999). For instance, how are humans able to determine who constitutes a reproductively healthy mate and thus worthy of pursuit? Some evolutionary psychologists have argued that there is an ideal female waist-to-hip-ratio (WHR) that reveals their reproductive health. Females like Sophia Loren, Jessica Alba, and Jennifer Lopez supposedly exhibit this ideal WHR (of having hips that are 1.5 times larger than the waist) and this explains their supposed universal appeal to men. Part of this mate selection story includes the claim that men must have a specialized evolved WHR module/detector designed to triangulate, size-up, and/or rule-out various WHRs as a way of resolving, in part, the problem of mate selection; presumably, this detector helps in choosing mates that approximate the Sophia Loren WHR. In this way, the closer a female is to the ideal WHR, the greater the likelihood that she is in tip-top reproductive health. So, according to some scholars (e.g., Singh, 1993) in the **SEP** camp, men most likely have a reproductive bias toward pursuing those women who exhibit a WHR that is close to the ideal WHR. No wonder why my friend's father was so keen on Sophia Loren!

Still, some argue that there is a fatal flaw in the reasoning of those who endorse this modularity of mind thesis. For instance, Buller (2006, pp. 199–200) responds that a woman who learns, by way of domain-general processors, that plump and juicy peaches constitute nutritious peaches would not use these same characteristics to choose a mate. Rather, still by way of domain-general social learning processors, she would draw upon cues from other women her age about what constitutes a good mate. Buller stresses that his peach-selection-and-mate-selection example reveals that it is possible for domain-general mechanisms to be receptive to domain-specific stimuli and respond with the appropriate behavioral specificity. Thus, even if there is a specific WHR that reveals female reproductive health, it could very well be the case that a flexible enough domain-general mental mechanism could assist in recognizing such reproductively beneficial stimuli. Similarly, the criteria my sister and her friends agreed upon regarding preferred male characteristics could very well be produced by this sort of domain-general mental mechanism. This suggests that domain-general mechanisms can do the work that **SEP** defenders insist can only be done by domain-specific mechanisms. Buller concludes that "the principal argument for modularity rests on a false premise" (Buller, 2006, p. 200). Thus, Buller would likely be quick to reply that **P3 of The Argument from Domain-specificity** is false, rendering the argument unsound.[5]

Buller's criticism is insightful, but it is not clear that it is as far-reaching as he suggests. For instance, **SEP** devotees could allow for such scenarios, but insist that they are anomalies compared to the vast number of domain-specific mental modules. So, although there may be occasional instances when a domain-general processor performs complex procedures, the vast majority of modules that produce complex and precise processes are domain-specific processors.[6] As evolved beings, **SEP** defenders think humans could not have continued as a species without this bag of adaptive psychological tricks. Thus, even granting Buller's social learning peach-mating scenario, the massive modularity thesis could very well remain intact. Thus, contrary to Buller's analysis, **P3 of The Argument from Domain-specificity** has not been revealed to be false.

Not so fast! To his credit, Buller cautions that the domain-general module alternative is also off the mark (Buller, 2006, p. 200). He stresses that evidence reveals that neurological development is best understood within a "proliferate and prune" perspective. What this means is that the brain produces a dispersed set of brain cells that later become reduced and further organized as a result of organism-environment interaction. Buller concludes from this that "the specialized brain structures we have are primarily environmentally induced, not 'genetically specified'" (Buller, 2006, p. 201). In terms of **P3 of The Argument from Domain-specificity**, Buller points out that domain-general processing is in keeping with humans as evolved entities. It just happens to be the case that proliferation of brain cells is what evolved, while the module construction and organization of the proliferated brain cells is the product of organism-environment interaction. From this perspective, brain development is "plastic" or malleable to local environments such that domain-general modules can assist in the production of precise behavioral repertoires. So, although brain cell proliferation is an evolved feature of the human organism, "the cognitively specialized brain structures that are the outcome of brain development have not been shaped by

---

5   On an interestingly related point, the journal *Cognitive Science* (vol. 34, 2010) invested an entire volume on the issue of domain-general modules. A number of articles were commissioned to offer domain-general module arguments for formally domain-specific module arguments. Although not necessarily conclusive, these essays were designed to show that domain-general modules can produce the kind of specificity-oriented results suggested by Buller.

6   With respect to their discussion on the evolution of language acquisition, Pinker and Bloom appear to concede that some mental modules may be general and flexible while still affording needed precision, but they think that these sorts of adaptations are the exceptions (Pinker and Bloom, 1992).

natural selection" (Buller, 2006, p. 201). For instance, even if the proliferation of neurons includes a bias for reproduction, environmental influences throughout an individual's lifetime can influence reproductive choices. So, in some social circumstances, my old man at the grocery store could very well be pursuing a praiseworthy course of action in his search for a young paramour, while in other social circumstances, he would be deemed "a dirty old man" or have no interest in such pursuits. Again, similarly, both the ordinal and cardinal structure of the preferred male characteristics suggested by my sister's friends would change in different social environments. What this means for our discussion is that, if Buller's analysis is on the mark, **P3** of **The Argument from Domain-specificity** is false, rendering **The Argument from Domain-specificity** unsound.[7]

REPLY TO P1 OF THE ARGUMENT FROM DOMAIN-SPECIFICITY

Recall that **P1** of **The Argument from Domain-specificity** offers two options with respect to the architecture of the mind: either the mind is predominantly composed of domain-general modules or it is predominantly composed of domain-specific modules. In the light of Buller's proliferate-and-prune thesis, the truth of **P1** appears to be in jeopardy. As a point of emphasis, Buller educates his audience about brain development. What has evolved is a small number of modules that are susceptible to a range of expression based on environmental influences. This range of expression is what Buller means by "plastic." For instance, the face recognition module in infants initially responds to "three high contrast blobs" and by way of a gradual process is molded to a full-blown face recognition module (Buller, 2006, p. 202). In a sense, Buller is arguing that, contrary to **P1** of **The Argument from Domain-specificity**, domain-specific modules developmentally unfold out of domain-general modules rather than both kinds of modules functioning independently of each other. To the extent that Buller has the science correct, **P1** is false; that is, it is not the case that the structure of the mind must be *primarily* either domain-general or domain-specific modules. Rather, domain-specific modules are either (1) embedded within domain-general modules or (2) developmentally emerge out of domain-general modules. Neither of these scenarios seems to be in keeping with **SEP**'s massive modularity thesis.

---

7   One may be unmoved by this reply and insist that this exchange is really one of semantics. Buller allows for specific modules as do defenders of **SEP**. Note, however, that the difference in the nature of the modules is substantial. Specifically, the major difference appears to be the role of natural selection. Buller's specific modules are not the product of natural selection, while those of **SEP** are present as a result of natural selection working at the genetic level. Additionally, Buller's modules emerge or are embedded within a few general purpose modules (see the very next section), while the modules of **SEP** are distinctly unique and encapsulated modules. These differences are enough to suggest that the battle is more than a mere play on words.

CHEATING-DETECTORS TO THE RESCUE! LONG LIVE SEP!

Now, defenders of **SEP** have insisted that there are dedicated domain-specific modules and the human cheating-detector module is a knockdown example in favor of rejecting both the traditional social science reliance on domain-general processors and Buller's embedded developmental alternative. The argument can be represented as follows:

> **P1.** Either there is a domain-specific cheating module or there is not a domain-specific cheating module.
>
> **P2.** If there is not a domain-specific cheating module, then experimental tests will reveal that this is the case.
>
> **P3.** Experimental tests do not reveal that this is the case.
>
> **P4.** It is not the case that there is not a domain-specific cheating module.
>
> ---
>
> **C1.** Thus, there is a domain-specific cheating module.

P3 is the crucial premise here. Defenders of **SEP** think that it is true because of the results of a number of experiments (e.g., Wason Selection Task). Subjects were asked to determine the truth of certain logical expressions and social scenarios to determine when rules were violated. Invariably, subjects were able to answer the social scenarios correctly, but failed pretty miserably on the logic problems—even though the logical structure of both kinds of conditionals were the same. Before turning to the sort of examples used by **SEP**, let us review a standard Wason Selection Task example. Consider the following cards:

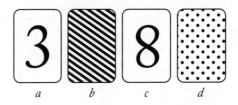

Given the above four cards, **card a**, **card b**, **card c**, and **card d**, here is the set-up:

> **The Rule of the Game:** "If a card has an even number on one side, then it has dots on the other side."
>
> **The Question of the Game:** Which card or cards *must* you turn over (without turning over any unnecessary cards) to determine if the rule *does not* hold?

No doubt, you could turn over all of the cards to determine if the rule is true, but we are looking for *the most efficient option*—this is why 'must' and the parenthetic qualifier are used in the question of the game. Given this efficiency demand, it turns out that only about 10 per cent of participants get the answer correct. The result of these findings have led to all sorts of interesting conclusions about both the quality of the test and the quality of human reasoning—conclusions that have included evolutionary considerations.

In terms of the solution, it should be fairly straightforward that **card a** should not be turned over because the rule of the game makes no reference to the status of odd numbered cards. Additionally, it should be pretty clear that the choice of turning over **card c** is a must; for it is an even numbered card and if it does not have dots on the other side, then the rule is invalidated. Now, you might be tempted to think that it is a must to turn over **card d** given that its face has dots. This would be a mistake. The reasoning is two-fold. First, if the opposite side of **card d** is an odd number, then this would not invalidate the rule because the rule does not prohibit odd numbers from having dots—it only requires that even numbered cards have dots. Second, if the opposite side of **card d** is even, then this validates the rule *but* does not invalidate it. You need to locate only those cards that invalidate the rule. So, there is no need to turn over **card d**. Finally, you should now see that **card b** must be turned over. If it turns out that the opposite side of **card b** is even, then the rule is invalidated. Thus, only **card b** and **card c** must be turned over to invalidate the rule of the game.

What is interesting about the Wason Selection Task is that, although most people do poorly on the even number/dots and logic versions of the test, participants do better on social interaction versions of the test. Notice that the rule of the game is in the form of a conditional (**if P, then Q**: If a card has an even number on one side, then it has dots on the other side). If the conditional is changed to a social interaction scenario ("If you drink in this bar, then you must be at least 21 years old") it turns out that people do much better—even though the social scenario is still in the form of a conditional. Here is a variation on a classic social scenario example related to alcohol consumption in a bar and age requirement:

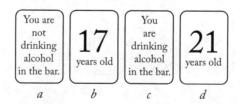

Given the above cards, **card a, card b, card c,** and **card d**, here is the set-up:

> **The Rule of the Game:** "If you drink alcohol in this bar, then you must be at least 21 years old."

> **The Question of the Game:** Which card or cards *must* you turn over (without turning over any unnecessary cards) to determine if the rule *does not* hold?

So, there are ages indicated on one side of the cards and presence or non-presence of an alcohol-drinking person in the bar indicated on the opposite side of the cards. Again, like the previous card example, what is the *most efficient* way of determining whether or not the rule is violated? Clearly, **card a** should not be turned over because you are not at the bar drinking alcohol. So the rule is not relevant in this instance (like the number 3 card is not relevant to the first scenario) because it makes no reference to those who are not at the bar drinking. Additionally, it should be obvious that **card c** must be turned over, because you have to know whether or not the person drinking is at least 21 years old. If he or she is not at least 21 years old, then the rule has been violated. Furthermore, you might be tempted to turn over **card d**, but like the dots card, this would be a mistake. The reason is two-fold. First, if the person in the bar is drinking alcohol, then this merely validates the rule (since the person is at least 21 years old—like the dots card in the previous example). Second, if this 21 year-old person is not at the bar drinking, then his/her status with respect to the rule of the game is irrelevant—the rule says nothing about a person who is allowed to drink at the bar, but does not drink at the bar. In either case, there is no requirement to turn over **card d** to determine if the rule has been violated. Finally, it should be clear that **card b** must be turned over; for if 'you are drinking alcohol in the bar' is on the opposite side, then the rule has been violated because this person is not at least 21 years old. Thus, like the even number/dots card example above, only card **b** and card **c** should be turned over in the drinking example.

Like most of my students, people do very well on the drinking example and very poorly on the even number/dots example. What are we to make of such results? Evolutionary psychologists think that such results reveal the truth about the modularity of mind thesis by way of the presence of a cheating-detector module. For example, as noted before, subjects could not correctly answer when a conditional in symbolic form is true (**if P, then Q**) or when the conditional related to the rule of the even number/dots card game scenario above is violated (if a card has an even number on one side, then it has dots on the other side), but they could determine when a social exchange scenario (like the drinking example) in the form of a conditional is true by revealing unwarranted benefits gained by one person as a result of "cheating"; that is, if you are able to secure a drink at a bar and you are under the age of 21, this signals that you were able to access benefits that are prohibited to you.

To connect the Wason Selection Task discussion directly to **SEP**, consider the following conditionals:

Ex. 1 → If a woman drinks a sexually stimulating beverage, then she must be married.

Ex. 2 → If a woman is married, then she drinks a sexually stimulating beverage.

Ex. 3 → If P, then Q.

In the above scenario, only married women are permitted (i.e., socially and legally allowed) to drink sexually stimulating beverages much like 21 year old individuals (or older) are allowed to drink alcohol. It is a rule-violation or an instance of cheating if a non-married women drinks a sexually stimulating beverage like it is a rule-violation or an instance of cheating if a person under the age of 21 consumes alcohol. So, for an under-aged person to consume alcohol and for an unmarried woman to consume a sexually stimulating beverage, these constitute instances in which undeserved benefits were secured and such individuals would be considered rule violators. Cosmides uses scenarios, like the ones above, on subjects and provides them with specific cards from which to choose (like the even number/dots and drinking examples above). Subjects were asked to determine which cards reveal that the conditionals (like the 'if ... then' scenarios in **Ex. 1** and **Ex. 2** above) are true or when a violation has occurred. For example, cards like "she is not married"; "she has numerous boyfriends"; "she studies a lot"; "she does not drink sexually stimulating beverages," etc. could be used. So, marriage status could be put on one side of the card and whether or not the sexually stimulating beverage was drunk could be put on the other side of the card. Putting to one side some of the details, Cosmides points out that the data reveal that people answer **Ex. 3** poorly, but answer **Ex. 1** and **Ex. 2** much better—even though all three examples have 'if P, then Q' as part of their formal structure—this is exactly what we saw in the even number/dots and drinking examples. Cosmides thinks that two conclusions can be reasonably drawn from all this:

> (1) When the content of the conditionals represent a social exchange, people are good at picking out those cards that reveal the truth of these sorts of conditionals because humans possess an evolved domain-specific cheating-detector that is sensitive to social scenarios in which people secure undeserved benefits.
>
> (2) Even when they pick the wrong card, participants frequently choose cards that are related to a rule violation. This gives further support to the claim that a cheating-detector module is filling in the social scenario in a way that recognizes/interprets ill-gotten gains. This explains, argues Cosmides, why participants get examples like **Ex. 1** correct more often than **Ex. 2**.

Evolution has created, thinks Cosmides, an adaptation in the form of a domain-specific cheating-detector module (not a general logic-probing algorithm) to discern when individuals defect from paying appropriate social exchange costs. (Recall my opening example of how distraught my friend was over his girlfriend's alleged infidelity in the light of the benefits she received from him.) Indeed, Cosmides stresses that people do more poorly on **Ex. 2** relative to **Ex. 1**. This further reveals a cheating-detector module; for people seek out a cheating solution even when a social exchange conditional does not require picking such a solution. This further reveals, thinks Cosmides, that people search for rule violators in social settings (like the drinking example). The overall upshot of these tests is that there can be no general processing mechanism that could account for all this; a domain-specific cheating-detector module must be present. From these conclusions,

Tooby and Cosmides infer that the mind is filled with such special modules designed by evolution. The upshot is that **P1** and **P3** of **The Argument from Domain-specificity** could still be salvaged, displaying the soundness of the argument.

Yet, Buller is quick to reveal that this analysis is a smoke-and-mirrors ploy. He offers a two-part response. Specifically, he attacks the set-up of the Wason tasks in both responses. First, he explains that the examples used by Tooby and Cosmides (a drinking-age scenario and a marriage example) are *deontic conditionals*, while the logic problems are *indicative conditionals*. That is to say, the former type of conditionals have built-in obligations, while the latter do not; that is, according to the example and the social context, it is built into the drinking of a sexually stimulating beverage by a woman that she be required to be married. In contrast, no such obligation is present in 'If it rains, then people get wet' or 'If a card has an even number on one side, then it has dots on the other side.' Because the Q-part (i.e., the consequent) of the conditional is required in deontic conditionals, Buller insists that the requirement immediately catches the attention of the participants in these tests and accounts for why they pick the correct cards so frequently (Buller, 2006, p. 205). Thus, contrary to Cosmides, Buller claims that it is a logic effect (as opposed to a cheating-detector effect) that accounts for the results of these experiments.

By itself, this response is not entirely persuasive. Defenders of **SEP** could agree with much of what Buller has to say, but insist that the reason that attention is drawn to the correct answers in the case of deontic conditionals (and not indicative conditionals) is because a cheating-detector mechanism is triggered to bias such a response. If the mechanism was a general processing logic mechanism, as opposed to a cheating-detector mechanism, then there should be no statistically significant difference in the answers given by participants. Indeed, if the detector is a general logic mechanism, then it should plausibly ignore the so-called attention-grabbing influence of the deontic conditionals and apply the truth-table values to all conditionals equally well (instructors should remind students about the truth-table details pertaining to conditionals and the relevance to this discussion). So, unless there is evidence to show this kind of logic-equality across conditionals, **SEP** supporters could withstand Buller's criticism and insist that a domain-specific cheating mechanism is the most reasonable account for the differences revealed by the Wason tasks.

Unfortunately for **SEP** wardens, like Tooby and Cosmides, Buller points out that other Wason tasks reveal that participants *do* answer both types of conditionals with the same level of success (Cheng and Holyoake, 1989; Manktelow and Over 1990; Sperber et al., 1995). The reason, according to Buller, is that the scenarios posed in the indicative conditionals of these other tests are more "real-world" than the number and symbol examples put forth by Tooby and Cosmides. So, Buller confidently concludes that subjects of these tests do not ignore the logic of conditionals in favor of ferreting out cheaters. Again, the implication is that humans possess a general logic-probing module rather than a domain-specific cheating-detector.[8]

---

8 As discussed earlier, a domain-specific response could be possible, but Buller would claim that such a specific response could emerge from the relevant domain-general module mechanism.

It appears that Buller has the upper hand in this exchange. This is further supported by Buller's second reply. He claims that Cosmides' background story of her deontic conditional (the switched one in **Ex. 2**) that she told to subjects did not make sense with respect to the actual scenarios posed. So, speculates Buller, subjects represented the scenarios in a more reasonable fashion in the light of the original background story; then they applied the relevant conditional logic to the deontic conditionals. The result, argues Buller, is that the participants did not pick the wrong cards. Rather, they picked the logically correct cards after modifying the posed scenarios to fit the background information initially provided. For instance, participants could think to themselves that the investigator must want them to treat **Ex. 2** like **Ex. 1**, given certain details about drinking this sort of sexually stimulating beverage. Buller's point is that, contrary to Cosmides' assessment, participants did not seek out cheaters in the switched **Ex. 2**-type scenarios. Rather, argues Buller, they employed the domain-general logic module after changing the scenario to fit correctly the initial background story given.

From all of this, we can conclude that **P3** of the above argument is false, rendering the **SEP** defense of the presence of a cheating-detector module dubious. We should not at all be surprised, then, to find Buller concluding that there is "no good evidence of a cheater-detection mechanism" (Buller, 2006, p. 205). Yet, **SEP**'s claim that there is a large bundle of domain-specific modules may still be on the mark, but more tests and data would be needed to substantiate the presence of this evolutionary bag-of-tricks. In the light of the present analysis, we must conclude as follows: "silly defenders of **SEP**, your bag of evolutionary tricks is for kids!"

2. The Environment of Evolutionary Adaptedness

Although the previous section casts some doubt on the ubiquity of domain-specific modules, let's assume that some combination of domain-general and domain-specific modules best makes sense of the structure of the human mind. Still, it is reasonable to ask whence did these modules arise? In effect, we need to re-examine **P2** of **The Argument from Domain-specificity**; that is, given that humans are evolutionary entities, we need to determine if this fact tips the scale in favor of **SEP**. Recall that the traditional social science camp claims that the human mind is more or less a "blank slate," suggesting that **P2** is true. In sharp contrast, as we have seen, the **SEP** camp thinks that the human mind comes "pre-packaged" with a set of domain-specific modules—a kick-ass evolutionary bag of tricks.[9] Given this perspective, **P2** of **The Argument from Domain-specificity** is considered false. How do defenders of **SEP** justify their account? The standard answer given by proponents of **SEP** is that there is a time period and place, as noted earlier, known as the

---

9   This claim is an over-simplification of various nuanced views of the mind's structure. There is little doubt that few scholars in either the social science camp or the psychology troupe ascribe to all the ideas put forth by the extremists of their respective units. Still, there are many adherents to these two extreme views.

Environment of Evolutionary Adaptedness (**EEA**), from which these domain-specific modules are derived. Upon explicating what the **EEA** is and what specific thesis **SEP** supporters defend with its help, this chapter addresses whether or not **SEP**'s reliance on the **EEA** is as powerful as it is made out to be. To make this manageable, focus will be given to the recent **EEA** defense of Starratt and Shackelford (2010).

What, then, precisely is the **EEA**? Starratt and Shackelford tell us that it is not a place or time; rather it is a kind of statistical amalgamation of a range of environments and the corresponding pressures these environments presented (Starratt and Shackelford, 2010, p. 232). The range of this statistical composite is drawn from approximately 1.8 million years ago to about 10,000 years ago (commonly referred to as the Pleistocene Epoch) because it is thought that this range contains "the **EEAs** of the majority of human-specific adaptations" (Starratt and Shackelford, 2010, p. 232; my bold).

With the above account of the **EEA** in mind, what specific thesis is **SEP** trying to defend? The answer offered by Starratt and Shackelford comes in two parts: (1) "The Pleistocene is the period of time hypothesized to contain the **EEAs** of the majority of human-specific adaptations and (2) that most of our psychological mechanisms evolved during the Pleistocene and continue to be expressed today" (Starratt and Shackelford, 2010, p. 232). What arguments do these authors give in defense of these two claims? They offer two arguments, each designed to support (1) and (2). Let us turn to each argument with an immediate response to each.

THE ARGUMENT FROM TIME

The standard argument offered by many within the **SEP** camp (e.g., Tooby and Cosmides, 2005, Buss, 1999, and Symons, 1979) is also one offered by Starratt and Shackelford. In general, they argue that post-Pleistocene influences on human psychology constitute "one half of 1% of 1.8 million years of human existence" (Starratt and Shackelford, 2010, p. 232). This fact combined with the fact that there is only 10 per cent genetic variation amongst humans across the globe and 80 per cent of this 10 per cent difference is among individuals within the same population, suggests that "most genetic variation occurred before modern humans migrated out of Africa roughly 100,000 years ago" (Starratt and Shackelford, 2010, p. 232). The authors then conclude that human psychological adaptations are pretty much as they were during the Pleistocene and "continue to be expressed today" (Starratt and Shackelford, p. 232). The argument is as follows:

The Argument from Time

P1. If most of human psychological adaptations are the product of approximately 2 million years of evolution and there is little genetic change amongst and between human populations, then it is reasonable to think that humans continue to express a Pleistocene adapted mind.

P2. Most of human psychological adaptations are the product of approximately 2 million years of evolution and there is little genetic change amongst and between human populations.

C1. It is reasonable to think that humans continue to express a Pleistocene adapted mind.

Both **P1** and **P2** appear to be problematic, rendering the above argument unsound. Let us take the time component of **P2** first. Why is it reasonable to think that most human psychological adaptations took place within this two million year Pleistocene time frame? The answer given by most defenders of **SEP** is that it seems implausible that complex psychological modules, which exhibit precise coordinated responses to specific environmental stimuli, could manifest the kind of genetic stability needed to ensure survival and reproductive success in a mere 10,000 years (or 400 generations). The implication is that most human psychological mental modules had to have been manifested and cemented as part of our "nature" during the hunter-gatherer time of the Pleistocene. In terms of understanding humans as Pleistocene hunters and gatherers, Symons claims just this: "The hunting and gathering way of life is the only stable, persistent adaptation humans have ever achieved.... Humans can thus be said to be genetically adapted to a hunting and gathering way of life" (Symons, 1979, p. 35). Thus, **P2** is true given the implausibility of substantive changes in mental modules in the post-Pleistocene time frame.

REPLY TO THE ARGUMENT FROM TIME

The initial problem with **P2** of **The Argument from Time** is related to the claim that the Pleistocene represents a composite of a range of time and environmental challenges. Although this claim is reasonable, the implication is that it is not clear *to what* environmental stimuli a given module would be adjudicating. Given that considerable changes are possible to have occurred during the two million years of the Pleistocene, talk of humans as hunters and gatherers does not help a great deal in locating specific mental modules. For example, it is reasonable to think that human life expectancy during the two-million-year period demarcated by **SEP** defenders fluctuated based on various sorts of resources, climate factors, human modification of local environments, and social bonding dynamics.

To the extent that this is correct, it is not obvious that there was, for example, a stable and uniform WHR (waist-to-hip ratio) present such that a male WHR detector could have evolved (Starratt and Shackelford, 2010, p. 239). For instance, when food was scarce, the overall population would likely have been on the emaciated end of the body fat spectrum—including women. If this is correct, then it is unlikely that enough (or any) women exhibited the so-called ideal WHR *such that a selective advantage would have been made manifest.* At other times, there could have been an over-abundance of high calorie foods, moderate climate change, and a strong social bonding in favor of WHRs that exceeded the putative ideal WHR. Sterelny (2012, p. 4) helps us to understand the unstable dynamism of variation with respect to human evolution. In summary, what we can take away from Sterelny's research is that humans evolved in environments that were constantly changing. This degree of change created much variability in human biological, psychological, and social domains. The upshot, thinks Sterelny, is that it is a serious error to think that there is a unique set of mental modules that can be created and then eventually maintained in the light of such unpredictable human and environmental changes. To the contrary, it is the ability of the human mind to contend with novelty that suggests more of a plastic or malleable mind indicative of a set of domain-general modules at work (Sterelny calls this ability *competence*.). This is why Sterelny is suspicious of **SEP**-type models. As he puts it, "to the extent that those models explain competence by appealing to preinstalled information, they are not well designed to explain competence in the face of the new" (Sterelny, 2012, p. 6).

Again, in the light of Sterelny's insights, the ideal WHR would not necessarily have a selective advantage under these complex varying conditions. Clearly, many other possibilities are plausible, even with this sample (and incomplete) criteria. Since the **EEA** is an amalgamation of complex systems within the Pleistocene and the fact that adaptations are adaptations to local environments, it is not at all clear how **SEP** supporters are able to locate a unique and stable environment that captures the putative ideal WHR. Therefore, since adaptations are adaptations to specific local environments and **EEA** is not a specific local environment, **SEP**'s reliance on **EEA** for its modularity of mind thesis lacks credibility. P2 of **The Argument from Time** is thus false. **The Argument from Time**, therefore, is unsound.

THE PHYSICAL-TO-PSYCHOLOGICAL ANALOGY

SEP holdouts could grant, to a certain extent, the argumentative turbulence related to the **EEA** and continue to champion their cause. To see this, remember that many defenders of **SEP** draw upon the following analogy: *In the same way that the body has specific physical traits or physical features (e.g., the eye) that have evolved to navigate specific environmental stimuli, the mind has specific psychological features (i.e., mental modules) that have evolved to overcome specific environmental obstacles.* Starratt and Shackelford press the point as follows:

> Our eyes respond to light waves, not smell, for much the same reason as our hearts pump blood instead of pumping blood *and* storing waste. One system performs one function.... The human mind works in much the same fashion. (Starratt and Shackelford, 2010, p. 235)

The Physical-to-Psychological Analogy can be illustrated as follows:

Physical Adaptation	Psychological Adaptation
Specific Physical Function(s)	Specific Psychological Function(s)

The analogy makes clear that, in the same way that most physical adaptations perform a unique task set, psychological adaptations also perform a distinctive task set. Given that the human body is littered with such precisely evolved physical features (like the eye and the heart), it is eminently reasonable to claim that psychological features also follow the same biological pattern of performance. Thus, Starratt and Shackelford think that the overwhelming majority of psychological mechanisms "attend to stimuli specific to the adaptive problems of our ancestors and motivate behaviors that function to solve those adaptive problems" (Starratt and Shackelford, 2010, p. 241). So, even if certain **SEP** arguments succumb to critical dismissal, **The Physical-to-Psychological Analogy** still vindicates the **SEP** project. Maybe the adaptation component of this analogy is the missing link?

REPLY TO THE PHYSICAL-TO-PSYCHOLOGICAL ANALOGY

It is not evident why **SEP** defenders think that the **EEA** represents a "stable" time period in which a vast majority of the set of domain-specific modules could be solidified. Even if it is granted that there is a large number of domain-specific mental modules, it could very well be the case (like human physical features) that many of these mental modules were formed and finely calibrated by natural selection both before and after the Pleistocene. Yet, many bodily physical features evolved before the Pleistocene (like the eyes) and some physical features evolved after the Pleistocene (e.g., lactose tolerance in adult humans evolved during the Halocene).[10] Indeed, in response to Starratt and Shackelford, Downes illustrates the worry as follows:

> The hand did not evolve in response to a particular environmental stimulus at any particular time. Rather, various selection pressures, including bipedalism, the occupation of niches with widely varying food resources, and our own niche construction, led to the musculature and bone structure that supports the range of activities for which human hands can be used. (Downes, 2010, p. 249)

Given the hand example above, we can appreciate the conclusion that Irons posits: "So, at best, this analogy shows that mental modules could very well have evolved like physical features, but it does not show that the majority of these psychological features could have only evolved during the Pleistocene" (Irons, 1998). Again, the reasonable inference is that **P2** of **The Argument from Time** is false, revealing an unsound

---

10 For more on the lactose tolerance example, see Ananth (2005), Richerson and Boyd (2005), and Downes (2010).

argument. Thus, even if we ignore the adaptation problem related to a composite **EEA** noted on p. 269 and Sterelny's concern about the complexity of change, the very physical adaptation/psychological adaptation analogy employed by **SEP** defenders is the very analogy that warns against taking seriously their Pleistocene-only defense and their insistence that the type of mental modules encompassing the human mind must be *predominantly* domain-specific modules. In fairness, **SEP** may yet be vindicated, but this physical-to-psychological analogy approach works for them *and* against them. At the very least, a more limpid defense in favor of **The Physical-to-Psychological Analogy** is necessary for **SEP**'s project.

3. Is Mother Goose in the House of Biology? SEP's Talk of Adaptations

Throughout this chapter, it has been assumed that the mental modules (e.g., cheating-detector) discussed by the **SEP** camp are evolutionary adaptations. In fact, **The Physical-to-Psychological Analogy** assumes as much. Starratt and Shackelford use the emotion of fear as a prime example of an evolved psychological adaptation. They offer the following four-part explanation:

(1) Fear is most likely to be associated with stimuli that would have been ancestrally dangerous, such as snakes, spiders, and heights.

(2) Fear occurs automatically, that is, without the need of conscious processing.

(3) Fear appears to be disconnected from higher-level conscious thought.

(4) Fear is located in the amygdala, an evolutionarily old part of the brain that is shared with other mammals. (Starratt and Shackelford, 2010, p. 236)

REPLY TO SEP'S TALK OF ADAPTATIONS: LETTING MOTHER GOOSE LOOSE!

The above account of fear seems quite reasonable, but closer inspection is required. First, 'would have been ancestrally dangerous' in (1) requires much more detail. Do we know that these particular organisms are the source of the fear mechanism? Does relying on contemporary fear responses vindicate that these stimuli were the relevant ancestral stimuli? Even if these stimuli are correct and that similar fear mechanisms are present in other mammals, does it follow that the selection and calibration of this fear mechanism is restricted to the Pleistocene? These questions need to be answered well in order to claim that the fear mechanism is the kind of evolutionary adaptive trait endorsed by **SEP**. To see this, briefly drawing upon the work of Richardson (2007) will prove instructive. He offers the following criteria of what constitutes an adaptation:

(i) *Selection* → Evidence must be offered to reveal that selection of a particular feature has occurred. For instance, the reduction of the visual apparatus and the improvement of other sense-modalities (e.g., larger antennae) in certain cave-dwelling fresh-water crustaceans (*Gammarus minus*) compared to its related surface-dwelling relatives.

(ii) *Ecological Factors* → Field studies reveal that the larger antennae, smaller eyes and larger body sizes of the cave-dwelling crustaceans have a significant reproductive advantage over those crustaceans with smaller antennae, larger eyes, and smaller bodies in the near-pitch dark caverns.

(iii) *Heritability* → Lab studies of the reproductive results of both land and cave-dwelling crustaceans have shown that the features under consideration have a genetic basis.

(iv) *Population Structure* → Field studies reveal that there is no gene flow between the land and cave-dwelling crustaceans. This gives further support that the populations are isolated from one another and that the features under consideration have evolved separately.

(v) *Trait Polarity* → Primitive and derived traits need to be determined and the relevant phylogenetic history needs to be made clear. In the case of the crustaceans, evidence suggests that glacial progressions and recessions provided the pressure cooker to begin the process of differences in the size of the antennae, eyes, and bodies (including pigmentation differences) between the land and cave-dwelling crustaceans.[11]

Notice that, in order to claim that X is an adaptation, one must provide the kind of analysis suggested above by Richardson. This is particularly important for **SEP** since **The Physical-to-Psychological Analogy** relies upon physical adaptation as part of its persuasive analogical thrust. Without the above kind of account, the risk of mother goose story-telling without evidence about psychological features looms large. In other words, even if **SEP** can provide a good account of a physical trait (like eyes or hearts) by using the criteria above (like the crustacean example), it would need to provide the same kind of analysis for a specific psychological trait in order to vindicate the persuasiveness of **The Physical-to-Psychological Analogy**. Yet, according to Richardson, this is precisely what they cannot do given the paucity of evidence pertaining to human evolution; that is, they cannot offer a genuine account of a psychological adaptation closely resembling the criteria above (Richardson, 2007, p. 136 and pp. 170–71). The fear example offered by Starratt

---

11 See chapters 3 and 4 of Richardson (2007) for a careful elucidation of these criteria and how they relate to the ambitions of **SEP**.

and Shackelford lacks just the kind of specificity that renders their analysis empty. It may be the case that there is a specialized mental mechanism for fear, but the requisite details and data that need to be offered (like the ones offered above) to move beyond reasonable story-telling are either not given or are unavailable. Rather, what we get is "unconstrained speculation"—namely, mother goose in the house of biology! This does not bode well for **SEP**'s reliance on **The Physical-to-Psychological Analogy** at the present time, even if an example like the fear mechanism proves to be correct. Each proposed mechanism would require the same sort of robust analysis that Richardson puts forth in the crustacean example. This is a mighty task to say the least.

III. CONCLUSION

SEP is an ambitious research program designed to elucidate the psychological nature of the human mind. Two integral parts of its methodology include a modularity of mind thesis and a time-and-location (the **EEA**) thesis. This chapter, however, has revealed that **SEP**'s ambitions have far exceeded the evidence proffered. Even granting that humans are part of a rich evolutionary history, it is fairly clear that much more evidence is required in defense of a mind that is predominantly modular. Additionally, given the complex and interrelated dynamics of the social environments and physical environments foisted upon and constructed by various human practices, the suggestion that the **EEA** must be mostly restricted to the Pleistocene time frame also seems suspect. Ultimately, if these struggles prove to be insurmountable for **SEP** (things do not look very good as of now), then its claims about human nature and the mere varnish of cultural influence on human psychology must correspondingly fall by the wayside as well. The overall *prima facie* conclusion to be drawn is that, although **SEP** might be championed by the Sophia Loren fan-base, it should leave the rest of us with very little about which to cheer—even prophet Darwin might blushingly ask **SEP** researchers to proceed with caution!

CHAPTER REVIEW: DISCUSSION AND QUESTIONS

1. Why does the author begin with the chapter with what appear to be an odd set of examples? [pp. 253-54]
2. Why is Darwin labeled as a sort of prophet? [p. 254]
3. Explain the difference between **WEP** and **SEP**. [p. 255]
4. What are the basic principles of **SEP** and how are they related to each other? [pp. 255-56]
5. Try to explain **The Argument from Domain-specificity** and domain-specificity. [pp. 256-58]
6. Drawing upon Symons' account, how might one go about offering a defense of the truth of P3 of **The Argument from Domain-specificity**? Use the waist-to-hip ratio example to illustrate this defense. [p. 258]

7. How does Buller show that **P3** of **The Argument from Domain-specificity** is false? (make sure your answer includes the "proliferate-and-prune" discussion) [pp. 258–60]
8. How is the cheating-detector discussion supposed to help **SEP**? Try to explain the Wason task scenarios to support a defense of **SEP**. [pp. 261–65]
9. How does Buller respond to the use of the Wason tasks? What does this reveal about **P3** of **The Argument from Domain-specificity**? [pp. 265–66]
10. What is the **EEA**? [p. 267]
11. Explain **The Argument from Time** and how it supports a "Pleistocene mind." [pp. 267–68]
12. Explain how Sterelny's comments tell against **P2** of **The Argument from Time**. [p. 269]
13. Explain **The Physical-to-Psychological Analogy** and how it could support of **P2** of **The Argument from Time**. [pp. 269–70]
14. How does **SEP** use physical adaptations to make sense of human psychology? Use the fear example to make the case. [p. 271]
15. What are Richardson's five-fold criteria of a genuine adaptation? Why does he think that **SEP** cannot meet these criteria? [pp. 272–73]
16. What conclusions are drawn by the author regarding **SEP**'s claims related to human nature and our so-called "surface cultural psychology"? [p. 273]

REFERENCES

Ananth, Mahesh. "Psychological Altruism vs. Biological Altruism: Narrowing the Gap with the Baldwin Effect." *Acta Biotheoretica* 53, 3 (2005): 217–39.

Buller, David J. "Varieties of Evolutionary Psychology." In *Cambridge Companion to the Philosophy of Biology*, ed. David L. Hull and Michael Ruse, 255–74. Cambridge: Cambridge University Press, 2007.

———. *Adapting Minds: Evolutionary Psychology and the Persistent Quest for Human Nature*. Cambridge, MA: MIT Press, 2005.

Buss, David M. *Evolutionary Psychology: The New Science of the Mind*. Boston: Allyn and Bacon, 1999.

Cosmides, Leda, and John Tooby. "Evolutionary Psychology: New Perspectives on Cognition and Motivation." *The Annual Review of Psychology* 64 (2013): 201–29.

Darwin, Charles. *The Origin of Species*. London: Penguin Classics, 1985 [1859].

Downes, Stephen M. "The Basic Components of the Human Mind Were Not Solidified during the Pleistocene Epoch." In *Contemporary Debates in the Philosophy of Biology*, ed. Francisco J. Ayala and Robert Arp, 243–54. West Sussex, UK: Wiley-Blackwell, 2010.

Fodor, Jerry. *The Modularity of Mind*. Cambridge, MA: MIT Press, 1985.

Gaulin, Steven J.C., and Donald H. McBurney. *Evolutionary Psychology*. 2nd ed. Upper Saddle River: Pearson Prentice Hall, 2004.

Irons, William. "Adaptively Relevant Environments versus the Environment of Evolutionary Adaptedness." *Evolutionary Anthropology* 6, 6 (1998): 194–203.

Pinker, Steven. *The Blank Slate*. New York: Viking, 2002.

———. *The Language Instinct*. New York: HarperPerennial, 1995.

Pinker, Steven, and Paul Bloom. "Natural Language and Natural Selection." In *The Adapted Mind*, ed. Jerome H. Barkow, Leda Cosmides, and John Tooby, 451–93. New York: Oxford University Press, 1992.

Richardson, Robert C. *Evolutionary Psychology as Maladapted Psychology*. Cambridge, MA: MIT Press, 2007.

Richerson, Peter, and Robert Boyd. *Not by Genes Alone: How Culture Transformed Human Evolution*. Chicago: University of Chicago Press, 2005.

Singh, Devendra. "Adaptive Significance of Female Physical Attractiveness: Role of Waist-to-Hip Ratio." *Journal of Personality and Social Psychology* 65, 2 (1993): 293–307.

Sloutsky, Vladimir M. "Mechanisms of Cognitive Development: Domain-General Learning or Domain-Specific Constraints?" *Cognitive Science* 34 (2010): 1125–30.

Starratt, Valerie G., and Todd K. Shackelford. "The Basic Components of the Human Mind Were Solidified during the Pleistocene Epoch." In *Contemporary Debates in the Philosophy of Biology*, ed. Francisco J. Ayala and Robert Arp, 231–42. West Sussex, UK: Wiley-Blackwell, 2010.

Sterelny, Kim. *The Evolved Apprentice: How Evolution Made Humans Unique*. Cambridge, MA: MIT Press, 2012.

Symons, Donald. *The Evolution of Human Sexuality*. Oxford: Oxford University Press, 1979.

Tooby, John, and Leda Cosmides. "Conceptual Foundations of Evolutionary Psychology." In *The Handbook of Evolutionary Psychology*, ed. David M. Buss, 5–67. New Jersey: Wiley, 2005.

———. "The Psychological Foundations of Culture." In *The Adapted Mind*, ed. Jerome H. Barkow, Leda Cosmides, and John Tooby, 19–136. New York: Oxford University Press, 1992.

Workman, Lance, and Will Reader. *Evolutionary Psychology: An Introduction*. 2nd ed. Cambridge: Cambridge University Press, 2008.

Chapter 8

OF GNATS AND MEN
Evolution, Religion, and Intelligent Design

If it could be demonstrated that any complex organ existed which could not possibly have been formed by numerous, successive, slight modifications, my theory would absolutely break down.

—Charles Darwin, Origin of Species *(1872, p. 175)*

If I have erred in giving natural selection great power, which I am very far from admitting, or in having exaggerated its power, which is in itself probable, I have at least, as I hope, done good service in aiding to overthrow the dogma of separate creations.

—Charles Darwin, The Descent of Man *(1871, p. 863)*

I. SCIENCE AND RELIGION: A TROUBLED ROMANCE

The relationship between religion and science has been and continues to be quite tumultuous. Like young love, when science offers hypotheses that are congruent with religious tenets, both feverishly rush to walk hand-in-hand, overlooking each other's obvious differences. And, much like youthful paramours, both camps turn into scorned lovers when either expresses views that violently shake their respective foundational core.[1] For instance, the Catholic church of the sixteenth and early seventeenth centuries was steeped in the biology and science put forth by the Aristotelian champions of the day. From this perspective, the Earth was both stationary and the

---

1   There are many ways to explore the science and religion relationship. For a glimpse into some of these distinct debates, see Garvey (2007, chapters 12 and 13) and Stamos (2008, chapters 8 and 9).

center of the universe, all biological entities had their species-specific teleological structures, and God could be viewed as the creator and care-taker of all. *Science and religion: a match made in heaven and love was in the air!*

Imagine the church's shock when Galileo presented convincing evidence (drawing upon findings from his self-made astronomical telescope), in 1632, in support of Copernicus' hypothesis (in 1543) that the Earth was not the center of the universe, but a mere orb circling the sun! Endorsing Copernicus' heliocentric hypothesis would not only jeopardize the privileged status of the Earth as the center of the universe, but would also open the floodgates for rejecting other religious doctrines that constitute part of the core of Catholicism. As a result of a perceived betrayal, Galileo was forced to recant his scientific verdicts and was put on an eight-year house arrest.[2] *Science and religion: the pain of sleeping on the couch!*

In much the same way that Galileo was maligned for his views about astronomy, Darwin was lambasted by Christian supporters for his views regarding the nature and origin of humans. (This should startle none of the readers of this text given the structure of the *Origin* presented in ch. 2.) In the same manner by which Galileo kicked humans off the center stage of their own solar system, Darwin removed humans off the center stage of their own planet. It must have been quite staggering to digest and accept Galileo's account and almost equally over-whelming and appalling to entertain seriously humans as merely one amongst the animals—or as Gould eloquently quipped: "Homo sapiens [are] a tiny twig on an improbable branch of a contingent limb on a fortunate tree" (Gould, 1989, p. 261). If religion must begrudgingly accede to the location and movement of the Earth relative to other orbs in the sky, it damn well was not going to budge regarding the origin and nature of its human flock. God would remain the creator of all things and an interested observer and intervener of the human condition. This was the view expressed by many of Darwin's contemporaries and it's the same position taken by some contemporary anti-Darwinian religious groups.[3] *Science and religion: Irreconcilable Differences—let the divorce proceedings begin!*

Recent exchanges between defenders of evolutionary biology and upholders of various religious groups reveal that the irreconcilable differences noted above have not been resolved. The most recent rendition of these religious critical anti-Darwinian groups is known as the Intelligent Design Camp (hereafter **IDC**).[4] Interestingly, the **IDC** has melded God into a being that is interested in its human creation, but without explicitly attaching any sort of religious framework/dogma to this being. In this way, the **IDC** can endorse the power and anonymity of Aristotle's Unmoved Mover

---

[2] Note that not all historians of science view the Galileo affair as a battle between religion and science. Galileo was also having internal disputes with his science colleagues. For more on this, see Langford (1992).

[3] For attempts to reconcile evolution and religion, see Johnson's (2001b) critique. Also, see Witham (2002), Numbers (1998), and Larson (1997).

[4] Before the **IDC** movement, there was the *creation science/scientific creationism* rendition. For a careful analysis of the creation science project, see Kitcher (1983) and Pennock (2002).

(see ch. 1) and can retain a modicum of the personality of the Judeo-Christian God.[5] As a way of exploring a portion of the evolution and religion debates, this chapter will focus on two of **IDC**'s prominent defenders, Phillip Johnson and Michael Behe, and the critical rejoinder put forth against their arguments. This analysis will show that there is very little of the **IDC** project that is worth salvaging from Johnson's and Behe's efforts. *Science and religion: "Consider this a one-thousand horsepower divorce, sweetheart"* (Daven Anderson, *Vampire Syndrome*).

II. DESIGN, DESIGN, DESIGN: PALEY'S CHALLENGE

Recall that Darwin wrote the *Origin*, in part, to refute the belief that God-as-divine-designer is both the creator and caretaker of all life forms (see ch. 2). William Paley (1802), a naturalist and theologian, did much to defend the God-as-Designer hypothesis and appears to be one of the targets at which Darwin was aiming. Drawing on his empirical study of human and non-human organisms, Paley was convinced that the complexity and organization that he observed could not occur by happenstance. Much like the intricate structure and interacting parts of a watch, he argued, there is an analogous set of intricate mechanisms at play with respect to biological systems and their parts. And, in the same way that a watch is the product of a designer/watchmaker, biological systems and their parts have a corresponding designer; namely, God. Here is a reconstruction of Paley's argument to which Darwin took issue:

> The Argument from Design: Paley's Incredulity
>
> P1. Both biological systems and their parts are the product of chance events or both biological systems and their parts are the product of a divine designer.
>
> P2. If biological systems and their parts are the product of chance events, then they must also be poorly organized and lack goal-directedness.
>
> P3. Yet, it is not the case that biological systems and their parts are poorly organized and lack goal-directedness.
>
> P4. It follows that it is not the case that biological systems and their parts are the product of chance events.
>
> ---
>
> C1. Thus, biological systems and their parts are the product of a divine designer.

---

5   Whether this is true of all of the members of the **IDC** is unclear. For a compelling account of the Christian agenda of some of the key members of the **IDC**, see Forrest and Gross (2004).

If we assume the truth of **P1**, it is **P2** and **P3** that are doing the heavy lifting for Paley. How does Paley justify the truth of **P2** and **P3**? Primarily, he displays his impressive attention to detail by explaining various parts of animals and parts of the human body, including both the function and interrelationship amongst the parts. As Ayala summarizes, "After detailing the precise organization and exquisite functionality of each biological entity, relationship, or process, Paley draws again and again the same conclusion: only an omniscient Deity could account for these marvels of mechanical perfection, purpose, and functionality, and for the enormous diversity of inventions that they entail" (Ayala, 2010, p. 370). So, by way of his empirical efforts, Paley can only fathom that an all-knowing God could have produced such precision, complexity, and goal-directedness. It is the result of these findings that reveals why Paley thinks that **P2** and **P3** are true.

It is not too much of a stretch to infer that a tinge of incredulity attaches to Paley's argument in terms of alternative naturalistic accounts of biological complexity and functionality. Weber and Depew stress this point in their explication of Paley's argumentative strategy. These scholars remind us that, after Paley offers his detailed functional complexity account of a body part, "natural explanations are rejected using the criterion of incredulity: how *possibly* could the vertebrate eye appear as a result of random events, even under natural processes and laws?" (2004, p. 173). They conclude that, ultimately, Paley "dared anyone to come up with a fully natural explanation of biological functional complexity" (2004, p. 173). Interestingly, we see Paley offering a challenge to those who would defend a naturalistic account against a divine designer account; *and* we see Darwin (in the opening quotation) throw down an alternative challenge to those who would defend a non-naturalistic account against his naturalistic (natural selection) hypothesis. *Paley's intelligent designer and scientific/naturalistic explanations: Looking for love in all the wrong places!*

III. DARWIN'S DISMAY AND PROPOSED RESOLUTION

Interestingly, Darwin offers his own argument from incredulity as a response to Paley's version. His line of reasoning is that if one accepts that God is the creator of all life, then God's influence must be present in all of life's unfolding events. This implication, thinks Darwin, is absurd as is evident in the random bird that eats a gnat or the accidental lightning strike death of a person. As he puts it, "If the death of neither man nor gnat are designed, I see no good reason to believe that their *first* birth or production should be necessarily designed" (Darwin, 1860). Thus, Darwin's concern is captured by the following sort of argument against divine design:

> **P1.** If one should believe that the origin of life is the result of divine creation, then one must also endorse the view that every event is the product of divine creation.
>
> **P2.** It is absurd to think that every event is the product of divine creation.

C1. Thus, it is not the case that one should believe that the origin of life is the result of divine creation.

If Darwin is accurate that most people believe that every event's unfolding is the product of divine influence, then it appears that he does have a legitimate gripe. It is difficult to fathom how every single event—from the seemingly mundane to the monumental—could have God's finger prints on it. Strikingly, our putative understanding of accidental events must be eliminated as well as our beliefs in self-directed morally good and bad deeds. Of course, none of these concerns will budge the truly devout, but this does not by itself take away from the legitimacy of Darwin's skepticism.

Still, one might venture to think that if this is all Darwin has to say on the matter, then there should be some cause for concern. Primarily, **P1** could be false depending upon how God's causal influence is understood. From the belief that there is a God that designed everything, it does not follow that a belief that God's influence need necessarily pervade over all the minutiae of every event. Thus, Darwin's argument is rendered unsound, leaving Paley's argument from incredulity intact.

The above reply would be unfair to Darwin because it diverts from his intended target. Darwin is criticizing only those who believe that the intricacies of every event are divinely caused. It is this position that he rejects in the light of the variety of events—intentional and otherwise—that he has observed. So long as this target audience remains the focus of the debate, Darwin's position, though not rigorously defended, is not entirely preposterous.

Still, as we know from chapter 2, Darwin has much more to say on the matter. Rather than rehearse all of that analysis, Table 8.1 on the next page is offered as a reminder from Chapter 2.

Given 1–4, Darwin can claim that he has met Paley's challenge posed by his argument from incredulity; that is, it is possible to make sense of the intricate structure of the body and its parts by way of natural selection in conjunction with geologic time and various sorts of environmental pressures. Thus, contrary to Paley's incredulity, Darwin shows that it is eminently reasonable to make sense of complex biological features and their corresponding functionalities.[6] From this perspective, Darwin could claim that **The Argument from Design** is unsound—not because **P3** is necessarily false, but because **P2** is false;[7] that is, it is possible for biological entities and

---

[6] For more on Darwin's insights and their implications with respect to God-as-a-designer, see Ayala (2004).

[7] It is worth noting that Darwin goes to much trouble in his *Descent of Man* (1871) to show there are poorly organized, imperfect, "useless," and rudimentary parts of the human body in order to hammer home his defense of descent with modification. These details also tell against Paley's **P3** because biological systems, understood as bundles of incomplete and sub-optimal evolutionary compromises, tell against (to a degree) the precision and near-perfection claims related to Paley's mechanical analogy.

PRINCIPLES	EXPLANATION
1. Geologic Time	With the help of Ramsay and Lyell, Darwin defends the view that the Earth is a few hundred million years old.
2. A Dynamic Earth	Flexing his own geologic muscles, Darwin displays for the reader that, not only is the Earth a steadily changing place, but also that organism and species change occurs and has occurred accordingly and to a degree.
3. Domestic Breeding	If humans can modify features of organisms for their own benefit over a relatively short span of time, we should expect that nature can do considerably more, like produce species-level changes, over geologic time.
4. Malthusian Pressure Cooker	So long as populations exceed resources, there will be competition for those resources. Only those individuals who possess the requisite traits to cope with resource scarcity will likely survive and reproduce.

Table 8.1 *Four Principles Supporting Darwin's Defense of Natural Selection*

their parts to be the product of chance events and still exhibit complex structures and goal-directed activities. *Darwin's evolution and Paley's divine designer*: "*The course of true love never did run smooth*" (Shakespeare, *A Midsummer Night's Dream*).

IV. PALEY'S NEW DEFENDERS:
THE RISE OF THE INTELLIGENT DESIGN CAMP

The Paley-Darwin account above is provided as a sort of reminder that history can repeat itself. Indeed, the recent emergence of the Paleyesque **IDC** arguments and the counter replies by contemporary Darwinians reveals just this sort of historical repetition. As we will see, many scholars within the **IDC** claim that Darwin and contemporary neo-Darwinians have not shown the falsity of **P2 and P3**, leaving much of **The Argument from Design** intact. Broadly, much like Paley's, the **IDC**'s arguments employ the following three-step process:

Step 1 → Offer a description of a straightforward theoretical point or a biological fact or set of facts.

Step 2 → Proclaim that no set of natural biological processes *by itself* could produce the straightforward theoretical point or biological fact or set of facts noted in **Step 1**.

Step 3 → Insist that the only alternative to the theoretical point or natural biological processes must be an intelligent designer; namely, God.

Keep in mind that **Steps 1–3** are a way of making sense of how the **IDC** defends **P2** and **P3** of **The Argument from Design**. What this amounts to is that, according to the **IDC**, the presence of various types of complexity necessitates a belief in a divine designer. The reader should keep in mind that **The Argument from Design** is the over-arching argument that governs this chapter. So, upon examining a few attempts by the **IDC** to defend its specific versions of the creationism/design argument, this analysis will provide a brief reminder of what is at stake with respect to **P2** and **P3** of **The Argument from Design** in relation to those **IDC** arguments. *Religion to Naturalism/Scientism/Evolution*: "Goodbye foolish one, time heals and reveals...."

Phillip Johnson: Metaphysical Darwinism and Pennock's Response

The recent incarnation of God-as-a-legitimate-alternative-to-science (hereafter called 'creationism') was spear-headed by Phillip Johnson, one of the leading members of the **IDC**. In summary, Johnson argues that the creationism voice has been muffled by the scientific community's deafening scream that knowledge concerning the nature of reality is strictly gained by the findings of the natural sciences. Johnson labels such a view as *scientism* (Johnson, 2001, p. 72; also see Gould, 1988, p. 84). This move *from evolutionary biology to metaphysics* deeply worries Johnson. In response, he proceeds to attack versions of Darwinism that are wedded to scientism (Johnson labels this *naturalism* in the quotation below). Let us, then, label the attempt to wed metaphysical claims to evolutionary claims *Metaphysical Darwinism*. As he puts it, "By 'Darwinism' I mean fully naturalistic evolution, involving chance mechanisms guided by natural selection" (Johnson, 1991, p. 4). The formal definition is as follows:

Metaphysical Darwinism → The view that the underlying history and nature of life is explained *entirely* by natural selection working on chance variations.

We can make sense of Johnson's reaction to **Metaphysical Darwinism** using the **IDC** three-step approach:

Step 1 → Evolutionary biology is committed to scientism, revealing that evolutionary biology is committed to **Metaphysical Darwinism**.

Step 2 → Evolutionary biology's commitment to **Metaphysical Darwinism** is unwarranted on the grounds that its discovery and endorsement of any set of biological facts does not exhaust the set of metaphysical entities causally related to the set of biological facts.

Step 3 → It could very well be the case that an intelligent designer, namely God, is the metaphysical entity causally related to the set of biological facts.[8]

Johnson is arguing that naturalism (also called *scientism* above) is a metaphysical doctrine that rules out any sort of interacting or personal God on the grounds that the natural world has both its origins and continued changes governed solely by various sorts of material/physical processes, including evolution (Johnson, 1991, pp. 114–15).[9] Put another way, **Metaphysical Darwinism** rules out the existence of a personal God that influences the natural world (Johnson, 2001b, p. 447). The general argument for this discussion (and another way of making sense of **Step 2** above) can be put forth as follows:

The Argument from God-as-Designer-Interactor

P1. Either **Metaphysical Darwinism** is true or **The Argument from God-as-Designer-Interactor** is true.

P2. If **Metaphysical Darwinism** is true, then it is because it provides a complete account of the causal forces involved in an account of the history and nature of life.

P3. It is not the case that **Metaphysical Darwinism** provides a complete account of the causal forces involved in an account of the history and nature of life.

P4. It is not the case that **Metaphysical Darwinism** is true.

C1. Thus, **The Argument from God-as-Designer-Interactor** is true.

Notice that **P1** of **The Argument from God-as-Designer-Interactor** is presented as an "either-or" proposition that excludes any other possibility. **P1** is, however, clearly false given that God could have used evolution as part of the overall plan (*theistic evolution* is the label frequently given for these sorts of proposals). From this perspective, God indirectly created all the complexity of life by setting the wheels of biology in motion via evolution (see Sober, 2011, pp. 189–91). Alternatively, God could have employed evolution as the primary cause of those things biological and occasionally intervened at whim as a way to ensure certain outcomes (see Pope, 2007). Yet another possibility is that God created the beginnings of the universe and then took a permanent vacation

---

8   A relaxed variation of this three-step argument can be found in Graham (2011), p. 67.
9   The luxury to expand on the scientism debate is not available. See Rosenberg (2011) for more on what it means to take scientism seriously and Kitcher's (2012) worry about doing just this.

(sometimes called *deism*). These alternative accounts are logical alternatives to either **Metaphysical Darwinism** or **The Argument from God-as-Designer-Interactor**. Thus, from the get-go, the strategy employed by the **IDC** leaves **P1** false, rendering the entire argument unsound.

It is worth noting that some defenders within the **IDC** would balk at the nature of God presented in the above critique. Their counter-reply has two parts. First, they would respond that viewing God as a being that merely "knocked over the first domino" or behaves whimsically gives little credence to the belief in a personal and interactive God (See Kitcher, 2001, pp. 251–57). On this view, **P1** of **The Argument from God-as-Designer-Interactor** is not rendered false because the above depiction of an alternative God relies on God being no more than another mechanical cause in the chain of causes or an impulsive/fanciful child rather than a rational being that shows genuine care for the creatures it created. So long as God is viewed as a personal and rational God, the logical possibility of a mechanical or capricious and puerile God can be dismissed. This first counter-reply would appear to reveal that **P1** is true and that **The Argument from God-as-Designer-Interactor** is sound.

The second part of the counter-reply relies on a supposed logical error made by those who support theistic evolution. Primarily, the argument is that evolution proceeds in a non-directional fashion, but God proceeds with a directional plan. Given the former, those life forms that are and have been on Earth are the product of chance events; these beings could just as well have not come into existence if other random events took place (see the progressive evolution discussion in ch. 6 for a reminder). Thus, critics charge that it would be odd for God to use evolution as part of its overall strategy because there is no way to ensure that humans would ever come into existence—unless God is given the interactive characteristic of ensuring that humans would come to be. Yet Johnson retorts that the non-directional feature of evolution "means that God neither programmed evolution in advance nor stepped in from time to time to pull it in the right direction.... Once this logical difficulty is recognized, the attempt to reconcile Darwinism and theism collapses" (Johnson, 2001, p. 443). Thus, if ensuring that a divine being intended to produce humans on Earth is an integral part of theism, then relying on evolution-as-a-non-directional-process to do so would be a mistake. The upshot is that **P1** is correct as stated from the perspective of the **IDC**, resulting in the soundness of **The Argument from God-as-Designer-Interactor**.

Additionally, recall that **P3** of **The Argument from Design** claims that it is not the case that biological systems and their parts are poorly organized and lack goal-directedness. Notice that the soundness of **The Argument from God-as-Designer-Interactor** contributes to the soundness of **The Argument from Design**; for if a powerful divine being creates and monitors life on Earth in a non-haphazard way, it stands to reason that its creations would be well-organized teleological entities. Thus, if **The Argument from God-as-Designer-Interactor** is sound, then the **IDC** has secured further support in defense of the soundness of **The Argument from Design**.

A REPLY TO THE ARGUMENT FROM GOD-AS-DESIGNER-INTERACTOR

Both of the above possible counter-replies suffer from a similar problem: both rely upon a particular knowledge of the nature of God that appears difficult to justify. Restated, many **IDC** defenders point out that there is more to the metaphysical nature of reality than the rules and practices of science can come to know. As Johnson puts it, "Once we allow God to enter the picture at all, there is no reason to be certain a priori that natural science has the power to discover the entire mechanism of creation" (Johnson, 2001, p. 444). Johnson is stressing that science, as a way of knowing, is not in a position to offer an exhaustive set of causal mechanisms for all that exists. It is possible, thinks Johnson, that a divine being could invoke causal forces beyond the purview of science to detect. In the same way that science cannot offer a sweeping list of the "things" that exist (like God), it also cannot provide an all-inclusive list of the causes of the "things" that exist.

The problem with this sort of knowledge claim is that it is a double-edged sword. One cannot claim that **X** is ignorant about **Y** based on a particular standard of knowledge **Z** and concomitantly insist that one's own epistemic claims about **Y** are on the mark given that one is also bound by **Z**. It may be correct to claim, as Johnson does, that science cannot know *with certainty* God's specific set of mechanisms employed to create life on Earth; yet, it correspondingly follows that the same certainty requirement placed upon scientific practices must also be placed upon the claims made by the **IDC**. The result is that the certainty requirement renders all knowledge claims moribund because the certainty requirement cannot be met by epistemically limited creatures like humans. From this perspective, neither science nor the **IDC** can be certain of its claims once God is allowed to enter the discussion.[10] Reasonably, then, if certainty is the standard—one that cannot be met—then some version of "we can't know" is all that can be said about the nature and role of God's creation mechanisms. So, the two possible counter-replies put forth by some members of the **IDC** do not provide the needed support to reveal the truth of **P1**. The implication is that the **IDC** criticism of **Metaphysical Darwinism** in **Step 2** does not secure any support for **Step 3**. Put another way, even if the **Metaphysical Darwinism** part of **P1** of **The Argument from God-as-Designer-Interactor** is false, it does not follow that the personal/interactive God of **P1** of **The Argument from God-as-Designer-Interactor** is true. And, given how **The Argument from God-as-Designer-Interactor** can be viewed to bolster **P3** of **The Argument from Design**, the soundness of **The Argument from Design** is, at the very least, uncertain.

---

10 See Sober (2011) for an in-depth discussion of this point and Scott (2009, ch. 1) on science and certainty.

PENNOCK'S REPLY TO JOHNSON

Is, however, Johnson's construction of **Metaphysical Darwinism** fair to the practitioners of biology? Pennock answers in the negative, pointing out that biologists need not be committed to **Metaphysical Darwinism**. (This suggests that Johnson and other members of the **IDC** have constructed a straw-person argument, pointing to the falsity of both the **Metaphysical Darwinism** portion of **P1** and **Step 1**.) Relying upon *Methodological Naturalism* as an alternative to **Metaphysical Darwinism**, Pennock stresses that the methods and practices of science help to make sense of whatever phenomenon is under investigation without being committed to claims about what exists in the world (Pennock, 2001, p. 84). From this perspective, for example, the methodological naturalist endeavors to discover and explain the eating habits of a particular grasshopper after exhaustive empirical study of the species under investigation. Notice that such a study need not at all be committed to the ontological status of grasshoppers. Rather, a methodological naturalist need only be cautiously committed to knowledge claims about phenomena that have been directly or indirectly observed and vetted by peer-reviewed critiques. Again, this means that knowledge claims about X need not commit the **Methodological Naturalism** to ontological claims about X. Scientists, at least Pennock's methodological naturalists, are committed to this sort of research program, but are also quick to distance themselves from the later sort of potentially speculative ontological claims associated with **Metaphysical Darwinism**.

Pennock continues his discussion in defense of **Methodological Naturalism** as the core framework of the scientific method by revealing the kind of evidentiary-based intellectual modesty tethered to **Methodological Naturalism** (Pennock, 2001, p. 84–85). The point that can be taken away from Pennock's account is that science is not in the business of knowing under the guise of certainty (unlike the certainty requirement set by Johnson and other proponents of the **IDC**). Rather, science will employ its methods of empirical inquiry and offer claims based on evidence that has been accrued—knowing full-well that its claims could require revision or abandonment based on additional evidence. **Methodological Naturalism** can be defined as follows:

> Methodological Naturalism → The view that the methods of science assist in understanding, in an intellectually cautious and tentative way, various sorts of phenomena in the world.

So, according to Pennock, although some Darwinians might endorse certain speculative **Metaphysical Darwinism** claims, most contemporary naturalists would pursue the kind of science that is wedded to **Methodological Naturalism**. Pennock can make this claim because of the defeasibility element that is already present in the scientific practices of most scientists. What this illustrates is that evolutionary thinking need not be committed to the dogmatic **Metaphysical Darwinism** worries presented by Johnson. Pennock attempts to put an end to this straw-person discussion generated by Johnson by concluding that the methods of science take on very little metaphysical baggage, but cautiously offer criteria about valid investigation procedures (Pennock, 2001, p. 88). If Pennock is correct about the role of **Methodological Naturalism** in the practices of science, then both the **Metaphysical Darwinism** portion of **P1** and **Steps 1** and **2** are false and ought to be rejected. Thus, given the knowledge criticism of **The Argument from God-as-Designer-Interactor** part of **P1** and the **Methodological Naturalism** criticism of the **Metaphysical Darwinism** part of **P1**, **P1** is, at first glance, false. *And* the further implication is that Johnson's **The Argument from God-as-Designer-Interactor** is unsound.[11] And this further suggests that the soundness of **The Argument from Design**, which is supposed to be bolstered by **The Argument from God-as-Designer-Interactor,** is questionable. *Intelligent Design and Evolutionary Biology*: *"Merely two ships passing in the night ..."*

Behe's Irreducible Complexity Defense of The Argument from Design and Miller's Critical Assessment

As mentioned earlier, the opening quotation by Darwin at the beginning of this chapter can be viewed as a challenge. Darwin is claiming that his theory of evolution by natural selection will not be worth the paper on which it is written, if it can be shown that a biological feature exists such that its presence, characteristics, and functionality are not the product of similarly and slightly modified descendants. The biochemist, Michael Behe (1996 and 2002), takes on this challenge offered by Darwin by arguing that there are some features and/or parts of entities that are *irreducibly complex* and thus not the product of a history of modified descendants. Behe's conclusion is that such irreducibly complex entities must have come into existence "in one fell swoop" (as opposed to a protracted geologically-paced process) by way of an intelligent designer rather than by natural selection (Behe, 2007).

What does Behe mean by 'irreducibly complex' and what are these irreducibly complex entities of which Behe speaks? The answer to the first question is as follows:

(i) **Irreducible Complexity** → X is irreducibly complex if and only if the functions of X's components cease to be actualized when X's components are no longer part of contributing to X's overall function(s). (Behe, 2002, p. 74)

For example, Behe thinks that the tiny structures of cells, called organelles (e.g., nucleus, mitochondria, endoplasmic reticulum, golgi apparatus, ribosomes, etc), have functions only within the framework of the cell. Remove any one of these organelles from the cell and they cease to function as they did within the cell. Remove, for instance, ribosomes from a cell and they no longer have the function to produce the

---

[11] This discussion could continue by offering additional worries about the legitimacy of **Methodological Naturalism** within science. For a glance into this interesting discussion, see the criticism of **Methodological Naturalism** put forth by Plantinga (2001a) and the counter-reply by Ruse (2001).

relevant proteins needed for cell repair and various other specific cellular activities. So, for Behe, the cell constitutes an irreducibly complex system because its component organelles cease to actualize their particular functions once they are removed from the cell. What follows, thinks Behe, is that "the existence in nature of irreducibly complex biological systems poses a powerful challenge to Darwinian theory ... perhaps molecular machines appear to look designed because they really are designed.... I am hopeful that the scientific community will eventually admit the possibility of *intelligent design*" (Behe, 2002, p. 74; my italics).

The answer to the second question, regarding the kinds of entities under consideration, is the biochemistry related to these particular biological systems:

(ii) **Instances of Irreducibly Complex Systems** → cells and organelles, blood clotting, immune system, flagellum of bacteria, and origin of life.

Keeping in mind (i) and (ii), here are the three steps related to Behe's irreducible complexity argument:

Step 1 → Evolutionary biology is committed to the idea that the existence of any biological entity can be traced back to many successive modified ancestral entities.

Step 2 → There are some biological systems that cannot be traced back to many successive modified ancestral entities [see (ii) above]. These are known as irreducibly complex systems.

Step 3 → The existence of irreducibly complex systems, which cannot be explained by way of Darwinian principles, suggests the presence of an intelligent designer as their creator.

It is not unreasonable to think that evolutionary biologists are committed to **Step 1**. Put in Darwinian terms, organisms and their parts are the products of natural selection operating on fortuitous chance variations. The cumulating effects of these beneficial variations are successively modified organisms. From this perspective, any contemporary species will represent the existing set of modified organisms of a varied, yet related, set of modified ancestral species. Thus, most contemporary biologists would be quick to endorse the truth of **Step 1**.

MILLER'S REJOINDER TO BEHE

Just as fast as contemporary biologists would be willing to embrace **Step 1**, they would be equally eager to denounce **Step 2**. To see this, focus will be given to Behe's flagellum example. As a reminder, a flagellum is a thin whip-like tail structure that helps bacteria, spermatozoa, and protozoa with propulsion, sensation, and chemical information exchange between a cell and its ambient environment. According to Behe, the flagellum is a classic case of an irreducibly complex system because there exists a core set of proteins, *which do not have unique functions independent of their activities to aid in the functions of the flagellum*, that produce propulsion, sensation, and information exchange. Behe concludes that such an ensemble of proteins constitutes an instance of irreducible complexity because they have no ancestral history to make sense of their functional integrity (Behe, 2002). If this is true, then the result is that Darwin's challenge will have been met and evolution cannot account for the functional activities of the flagellum. Behe thinks that an intelligent designer proposal of such functional activity is an eminently reasonable alternative.

Kenneth Miller takes serious umbrage at Behe's handling of the facts.[12] To the contrary, he insists that the flagellum does not constitute an instance of irreducible complexity. Drawing upon the biochemistry related to the poison-injection mechanism of some bacteria, Miller points out that this mechanism (known as the type III secretory system or TTSS) remains fully functional even after it is separated from most of the parts of the flagellum. The important element here is that the proteins that are present in the TTSS are *homologous* to the proteins in the base layer of the flagellum. In evolutionary terms, what this means is that, in the same way that the flipper of a dolphin is similar in structure and location, but distinct in function, to the arm of a human, the proteins of the TTSS are similar in structure and location, but distinct in function, to the proteins in the base of the flagellum.

What Miller has shown is that a component TTSS of the flagellum can retain its unique function even after it has been separated from the so-called irreducible system of which it is a part (Miller, 2004, pp. 86–87). This directly contradicts that claim offered by Behe about the flagellum and its parts *and* his own definition of irreducible complexity. Moreover, given that the TTSS is homologous to proteins at the base of the flagellum, it is also the case that the TTSS is the product of a line of descent and modification. The conclusion is clear: one had better look for irreducible complexity elsewhere because it cannot be found with respect to the flagellum. **Step 2** is false, given Miller's account of the empirical findings, rendering dubious the soundness of Behe's irreducible complexity argument. Furthermore, wrapping-up where this discussion started, Behe has not offered an account that constitutes a powerful response to Darwin's challenge—**Step 3** is also false.

A cautionary point is worth stressing at this juncture: although there are a large number of unanswered concerns regarding the details related to many biological phenomena, it would be prudent to proceed with humility regarding what science can and cannot explain. The history of science is replete with instances of "unanswerable gaps" being filled in by a future scientific community—even if the time period

---

12 For a very illuminating discussion on this biochemistry debate, see Shanks and Joplin (1999) and Liu and Ochman (2007).

between discovery of the problem and the corresponding resolution is protracted.[13] This is not to say that such answers are impermeable to error; rather, the point is that additional evidence can yield compelling support for a particular hypothesis from a methodological naturalistic perspective where either ignorance or divine intelligence were the prevailing alternatives. This irreducible complexity argument put forth by Behe and others in the **IDC** illustrates this lack of humility and prudence.

No doubt, Behe or some other member of the **IDC** could contend that the TTSS is itself an irreducibly complex system within the flagellum. This possibility, it could be argued, suggests that now we have two irreducibly complex systems—the flagellum and the TTSS—that require an explanation (Miller, 2004, p. 87). With no such explanation forthcoming, this quick rejoinder would suggest that the truth of **Step 2** remains in place. The kicker here is that the possibility of an intelligent designer is still a legitimate candidate for irreducibly complex phenomena.

This possible objection is flawed based on Behe's own definition of irreducible complexity. Recall, if X is irreducibly complex, then it is composed of reducible parts that have functions only so long as these parts contribute to the overall function(s) of X. Remember, X was brought to exist (by a designer) as a whole with integrated parts, all of which cannot be accounted for by the evolutionary biologist. Yet, if there is an irreducible Y within an irreducible X, this leaves open the possibility that Y does not contribute to the overall function of X since Y, as an irreducibly complex entity, will have a unique function of its own. This unique function could be made sense of from an evolutionary perspective, rendering Behe's definition of irreducible complexity dead in the water. As Miller puts it, "However, if the flagellum contains within it a smaller functional set of components such as the TTSS, then the flagellum itself cannot be irreducibly complex—by definition" (Miller, 2004, p. 88). Thus, it would be best to avoid drawing upon a hierarchy of irreducibly complex entities as a counter reply to any empirical findings that tell against proposed instances of irreducibly complex entities.

As a last gasp, the **IDC** could insist that its intelligent design account remains intact because a detailed account of the evolution of the TTSS has not been offered by the evolutionist (Miller, 2004, p. 88). Even if a set of proteins of the TTSS is homologous with other proteins within the flagellum, all this does is force us to wonder not only what the evolution of the flagellum is (in detail), but also what the evolution of the TTSS is (in detail). Given that no answer is on the horizon, the **IDC** can claim

---

13 Ratzsch suggests that there may be a time for science to "call it quits" in its pursuit of some lines of inquiry, but that methodological naturalism would not allow for the possibility of futile scientific activity (Ratzsch, 2010, pp. 346–47). Yet, imagine if Kepler (circa 1600 CE) quit on trying to make sense of Ptolemy's (circa 200 CE) erroneous account of retrograde motion! Now, this does not settle the matter regarding when science should stop its pursuit of a specific topic, but it also does not suggest that methodological naturalism should be abandoned when no resolution of a perplexing concern appears forthcoming. Some puzzles with which science grapples are rather complex and require great ingenuity and technological advancement to solve. The detours and short-term dead-ends can be viewed as a kind of motivation for another generation of researchers to continue to engage unresolved puzzles and anomalies.

that it is reasonable to conclude that an intelligent designer is a plausible position to maintain (Behe, 2004, p. 355). This is the position recently taken by Jonathan Marks: "If the flagellum gives the overwhelming *appearance* of having been designed by an intelligent agent, are we not justified in inferring design until a more compelling candidate explanation (which better explains the data) is offered?" (Marks, 2011).

Unfortunately, this alternative-candidate-move fares no better than the previous possible reply. There is a two-part response. First, the work of Shin-Ichi Aizawa (2001) and Ariel Blocker et al. (2003) and the more recent efforts of Diepold and Wagner (2014), suggest that it is only a matter of time and technology-upgrade before more of the subtle details of the flagellum and TTSS are understood. Aizawa, for instance, has shown that both the flagella and TTSS share many proteins and amino acids. Aizawa concludes that "flagella and the type III secretion system [TTSS] consist of homologous component proteins with common physico-chemical properties, suggesting that these two systems could have evolved in parallel" (Aizawa, 2001, pp. 163–64; bracketed addition mine). Blocker et al. have pushed further and shown that the machinery governing TTSS and flagella share a great many characteristics when their genetic structure, morphology, regulation and function, and integration of structural information are compared. Finally, Deipold and Wagner have offered a painstaking account of the parts of the injection mechanism of a specific TTSS and have provided the proteins that are homologous to those of the flagellum. The point to stress here is that the details of the relationship between the TTSS and the flagellum demanded by members of the **IDC** are the very details that are slowly being offered by the scientific community. Are *all* the details available yet? No, but the reasonable inference is that time and improved technology will assist in the production of further details in terms of comparative biochemical structure and evolutionary relationships. Thus, given the direction of the science at present, there is neither compelling reason to think of the flagellum as irreducibly complex nor to invoke an intelligent designer regarding the intricate structure of the flagellum.

Second, putting to one side the current state of the science, the reasoning employed by some members of the **IDC** is incompatible with reliance on irreducible complexity. As noted above, Marks thinks that the appearance of organization warrants an intelligent designer defense until a superior alternative account can be proffered. As Miller puts it, "until we have produced a step-by-step account of the evolutionary derivation of the flagellum, one may indeed invoke the argument from ignorance for this and every other complex biochemical machine" (Miller, 2004, p. 88). Yet, as Miller stresses, relying on scientific ignorance does not help the defenders of irreducible complexity; for they are committed to the view that no amount of scientific data can provide an evolutionary explanation of a system that is already deemed to be irreducibly complex. This is because **IDC** defenders, like Behe (2004, p. 353), have denied that a Darwinian gradualism can build (on evolutionary time scales and in a step-by-step fashion) such systems. Thus, invoking intelligent design based on scientific ignorance of the step-by-step details is yet another path not available to the champions of irreducible complexity.

It should come as no surprise, then, to learn that Miller concludes that "the claim of irreducible complexity has collapsed, and with it any 'evidence' that the flagellum was designed" (Miller, 2004, p. 88). Additionally, given the counter-replies to both of the possible objections that the **IDC** could offer with respect to irreducible complexity and the flagellum case, the truth of **Step 2** is in serious jeopardy; at least to the extent that the truth of **Step 2** relies upon the legitimacy of the flagellum case. The overall thrust is that **P3** of **The Argument from Design** is not supported by Behe's irreducible complexity argument.

To reiterate, this sort of Behe/Johnson-type challenge is being taken seriously by the scientific community and I strongly encourage the reader to explore some of these other debates. For instance, regarding the evolution of the blood clotting process, see Doolittle (2009); for the evolution of the immune system, see Janeway et al. (2001), Flajnik and Kasahara (2010), and Cooper and Herrin (2010); and for the evolution of life, see Pross and Pascal (2013) and their citations. The reader will learn that, in these other cases, the findings will be similar to those put forth in the flagellum case; that is, some of the gaps in our knowledge will be or have been filled, while others will be works-in-progress, and others have yet to be determined or explored. None of this suggests that the **IDC** has met Darwin's challenge in the opening quotation of this chapter. Again, ignorance regarding a specific issue does not in itself justify the truth of an alternative account. *Evolutionary Biology and Intelligent Design:* "*Why don't you just pretend that the asshole dropped dead? You can't call or write to a dead man. Put a couple of candles in front of his picture, say a few Hail Marys, and get it over with*" (Isabel Lopez, *Isabel's Hand-Me Down Dreams*).

V. THE PINK DEMARCATION ELEPHANT IN THE ROOM

Usually, at this point in the discussion, students wonder why **IDC** defenders ever thought that they really had a chance of infiltrating the field of science with their set of theistic views as a genuine alternative to scientific accounts. There is a two-part answer to this obvious query. First, **IDC** advocates push their theistic agenda in hopes of influencing impressionable faculty, parents, and students in the high school arena. Because high school curricula can be influenced at the PTA and district levels—the very places where fellow theistic "groupies" hold sway—**IDC** can and has imperceptibly pushed for "equal time" in the biology classroom. Quite a bit has been written on this religion-in-the-classroom debate and will not be explored directly here.[14]

---

14   For a gripping discussion of **IDC**'s attempt to influence high school biology curricula, see Pennock (2001a, 2001b, and 2003) and Plantinga (2001b). For a legal perspective regarding **IDC**'s attempt to gain footing in the high school science curricula by way of the interpretation of the Establishment Clause, see Nagel (2008). Additionally, for a defense of creationism/intelligent design as a legitimate "worldview" that should be included in high school curricula, see Reiss (2011).

Second—and this is the focus of this section—**IDC** defenders claim that it is eminently reasonable to view the notion of an intelligent designer as an alternative to and an equal participant with any sort of non-designer scientific/naturalistic account because *they claim that there is no substantive way of demarcating science from non-science* (this is known as **the demarcation problem**). From this standpoint, **IDC** champions can either claim that their theistic propositions are "scientific enough" or that the inability to demarcate science from other non-scientific theoretical perspectives allows for the legitimacy of other non-scientific theoretical perspectives, namely their intelligent design hypothesis. It follows, think the **IDC** proponents, that science is *just* another theoretical perspective, like the intelligent design perspective, because science cannot demarcate itself as legitimately peerless.

Why do defenders of the **IDC** think that there is no legitimate way to circumscribe science from other non-science practices? **IDC** proponents have latched onto the works of Larry Laudan (1983) and Philip Quinn (1984) as part of their answer to this question. As Laudan explains, philosophers have taken up the task of locating the epistemic (i.e., knowledge) elements that distinguish science from non-science. Philosophy's track record, according to Laudan and others (e.g., Resnik, 2000), has been pretty dismal. This leads Laudan to conclude that "it is probably fair to say that there is no demarcation line between science and non-science, or between science and pseudo-science, which would win assent from a majority of philosophers. Nor is there one which *should* win acceptance from philosophers or anyone else ..." (Laudan, 1983, p. 112).

To make sense of all this, we must remind ourselves that philosophers have been trying to tackle this demarcation problem since at least the beginning of the twentieth century and likely as far back as ancient times, when Greek philosophers were pondering the nature of the universe. For our purposes let us focus on the history relevant to the **IDC**'s position. Specifically, this history (1) weaves through the work of the logical positivists' verifiability criterion of meaning and then more specifically (2) through Popper's falsification standard, and ultimately moving forward to (3) Lauden's strong proclamation that there is no way to distinguish science from non-science. Let us briefly turn to each of these.

In the 1920s, a group of philosophers, mathematicians, and scientists known as the "Vienna Circle" decided it was high time that philosophy should start mirroring the ways of science. According to many members of this group, (e.g., Carnap, Schlick, Ayer, and Neurath), a criterion for the meaningfulness of statements needed to be constructed to fend off what they perceived to be a burgeoning and obscure set of ideas developing within metaphysics, phenomenology, and psychology that was starting to take hold in philosophy.[15] In response, these scholars spearheaded a movement known as *logical positivism*. According to these logical positivists, the only

---

15 No doubt, it is a bit unfair to ignore the subtle differences amongst the logical positivists. I would direct the reader to Uebel's (2016) discussion where these differences are more carefully explicated.

kinds of statements that are meaningful are those that could be empirically verified (e.g., All swans are white) or those that are analytic (e.g., All triangles have three sides). In brief, the *verification principle* is the view that a statement X is meaningful, if and only if, X can be either empirically verified or is analytic. In terms of science vs. non-science, statements that can satisfy the verification principle are both scientific and meaningful, while statements that cannot satisfy this principle are deemed non-scientific and meaningless.[16]

In practice, the verification principle was expanded to a formal logical model known as "the covering law" model or "the deductive-nomological" model. This model basically captures the belief that in order to explain meaningfully an event, it must be governed by a set of laws and initial conditions. The schema looks like this:

There is a set of laws **L** that make sense of event **E**.

↓

There are a set of initial conditions **C** that are empirically related to **E**.

↓

If **C**, then it follows that **E** occurs.

From the above set-up, the logical structure of the positivists covering law model is in the form of *modus ponens* (If p, then q) as follows:

P1. If **C** occurs, then it follows that **E** occurs *and*, in general, this holds for any instance of **E** given that any corresponding instance of **C** occurs.

P2. **C** occurs.

C1. So, **E** occurs *and*, in general, this holds for any instance of **E** given that any corresponding instance **C** occurs.

For example, consider the statement: "If a pen is released from your hand, then the pen moves downward." The event, **E**, under consideration is "the pen moves downward"; the initial conditions, **C**, are "pen released from your hand"; finally, the physical laws, **L**, governing gravity and any other related physical laws are included to account for the event necessarily occurring. The more general thesis that emerged from this sort of reasoning is that it is possible to defend the truth of particular

---

16 For further discussions regarding the history and philosophy of logical positivism, see Beckwith (2011), Richardson and Uebel (2007), Friedman (1999), and their citations.

generalized statements by way of a set of empirical data. In our example, instances of pens released from hands would count as the set of empirical data from which follows that pens, in general, move downward. It is this sort of deductive logical structure that the verification principle of logical positivism included and accounts for what logical positivists think is the correct form of meaningful and scientific reasoning.

At this point, one might be wondering: what is the connection between logical positivism, the covering law model, and the demarcation problem? Also, one might be trying to understand the connection between all of this and the intelligent design discussion. Briefly, the demarcation problem started to be treated as a problem once it was clear that what counts as genuine scientific explanations from those that are not genuine scientific explanations was based on the covering law version of the verification principle. So, if the verification principle could be shown to be defective, then the positivists' covering law verification principle, as a justification of what counts as genuine scientific queries (from those that are not), could be rejected. This sort of critical reply partly captures the response of the IDC defenders in their attempt to discredit the view that there is a distinctly genuine theoretical scientific perspective. All of this will become clearer in the critical reply to logical positivism.

Logical positivism met with a barrage of criticisms.[17] One glaring concern is that any number of finite observations (however abundant!) cannot validate any general conclusion. For instance, it may very well be the case that every hand-dropped pen I have observed falls downward (let us say I have seen this occur a million times), but I cannot conclude that it is necessarily the case that all such objects in similar circumstances fall downward. Similarly, you might observe a million white swans, but it does not necessarily follow that all swans are white (go check the southwest regions of Australia!). *The point is, and this is what tells against the verifiability principle, that generalized conclusions from a particular data set cannot be verified, rendering such conclusions meaningless.*[18] So, the very principle that governs logical positivism is the very principle that led, in part, to its demise.

Partly in response to the verification principle put forth by the logical positivists, Karl Popper (1963) developed his *criterion of falsifiability* to assist in resolving the demarcation problem.[19] According to Popper, a theory, statement, or hypothesis X is scientific, if and only if, there is, in principle, empirical evidence that could refute

---

17 See Bechtel (1988, chaps. 2 and 3) for a good introduction to logical positivism and a corresponding set of criticisms.

18 For more on this difficulty of producing a workable principle of verifiability to distinguish sense from nonsense, see Ayer (1946).

19 Popper rejected the positivists' belief that non-falsifiable statements or hypotheses are meaningless. Rather, he was adamant that his criterion of falsifiability did not imply the meaninglessness of non-falsifiable statements or hypotheses. See Preston (1994) for more on Popper's reply to the positivists and for a more comprehensive account and critique of Popper's epistemological views, see Naydler (1982).

the truth of X. Put in a way to capture Popper's sense of whether or not a hypothesis is scientific,

> If hypothesis **H** is scientific, then it must be empirically possible for the purported set of predictions **P** of **H** to be false based on a countervailing data set **D**.

For example, 'No zebras are entirely white' is a scientific statement since a single instance of an entirely white zebra could refute this statement. In contrast, consider some of the statements generated within Freudian psychological theory. For instance, 'Repression of the Id explains why the man jumped into the water to save the child' is a statement that cannot be falsified because, on the assumption that Freudian concepts are being employed, any attempt to refute the statement results in a counter Freudian reply—a reply that relies on an alternative Freudian concept from within the theory. To illustrate, if evidence is put forward that repression of the Id is not likely relevant to the man who saved his child, then a Freudian-type response would be that the man may have been expressing some sort of *reaction formation* such that he was conveying bravery as a way to combat his life-long struggle with cowardice. Notice that there is no way to falsify the Freudian theory because there is always an alternative rendering to save the theory. Thus, on the Popperian criterion of falsifiability, many (if not all) hypotheses derived from Freudian psychoanalysis are not scientific.

As insightful as Popper's own resolution might seem, it too runs into difficulties. Two of these road blocks are worth mentioning here. First, his *criterion of falsifiability* rules out many theories or statements which seem intuitively to be scientific despite the fact that they cannot be directly falsified. Classic examples include theoretical statements in evolutionary biology (Resnik, 2000, p. 254) and physics (Sarkar, 2011, pp. 295–96). The former relies on events that occur on geological time scales/history that we cannot directly disconfirm, while the former might invoke entities (e.g., strings or parallel universes) that we cannot directly controvert. This suggests that Popper's *criterion of falsifiability* does not provide a necessary condition for what constitutes a scientific statement or theory.[20]

Second, recall that Popper's criterion of falsification requires that each statement or hypothesis (or theory or law) has to meet the standard of observational/empirical evidence and that this is done on a one-on-one basis. In other words, observational evidence **O** tests only the falsifiability of a particular law, theory, statement, or hypothesis **H**. Now, in response to this sort of *atomic view* of the relationship between evidence and hypothesis, Quine (1951) argued that it is actually a web of

---

20 Because of reliance on "deep/geological history," a history that does not allow for the empirical testability of many of its claims, some have argued that evolutionary theory is not really a science. Indeed, it seems that Popper struggled with evolutionary theory for this and other reasons (see Stamos, 1996, Bradie, 2016, and Elgin and Sober, 2017). A full account and review of Popper's philosophy of science cannot be offered here. For further discussions, see Johannson (1975), Godfrey-Smith (2016), and their citations.

beliefs—that includes H and its auxiliary assumptions and hypotheses—that is being tested by O. This is known as *holism*. To the extent that holism is correct, one further product is that tests are actually testing not only H, but also H's related auxiliary assumptions and hypotheses *as a whole web of beliefs*—not individual hypotheses (this is called the Quine-Duhem thesis). From this holism claim, Quine went on to explain that it cannot be fully determined if O actually confirms or refutes H or H's auxiliary assumptions or hypotheses. It is this indetermination that allows modifications to H's auxiliary assumptions that allows for the salvation of any theory from refutation. For example, consider the hypothesis "Dogs originally evolved from wolves." A population geneticist might think that this hypothesis can be tested by comparing specific genetic markers at specific loci of both wolf DNA and dog DNA. According to Quine, however, there is a whole host of beliefs held by the population geneticist that are also under scrutiny beyond this explicit hypothesis. For instance, other ideas related to evolutionary biology, like genetic drift, random mutation, epigenetic rules, etc., which are part of the population geneticist's conceptual tool kit, are all part of the web of beliefs that are brought under the "testability microscope." Additionally, if Quine is correct, any of these auxiliary beliefs could be the relevant target of the collected data regarding the evolution of dogs from wolves—not the primary hypothesis. The implication of all of this, to the extent that it is true, is that Popper's atomic view of falsifiability is dashed by Quine's holism because no single H can ever really be falsified.

It is in response to this history that we can begin to make sense of Laudan's critical pessimism regarding the possibility of resolving the demarcation problem and why the IDC defenders latched onto Laudan's critical assessment. Initially, Laudan observes that attempts to settle the demarcation problem rely upon accepting that a legitimate resolution must satisfy both necessary and sufficient conditions (p. 118)—this is the standard employed by both the logical positivists and Popper. Yet, Laudan shows that this standard is near-impossible to meet. For instance, the painstaking work by physicists in the area of string theory might not count as science because, although their efforts might meet some set of sufficient criteria to count as a science, they may not have adequate empirical evidence as part of their various hypotheses; that is, their hypotheses are currently not empirically testable even if they are mathematically plausible. From this perspective, if string theory is allowed to count as a science, then one could imagine that palm reading and astrology (both of which also lack empirical evidence) should get to count as sciences as well. (Imagine telling your loved one that "the movements of Venus or the meandering angles of the beauty line on your palm are the causes of the details of your future loving relationships"!) So, it is possible for a theory to have met a set of sufficient conditions to count as a science, but it may not meet all the necessary conditions to count as a science (and vice-versa). The upshot is that attempts to satisfy the necessary-and-sufficient approach to the demarcation problem will almost always run the risk of either letting too many theories count as scientific or not letting enough theories count as scientific depending upon the criteria involved. It is this difficulty that motivates Laudan to conclude that there is no rational way to distinguish between scientific and non-scientific theories (Laudan, 1983, p. 349).

My students are quick to reply that we need not be wedded to this broad necessary-and-sufficient-conditions approach to resolve this matter (i.e., they think it is better not to look for necessary and sufficient conditions). Surely, they insist, there must be a better framework to advance the view that there is a clear difference between scientific and non-scientific theories. The reply to the students is this is precisely what has happened, but the results do not appear much better. Briefly, in what can be viewed as further vindication of Laudan's pessimism regarding the resolution of the demarcation problem, Resnik (2000) has examined and critiqued four alternative approaches to the general necessary-and-sufficient-conditions tactic: (1) Historical, (2) Sociological, (3) Psychological, and (4) Epistemological. His conclusion is that all of these approaches also suffer from not being able to meet necessary and/or sufficient conditions for what counts as scientific.[21]

In terms of what counts as a legitimate explanation of the origin and development of life on Earth, we should start to see why the **IDC** thinks that it can claim a place at the "legitimate-explanation-table." Basically, following the likes of Laudan, Quinn, Plantinga (1991), Resnik and others, **IDC** proponents (e.g., Dembski, 1994) have argued that the naturalism of science does not get to rule-out the supernaturalism of **ID** because there is no genuine way to distinguish so-called science-type explanations from non-science explanations. So, even if it turns out that the empirical testability arguments of young-Earth creationists, Behe, Dembski, and others falter in the light of scientific scrutiny (and they have!), **IDC** stalwarts can merely shift the discussion away from the empirical testability standard and focus on some other aspect of **IDC**'s project regarding the origin and development of life on Earth *and* still insist that their accounts are just as reasonable as those put forth by the scientific community.

Re-Enter Pennock: Methodological Naturalism and Epistemic Values

In an attempt to re-establish the importance of scientific practices in the wake of this assault by the **IDC**, Pennock (2011) begins by rejecting the necessary-and-sufficient-conditions requirement embraced by the logical positivists, Popper, Laudan, and members of the **IDC**. Having made clear that the necessary-and-sufficient-conditions standard need not be the benchmark by which science and non-science practices are demarcated, Pennock moves on to re-institute the derived foundational indispensability of methodological naturalism (**MN**). Much like his response to Johnson, Pennock here stresses that the naturalism assumption within science need not be understood as some dogmatic metaphysical assumption (see Plantinga 1991). Rather, he claims

---

21 It is worth noting that Resnik (2000, pp. 260–65) does offer a "practical" resolution to the "deep" demarcation problem. This alternative will not be explored here, but his attempt is surely worthy of a critical essay by a curious student.

that the brand of naturalism embraced or assumed within science is of the **MN** sort and this commitment has its origin in the basic value of embracing the authority of empirical evidence (Pennock, 2011, p. 199). What this means is that, either implicitly or explicitly, **MN** is accepted by the scientific community as a result of accepting a more fundamental (epistemic) value associated with the authority of empirical evidence. Importantly, and here is where Pennock thinks he avoids the charge of dogmatism (e.g., see Plantinga, 1991), the results of embracing **MN** have proven successful. So, in effect, Pennock's account, as a reply to the **IDC**, looks like this:

P1. If theory **X** is to count as scientific, then it must endorse the epistemic values of empirical evidence and **MN**.

P2. The theory of intelligent design does not endorse the epistemic values of empirical evidence and **MN**.

C1. Thus, intelligent design does not count as scientific.

Why does Pennock think **P1** is true? His answer is in terms of success. He thinks that an integral part of a defense of endorsing any set of epistemic values is the results that such a set produce regarding the likely truth of a given hypothesis. From this perspective, endorsing the epistemic value of the need to provide empirical evidence for a given hypothesis is the product of proven success. As Pennock puts it, "it works." Correspondingly, the epistemic value of **MN** is and has been very successful and is why it is a near-inexorable value within the field of science. It further follows that if Pennock's explanation of what is part of the core values of the scientific enterprise is on the mark, then we have a way of demarcating science from non-science that need not leave us stuck with Laudan's pessimism. This is not to suggest that Pennock has offered a knock-down way of resolving the demarcation problem, but that certain core values of science must be present to be in the-let's-play-science-game (Pennock, 2011, p. 184).[22]

Given Pennock's account, since the theory of intelligent design does not put forth a theory or set of hypotheses that can be corroborated by the epistemic values of empirical evidence it is not a scientific theory. The upshot of Pennock's account is that the combination of both the epistemic values of empirical evidence and **MN** prohibit the **IDC** from bringing its theory of intelligent design to the "legitimate-scientific-explanation-table" because such accounts cannot be substantiated by empirical evidence.

---

22 Bradie (2009, p. 135) appears to be less confident about such moves as the one made by Pennock here. Although he is sympathetic to **MN**, he hints at a potential circularity problem in that if the value of empirical evidence is already built into **MN**, then **MN** cannot really be derived from the value of empirical evidence as suggested by Pennock. This suggests that Pennock has not evaded the dogmatism worry. I suspect that there is an interesting essay here for the relentless student.

Without hesitation, **IDC** members could respond that Behe and Dembski have offered hypotheses that have come under scientific scrutiny. This would suggest that **P2** of the reconstructed argument on the previous page is false. So, contrary to Pennock's claims, it is possible to offer an intelligent design account that could satisfy his core values of science. Such accounts might be deemed bad scientific theories, but they are scientific nonetheless. Thus, even if **P1** of the above argument is true, **P2** is false. The pressing implication is that Pennock's argument is rendered unsound.

The problem with this sort of response is that it is really a slight of hand move. Recall that Behe's irreducible complexity (and the same goes with Dembski's "complex specified information") relies upon traditional scientific accounts not being able to explain certain phenomena (recall the flagellum example). When an answer is not immediately forthcoming, Behe (and Dembski) tries to fill in this gap with an intelligent designer. The problem, however, is that there is no independent empirical evidence to substantiate this positive account; that is, everyone could agree that current scientific research cannot explain phenomenon X. Still, it does not follow that the lack of a scientific explanation regarding X warrants the vindication of an intelligent design explanation. Indeed, others have pointed out (e.g., Sarkar, 2011) that such moves by the **IDC** are not really explanations at all—they are merely gap fillers and do not count as playing the game of science at all. As Bradie eloquently makes the point, appealing to supernatural interventions "contributes nothing to our understanding of the phenomenon in the sense that our expectations with respect to experiential results will not be changed in any way." (Bradie, 2009, p. 136)

It is worth pointing out that another epistemic value embraced by the scientific community that is not endorsed by the **IDC** is: *epistemic humility*. This is the value that, *when a particular aspect about phenomenon X cannot be explained after existing scientific techniques and empirical resources have been exhausted, it is eminently reasonable to claim ignorance about this aspect of X*. Notice that epistemic humility should be viewed as an integral part of **MN**, eschewing any metaphysical or poorly substantiated or speculative alternatives. Rather, work on understanding this aspect of X should continue in the hopes that new insights, evidence, and/or instrumentation will be forthcoming. So, along with empirical evidence and **MN**, an integral epistemic value that points toward a theory being scientific is epistemic humility. This endorsement of epistemic humility stands in sharp contrast to Dembski's radical proclamation that "*science cannot plead momentary ignorance which it hopes someday to redress. When its empirical resources are exhausted, science is in no position to distribute promissory notes*" (Dembski, 1994, p. 91; my italics).

The italicized passage above reveals Dembski's flat-out rejection of epistemic humility. He goes on to proclaim boldly not only that intelligent design is preferable when science has exhausted all of its resources on a particular problem, but also that the intelligent design alternative is scientific (Dembski, 1994, p. 92). Yet, if this analysis is correct, there is nothing scientific about Dembski's repositioning here. It is this type of comment that leads scholars like Pennock to insist that **P2**—the theory of intelligent design does not endorse the epistemic values of empirical evidence and

MN—is true. Given his rejection of the epistemic values of empirical evidence, **MN**, and humility, there is very little that is scientific about his alternative offering of intelligent design—indeed, it seems he is not even in the science ballpark! Although Pennock may likely not have the last word on the demarcation problem initiated by the logical positivists, Popper, and Laudan, he thinks he has moved the discussion in a direction that reveals that this problem may not need to be resolved in order to address the claims made by many of those in the **IDC**. *Science to Intelligent Design Movement*: "*Hit the road Jack, and don't ya come back no more, no more, no more, no more!*"

VI. CONCLUSION

The evolution and intelligent design debate is one of many topics in the evolution and religion discussion. Like the previous chapters of this book, I only scratch the surface of the many sub-topics related to the evolution and intelligent design debate in this chapter.[23] Specifically, a glimpse into the efforts of both Johnson's and Behe's accounts and some of the counter-replies to their arguments has been provided. The preliminary findings here are that the arguments put forth by Johnson and Behe are not very persuasive. Students are strongly encouraged to explore further these and related discussions. Additionally, the demarcation problem was broached, suggesting that Pennock offers a *prima facie* cogent reply to some of the **IDC** proponents' belief that intelligent design can be part of scientific discussions. Although one might not endorse Kitcher's contentious conclusion that the **IDC** "is out of new resources. For all the fancy rhetoric, all the academic respectability, all the accusations and gesticulations, born-again creationism is just what its country cousin was. A sham" (Kitcher, 2001, p. 287), one would be quite foolish to bet the farm on the success of its defense of **The Argument from Design**. We might be better off trying to learn more about gnats and men ... *Intelligent Design and Evolutionary Biology*: "*And he hated himself and hated her, too, for the ruin they'd made of each other*" (Dennis Lehane, *The Given Day*).

CHAPTER REVIEW: DISCUSSION AND QUESTIONS

1. Drawing on the introduction and Gould's "twig" comment on p. 278, speculate as to why contemporary religion is able to tolerate the Earth not being the center of the universe, but is unable to accept that humans share an evolutionary history with other animals.
2. Explain Paley's **The Argument from Design** argument. How do you make sense of "Paley's Incredulity"? [pp. 279–80]

---

23 For example, I have not tackled the "complex specified information" argument put forth by Dembski (1995 and 2002). See Perakh (2004a and 2004b) and Sarkar (2011) for replies to Dembski's account.

3. In section III, "Darwin's dismay" is discussed. [pp. 280–81] What sort of reasoning does Darwin employ in the gnats and men quotation to try and refute Paley-like design arguments? How is Table 8.1 used to support Darwin's position? [pp. 281–82]
4. What is **Metaphysical Darwinism** and why does Johnson reject it? Make sure your answer includes the difference between knowledge claims and metaphysical claims. [p. 283]
5. Why does Johnson oppose naturalism as a "closed system" theoretical framework? How does he use **The Argument from God-as-Designer-Interactor** to refute naturalism/ Metaphysical Darwinism? Use **steps 1–3** as part of your answer. [pp. 283–85]
6. On p. 286, a reply to **The Argument from God-as-Designer-Interactor** is offered in the form of "knowledge as a double-edged sword." What is this objection and are you persuaded by it?
7. Explain how Pennock uses **Methodological Naturalism** to refute **Metaphysical Darwinism**. Then, explain how this refutation is used to refute both **The Argument from God-as-Designer-Interactor** and **The Argument from Design**. How does Pennock's reliance upon the fallibility of science assist him in his overall critique? [pp. 287–88]
8. What does Behe mean by 'irreducible complexity'? Give an example he uses to help make your explanation clear. [pp. 288–89]
9. Why does Behe think that irreducibly complex entities count as a legitimate answer to Darwin's challenge and support the reasonableness of intelligent design (and **The Argument from Design**)? Use **steps 1–3** as part of your answer. [p. 289]
10. Use the flagellum example to show the reasonableness of Behe's argument. Make sure your answer makes clear why evolution by natural selection should be rejected. [pp. 289–90]
11. How does Miller use the TTSS to argue against Behe's use of the flagellum example? Why couldn't a hierarchy of irreducibly complex entities existing together be a reasonable reply to Miller? [Make sure your answer reveals the contradiction that awaits Behe if he were to make this hierarchy move.] [pp. 290–91]
12. Why doesn't a lack of "step-by-step" details of evolutionary features support Behe's alternative irreducible complexity account? Your answer should include the strides made by science regarding the evolutionary details of biological features. [pp. 291–92]
13. Upon reflection, what additional points could you make to suggest a more sympathetic view of Johnson's and/or Behe's arguments in defense of **The Argument from Design**? What points could you offer to suggest a more skeptical view of Johnson's and/or Behe's arguments in defense of **The Argument from Design**?
14. What is the demarcation problem? [p. 294]
15. Explain the historical transition from (1) the positivists to (2) Popper and then to (3) Laudan. [pp. 294–99]
16. Why does the **IDC** think it can use the demarcation problem to its benefit? [p. 299]
17. What is Pennock's response to the **IDC**'s attempt to get into the science game? How does Pennock use methodological naturalism and epistemic values to do this? [Make sure your answer includes the epistemic values of empirical evidence, **MN**, and humility.] [pp. 299–302]
18. To what extent are you persuaded by Pennock's reply to the **IDC**?

REFERENCES

Ayala, Francisco J. "There Is No Place for Intelligent Design in the Philosophy of Biology: Intelligent Design Is Not Science." In *Contemporary Debates in the Philosophy of Biology*, ed. Francisco J. Ayala and Robert Arp, 364–90. West Sussex, UK: Wiley-Blackwell, 2010.

———. "Design without Designer: Darwin's Greatest Discovery." In *Debating Design: From Darwin to DNA*, ed. Michael Ruse and William H. Dembski, 55–80. Cambridge: Cambridge University Press, 2004.

Ayer, A.J. *Language, Truth, and Logic*. New York: Dover Press, 1946.

Bechtel, William. *Philosophy of Science: An Overview for Cognitive Science*. London: Psychology Press, 1988.

Beckwith, Burnham P. *Religion, Philosophy and Science: An Introduction to Logical Positivism*. Whitefish, MT: Literary Licensing, 2011.

Behe, Michael J. *The Edge of Evolution: The Search for the Limits of Darwinism*. New York: Free Press, 2007.

———. "Irreducible Complexity: Obstacle to a Darwinian Evolution." In *Debating Design: From Darwin to DNA*, ed. Michael Ruse and William H. Dembski, 352–70. Cambridge: Cambridge University Press, 2004.

———. "The Challenge of Irreducible Complexity." *Natural History* 111 (April 2002): 74.

———. *Darwin's Black Box: The Biochemical Challenge to Evolution*. New York: Free Press, 1996.

Blocker, Ariel et al. "Type III Secretion Systems and Bacterial Flagella: Insights into Their Function from Structural Similarities." *Proceedings of the National Academy of the Sciences of the United States of America* 100, 6 (2003): 3027–30.

Bradie, Michael. "Karl Popper's Evolutionary Philosophy." In *Cambridge Companion to Popper*, ed. Jeremy Shearmur and Geoffrey Stokes, 125–42. Cambridge: Cambridge University Press, 2016.

Cooper, Max D., and Brantley R. Herrin. "How Did Our Complex Immune System Evolve?" *Nature Reviews Immunology* 10, 1 (Jan. 2010): 2–3.

Darwin, Charles. *The Descent of Man*. 2nd ed. In *From so Simple a Beginning: The Four Great Books of Charles Darwin*, ed. Edward O. Wilson. New York: W.W. Norton & Co, 2006 [1871].

———. *Autobiography of Charles Darwin*. New Delhi: Rupa & Co., 2003.

———. *The Origin of Species*. 6th ed. New York: Penguin Books, 1958 [1872].

———. "Letter to Asa Gray." July 1860. http://darwiniana.org/religion.htm.

Dembski, William. *No Free Lunch: Why Specified Complexity Cannot Be Purchased without Intelligence*. Lanham: Rowman & Littlefield, 2002.

———. *The Design Inference: Eliminating Chance through Small Probabilities*. Cambridge: Cambridge University Press, 1995.

———. "The Incompleteness of Scientific Naturalism." In *Darwinism, Science or Philosophy?*, ed. John Buell and Virginia Hern, 79–98. Richardson: Foundation for Thought and Ethics, 1994.

Diepold, Andreas, and Samuel Wagner. "Assembly of the Bacterial Type III Secretion Machinery." *FEMS Microbio Rev* 38, 4 (2014): 802–22.

Doolittle, R.F. "Step-by-Step Evolution of Vertebrate Blood Coagulation." *Cold Spring Harbor Symposia on Quantitative Biology* 74 (2009): 35–40.

Elgin, Mehmet, and Elliott Sober. "Popper's Shifting Appraisal of Evolutionary Theory." *HOPOS* 7 (Spring 2017).

Flajnik, Martin F., and Masanori Kasahara. "Origin and Evolution of the Adaptive Immune System: Genetic Events and Selective Pressures." *Nature Reviews Genetics* 11, 1 (Jan. 2010): 47–59.

Friedman, Michael. *Reconsidering Logical Positivism*. Cambridge: Cambridge University Press, 1999.

Garvey, Brian. *Philosophy of Biology*. Montreal and Kingston: McGill-Queen's University Press, 2007.

Godfrey-Smith, Peter. "Popper's Philosophy of Science: Looking Ahead." In *Cambridge Companion to Popper*, ed. Jeremy Shearmur and Geoffrey Stokes, 69–103. Cambridge: Cambridge University Press, 2016.

Gould, Stephen Jay. *Wonderful Life: The Burgess Shale and the Nature of History*. New York: W.W. Norton, 1989.

———. "The Meaning of Life." *Life*, December 1988, p. 84.

Graham, Peter J. "Intelligent Design and Selective Purpose: Two Sources of Purpose and Plan." In *Oxford Studies in Philosophy of Religion*, vol. 3, ed. Jonathan L. Kvanvig, 67–88. Oxford: Oxford University Press, 2011.

Janeway, Charles A., Paul Travers, Mark Walport, and Mark Shlomchik. *Immunobiology*. 5th ed. New York: Garland Publishers, 2001.

Johannson, I. *A Critique of Karl Popper's Methodology*. Stockholm: Scandinavian University Books, 1975.

Johnson, Phillip E. "Evolution as Dogma: The Establishment of Naturalism." In *Intelligent Design Creationism and Its Critics*, ed. Robert T. Pennock, 59–76. Cambridge, MA: MIT Press, 2001a.

———. "Creator or Blind Watchmaker." In *Intelligent Design Creationism and Its Critics*, ed. Robert T. Pennock, 435–49. Cambridge, MA: MIT Press, 2001b.

———. *Darwin on Trial*. Washington, DC: Regnery Gateway, 1991.

Kitcher, Philip. "The Trouble with Scientism." *The New Republic* 243, 8 (May 24, 2012): 20.

———. "Born-Again Creationism." In *Intelligent Design Creationism and Its Critics*, ed. Robert T. Pennock, 257–87. Cambridge, MA: MIT Press, 2001.

———. *Abusing Science: The Case against Creationism*. Cambridge, MA: MIT Press, 1983.

Langford, Jerome J. *Galileo, Science and the Church*. 3rd ed. Ann Arbor: University of Michigan Press, 1992.

Larson, Edward J. *Summer for the Gods*. Cambridge, MA: Harvard University Press, 1997.

Laudan, Larry. "The Demise of the Demarcation Problem." In *Physics, Philosophy, and Psychoanalysis*, ed. R.S. Cohen and L. Lauden, 111–27. Dordrecht: D. Reidel Publishing Company, 1983.

Liu, Renyi, and Howard Ochman. "Stepwise Formation of the Bacterial Flagellar System." *Proceedings of the National Academy of the Sciences of the United States of America* 104, 17 (2007): 7116–21.

Marks, Jonathan. "The Awe-Inspiring 'Divine Beauty' of Flagellar Assembly." http://www.evolutionnews.org/2011/08/the_awe-inspiring_beaut049211.html.

Miller, Kenneth R. "The Flagellum Unspun." In *Debating Design: From Darwin to DNA*, ed. Michael Ruse and William H. Dembski, 81–97. Cambridge: Cambridge University Press, 2004.

Nagel, Thomas. "Public Education and Intelligent Design." *Philosophy & Public Affairs* 36, 2 (2008): 187–205.

Naydler, Jeremy. "The Poverty of Popperism." *Thomist* 46, 1 (1982): 92–107.

Numbers, Ronald L. *Darwinism Comes to America*. Cambridge, MA: Harvard University Press, 1998.

Pennock, Robert T. *The Tower of Babel: The Evidence against the New Creationism*. Cambridge, MA: MIT Press, 2003.

———. "Why Creationism Should Not Be Taught in the Public Schools." In *Intelligent Design Creationism and Its Critics*, ed. Robert T. Pennock, 755–77. Cambridge, MA: MIT Press, 2001a.

———. "Reply to Plantinga's 'Modest Proposal.'" In *Intelligent Design Creationism and Its Critics*, ed. Robert T. Pennock, 793–97. Cambridge, MA: MIT Press, 2001b.

Perakh, Mark. *Unintelligent Design*. New York: Prometheus Books, 2004a.

———. "There Is a Free Lunch After All: Dembski's Wrong Answers to Irrelevant Questions." In *Why Intelligent Design Fails: A Scientific Critique of the New Creationism*, ed. Matt Young and Taner Edis, 153–71. New Brunswick: Rutgers University Press, 2004b.

Plantinga, Alvin. "Methodological Naturalism?" In *Intelligent Design Creationism and Its Critics*, ed. Robert T. Pennock, 339–61. Cambridge, MA: MIT Press, 2001a.

———. "Creation and Evolution: A Modest Proposal." In *Intelligent Design Creationism and Its Critics*, ed. Robert T. Pennock, 779–91. Cambridge, MA: MIT Press, 2001b.

———. "When Faith and Reason Clash." *Christian Scholar's Review* 21, 8 (1991): 8–32.

Popper, Karl. *Conjectures and Refutations*. New York: Harper and Row, 1963.

Preston, John. "Methodology, Epistemology and Conventions: Popper's Bad Start." *PSA: Proceedings of the Biennial Meeting of the Philosophy of Science Association* 1 (1994): 314–22.

Pross, Addy, and Robert Pascal. "The Origin of Life: What We Know, What We Can Know and What We Will Never Know." *Open Biology* (March 2013). http://rsob.royalsocietypublishing.org/content/3/3/120190.

Quine, W.V.O. "Two Dogmas of Empiricism." *The Philosophical Review* 60, 1 (1951): 20–43.

Quinn, P.L. "The Philosopher of Science as Expert Witness." In *Science and Reality: Recent Work in the Philosophy of Science*, ed. J.T. Cushing, C.F. Delaney, and G.M. Gutting. South Bend, IN: Notre Dame University Press, 1984.

Ratzsch, Del. "There Is a Place for Intelligent Design in the Philosophy of Biology." In *Contemporary Debates in the Philosophy of Biology*, ed. Francisco J. Ayala and Robert Arp, 343–63. West Sussex, UK: Wiley-Blackwell, 2010.

Reiss, Michael J. "How Should Creationism and Intelligent Design Be Dealt with in the Classroom?" *Journal of Philosophy of Education* 45, 3 (2011): 399–415.

Richardson, Alan, and Thomas Uebel. *The Cambridge Companion to Logical Empiricism*. Cambridge: Cambridge University Press, 2007.

Rosenberg, Alex. *The Atheist's Guide to Reality.* New York: W.W. Norton & Co., 2011.
Ruse, Michael. "Methodological Naturalism under Attack." In *Intelligent Design Creationism and Its Critics*, ed. Robert T. Pennock, 363–85. Cambridge, MA: MIT Press, 2001b.
Sarkar, Sahotra. "The Science Question in Intelligent Design." *Synthese* 178, 2 (2011): 291–305.
Scott, Eugenie C. *Evolution vs Creationism: An Introduction.* 2nd ed. Westport: Greenwood Press, 2009.
Shanks, Niall, and Karl Joplin. "Redundant Complexity: A Critical Analysis of Intelligent Design in Biochemistry." *Philosophy of Science* 66, 2 (1999): 268–82.
Sober, Elliott. "Evolution without Naturalism." In *Oxford Studies in Philosophy of Religion*, vol. 3, ed. Jonathan L. Kvanvig, 187–221. Oxford: Oxford University Press, 2011.
Stamos, David. *Evolution and the Big Questions: Sex, Race, Religion, and Other Matters.* Oxford: Wiley-Blackwell, 2008.
——. "Popper, Falsifiability, and Evolutionary Biology." *Biology and Philosophy* 11, 2 (1996): 161–91.
Uebel, Thomas. "The Vienna Circle." *Stanford Encyclopedia of Philosophy*. 2016. https://plato.stanford.edu/entries/vienna-circle/#VieCirHis.
Weber, Bruce H., and David J. Depew. "Darwinism, Design, and Complex Systems Dynamics." In *Debating Design: From Darwin to DNA*, ed. Michael Ruse and William H. Dembski, 173–90. Cambridge: Cambridge University Press, 2004.
Witham, Larry A. *Where Darwin Meets the Bible.* Oxford: Oxford University Press, 2002.

INDEX

aboutness, 114–15
adaptation, xiv, 87, 108, 143, 272
Adaptation and Natural Selection (Williams), 86
agency
  argument from, 235
  objection from, xvii, 220, 224–26
Aizawa, Shin-Ichi, 292
allele combinations, 86
alleles (replicators), 100
altruism, 88–89, 91–92, 211–12, 217, 219, 233. *See also* other-regarding behavior
  reciprocal altruism, 218–20, 234–35, 247
altruism *vs.* selfishness, 106
altruistic motives, 236–37
analogy from artifact functions to organism functions, 126–27
Animal Dispersion in Relation to Social Behavior (Wynne-Edwards), 86
animal other-regarding and self-regarding behaviors, 211
animal social behavior, 217
animal-social-behavior argument, 226
animals other than man, 21
anthropocentric objection to final causation, 22–25
anti-evolutionary naturalistic concept of function, 130
Argument from Arbitrariness, 170, 177

Argument from Artificial Selection. *See* domestic breeding analogy in defense of natural selection
Argument from Causal Complexity, 101–02
Argument from Causal Efficacy, 169–70, 177
Argument from Design, 279–83, 285–86, 288, 293, 302
Argument from Domain-specificity, 256–60, 265–66
Argument from Facts and Values, 243–45
Argument from God-as-Designer-Interactor, 284–86, 288
Argument from Individuality, 185
  genetic-individual analogy, 186
  part-whole analogy, 186–87
Argument from Naturally Good Virtues, 241, 243
Argument from Progress, 227–28, 230–32
Argument from Time, 267–69, 270
Aristotle, xv–xvi, 3–5, 18, 126–27, 129
  artifact analogy, 21, 28–29
  as biologist and philosopher of biology, 1–5, 9, 11, 14–15, 22, 33
  causal analysis, 10–12
  distinguishing human animal from other animals, 211
  on divine thinking, 24
  four causes, 12–14
  four causes (table), 14
  Generation of Animals, 26

INDEX

Hippocratics' influence on, 4–5
History of Animals, 2–3, 8–9
Metaphysics, 3
mixed causal-empiricist methodology, 5
Parts of Animals, 9
philosophy of science, 5
Politics, 22–23
Posterior Analytics, 8
Prior Analytics, 8
theory of soul, 5–6
work in zoology, 1
Aristotle's conception of male semen, 30–31
Aristotle's final causation (teleology), 3, 15–33
 objections to (table), 32
Aristotle's God (The Unmoved Mover), 25
Aristotle's philosophy of biology (table), 14
Aristotle's Unmoved Mover, 278
Aristotle's virtue theory, 240
Arnhart, Larry, 240, 243–45
Arnhart's EE3 Project, 241–46
Arnhart's EE3 Project (table), 242
art-object analogy, 19
art-objects, goal-directedness in, 18
artifact functions, 126–27
artificial selection. *See* domestic breeding analogy in defense of natural selection

backward causation, principle of, 26
Barnes, Jonathan, 2
Beagle Voyage, 43, 59–60, 73
Bechtel, William, 150–51
Behe, Michael, xviii, 279, 288–92, 299, 301
Bigelow, John, 141, 146
biological altruism, 88–89, 92
biological classification, 162–63
biological function, xiv–xv, 15
 excludes conscious design or intention, 138
 goal of survival-and-reproductive success, 128
 should not include conscious design, 130

biological function, comparison of various theories (table), 156
biological function, four-fold criteria (table), 131
biological hierarchy (nested hierarchy of biological life), 108
Biological Species Concept, xvii, 165, 172–73, 182–85, 195
 nominalist challenge, 184–85
 reproductive isolation, 184
Biological Species Concept and the Four-Fold Species Criteria (table), 185
biotechnology, xiv
blade of grass, 40
Blocker, Ariel, 292
body as *primary* unit of selection, 100
Boorse, Christopher, 136–37
brain development, 259–60
Brandon, Robert, 104, 106–07
Buller, David, 255, 258, 260, 265–66
Buller's proliferate-and-prune thesis, 259–60
Burma, Benjamin, 170
Butch Cassidy and the Sundance Kid, 214, 225–26, 241, 244–45

Cambrian explosion, 67
Casebeer, William C., 240
categorization issue, 163
categorization of observed phenomenon, 10
Causal Complexity, Argument from, 101–02
Causal Efficacy, Argument from, 169–70, 177
causal efficacy condition, 174, 177, 179
cheating-detector mechanisms, 265, 271
cheating-detector mental modules, 257, 261, 264
common descent and natural selection, 59, 67–68
common sense function attributions, 132–34
"community" benefit/altruism, 91
compatibilism, 246

INDEX

competence, 169
completeness sense of priority, 26–27
connectedness of life via natural selection, 75
conscious design concept of function, 130
conservation biology, 166
contractual regulations, 234
conventional interactionist view, 111–15, 119
convergence argument, 230
cooperation, 76
Copernicus, 40, 278
Cosmides, Leda, 254–55, 264–66
covering law model, 295–96
Coyne, John, 196–97
creationism, 53, 283. *See also* God; Intelligent Design Camp (IDC)
critical thinking, xiii
culture, xiv
Cummins, Robert, 130, 132–35, 137–38, 142

Darwin, Charles, xvi, 128, 167, 254, 289–90
   analogy of war, 65
   argument from incredulity as response to Paley, 280–81
   Beagle Voyage, 43, 59–60, 73
   calculations about the age of the Earth, 56–57
   on community/group selection, 91
   defends individual as *primary* unit of selection, 89, 91–93
   defense of natural selection as alternative to a divine artisan, 53
   early life and bio-philosophical development, 41–43
   mixed causal-empiricist methodology, 42, 45, 68
   Origin of Species, xv, xvii, 40–45, 49–50, 54–55, 60, 73, 76, 78, 91, 130, 138, 278–79
   philosophy of science, 45–48
   praise for Aristotle's work in biology, 2–3
   on progress, 226
   on relationship between evolution and human morality, 212
   Strong Argument for Natural Selection, 87 (*See also* natural selection)
Darwinian perspective, 172
   other-regarding behavior is costly to the individual, 218
   on species as essentialist-Natural Kind, 177
Darwinian selectionist theory, 116
Darwinian-type teleology, 240
Darwin's "Argument from Artificial Selection," 63
Darwin's Defense of Natural Selection, Four Principles Supporting (table), 66, 282
Darwin's "highest-degree-probable" conclusion, 64
Dawkins, Richard, 86, 88–89, 93, 95, 97, 101–02, 104, 110
   as "Keeper of Genes," 87
   The Selfish Gene, 96
deductive-nomological model, 295–96
Dembski, William H., 299, 301
deontic conditional, 265–66
Design, Argument from, 279–83, 285–86, 288, 293, 303
determinism, 245–46
development, xiv
development sense of priority, 26–27
development systems theory (DST), 111–20
Developmental System and the Unit(s) of Selection (table), 115
Distortion Argument, 98–100
divine challenge to natural design, xvii, 52–53. *See also* creationism; God
DNA, 31, 85, 93–94, 114
DNA-sequence research, 133
domain-general modules, 169, 259–60
domain-general social learning processors, 258

domain-specific cheating-detector modules, 264–65
domain-specific mechanisms, 259. *See also* Argument from Domain-specificity
domain-specific mental modules, 259, 270
domain-specific modules, 260
domestic breeding analogy in defense of natural selection, xi, xvi, 60–65, 72–75
dual functionalism, 128, 140, 156

early-stage-to-late-stage-purposefulness. *See* sequential and developmental organization
Earth as center of the universe, 278
Earthly World, 23–25
efficient cause, 12–13
elements (humours) of the body, 4
embryology, 68–69
emotions, xiv
empirical evidence, 300
empirical testability standard, 299
entwined/fused levels of selection, 110
entwinement aspect of biological systems, 116
entwinement or fusion of properties, 111
environmental influences, xiv, 101, 169, 260
epistemic humility, 301–02
Ereshefsky, Marc, 177–78
Essay on the Principles of Population (Malthus), 43
essentialism, 172, 174
Essentialism/Natural Kind, 173
ethical theories and evolutionary ethics projects (table), 217
ethical theory, 214–16. *See also* moral theory
Etiological Evolutionary Functionalism, xvi, 140–41
  circularity charge against, 143–44
  replies to, 142–47
Etiological Evolutionary Functionalism, Four-Fold Criteria (table), 147
everyday ethics, 214–16, 232
evolution and intelligent design debate, 302

evolution and religion discussion, 302
evolution as inherently progressive, 230, 232
evolutionary adaptations, 271
evolutionary biology, xvii, 116, 254
evolutionary developmental biology, 117
evolutionary environment of adaptedness (EEA), 255, 270–71, 273
evolutionary ethics, xvii, 212–13, 217, 224–26, 232
evolutionary "facts," 216
evolutionary function attributions, 132–33
evolutionary functionalism, 139–40
evolutionary history of *Homo Sapiens*, 215
evolutionary progress, 226–28
evolutionary psychology, xvii, 254–55
"extinction by competition," 70

fear, 271, 273
fear mechanism as evolutionary adaptive trait, 271
fecundity, 47, 60
final causation. *See* Aristotle's final causation (teleology)
final cause, 12–13, 15
final/formal cause, 30
formal cause, 12–13
fossil record, 69–71
free choices, 236
free will, xiv
Freudian psychological theory, 297
functional analysis, criteria for, 129–30
functionalism, xvi, 130–39. *See also* mixed evolutionary functionalism
  dual, 128, 140, 156
  evolutionary functionalism, 139–40
functions, 125–28
  biological (*See* biological function)
  genuine functions and features that no longer possess genuine functions, 145–46

genuine functions and mere side effects, 151–52, 154
fusion or coalescence of forces/information, 111

Galileo, 278
Garvey, Brian, 197–98, 229
gene as *primary* unit of selection, xvi, 89–90, 93, 96–98, 110
  Gould's argument against, 101–02
general predictions (GP), 117–19
Generation of Animals (Aristotle), 26
genes, 87, 94, 98–99
genes (as replicators), 93, 100, 102
gene's-eye-view, 93, 95–96, 103, 117
gene's-eye-view and the unit(s) of selection (table), 97
genetic and epigenetic causal factors, 114, 116
genetic causal account, 103
genetic-individual analogy, 186
genetic level, natural selection at, 88
genetic sequencing, 85
genetic survivorship and replication natural selection favoring, 87
genetic variation amongst humans across the globe, 267
geographic isolation, 73–74
geologic time, Darwin's commitment to, xvi
Ghiselin, Michael, 88, 167, 169–70
goal-directedness, 127–28
  in nature, 18
  without referencing a designer, 19
God, xviii, 52, 278–79, 284–85
God-as-Designer hypothesis, 279. *See also* creationism
God-as-Designer-Interactor, Argument from, 284–86, 288
Godfrey-Smith, Peter, 149–50
Godly World, 23–25
'goodness' definition, 223
Gould, Stephen Jay, 100–02, 230, 278
  argument from mass extinction, 230–31, 236
  objection from progress, 238
  rejection of inherent progress, 229
Gould's Guillotine, 226–32, 238, 242
Gould's kaleidoscope causal complexity account, 102–03, 110, 113
Gray, R.D., 111, 116
Griffiths, Paul E., 99, 111–12, 116–17, 119
group as unit of selection, xvi, 86–88, 103, 109–10
group selection, 104, 218, 220, 242
  argument from heritability against, 105–07
  argument from speed against, 105–06
  group-level altruism that gives advantage over other groups, 219
  in Richards' Kantian/intention based evolutionary ethic, 237
group selection and the unit(s) of selection (table), 104

Hamilton, W.D., 86–87
haplodiploidy, 86
Hardcastle, Valerie, 141–44, 150
Henslow, John, 42
hereditary material, 85, 92. *See also* inheritance, strong principle of
heritability, 106–07
heritability element, 107, 272
heritable variations, 176
Herschel, John, 45–48
Herschel-Whewell criteria for a good philosophy of science, 49
Herschel-Whewell criteria for a good philosophy of science (table), 48
Herschel-Whewell philosophy of science, 45–48
Hippocratics, 4
History of Animals (Aristotle), 2–3, 8–9
Hochman, Adam, 112, 117, 119
Homo Sapiens, 215
Hull, David, 90, 185–90

human agency and control, 225. *See also* moral responsibility
human flourishing, 241
human genome project, xiv, 86
human mind, "blank state," 266
human mind as computer, 255
human mind comes "pre-packaged" with domain-specific modules, 266
human moral sense faculty, 233
human moral sense of justice, 234
human moral sensibility, 211
human other-regarding behavior. *See* other-regarding and self-regarding behaviors
human psychology, xvii
humans as merely one amongst the animals, 278
Hume, David, 221
Hume's Hatchet, 222–23, 235, 240, 245
Hume's is-ought objection, 221–22, 231–32

indetermination, 298
indicative conditionals, 265
indispensability condition, 174, 177, 179
indispensability features
 multiply realized in different species, 178
individual and the unit(s) of selection (table), 92
individual as *primary* unit of selection, 89, 91, 100, 107, 109–10
 Gould's arguments to vindicate, 101–02
individual organisms as unit of selection, xvi
Individuality, Argument from, 185–87
induction, 46
inductive reasoning (Darwin's), 49–50
inference to best explanation, 50, 137
Information Argument, 111–15
informational information, 114
inheritance, strong principle of, 54. *See also* hereditary material
inheritance, underlying biology of, 77
innate mental dispositions (IMDs), 233, 235–36

"innate" problem solving information, 258
"innate" universal nature specific to human beings, 256
Intelligent Design Camp (IDC), xviii, 278–79, 282–87, 291–94, 296, 299, 301–02
"intelligent design" creationism, 53
intelligent design discussion, 296
intentional causal information (i.e., "aboutness"), 115
intentional information, 114
intentional states, 19, 21
intentional value in biological processes, 116
intentionality, 19
interactor, 90, 95, 100
intrinsic teleology, 23
intuition, 134–35
irreducible complexity, 288, 292–93
Irreducibly Complex Systems, 289–91
is-ought fallacy, xvii, 221–22, 232
is-ought objection (naturalistic fallacy), 220
"is" to "ought," 235, 239–40
isomorphism, 200

Johnson, Phillip, xviii, 279, 283–88
justice, 233–35

kaleidoscope causal complexity account, 102–03, 110, 113
Kant, Immanuel, 40, 128–29
Kantianism, 215
kin and/or group selection account of human evolution, 239
kin selection, 218, 220, 237, 242, 247
Kitcher, Philip, 103, 213, 216–17, 246–47
Korsgaard, Christine, 225
Korsgaard's Katana, 224–26, 236, 240, 246

Lamarck, Jean Baptiste, 43, 58, 60, 77
language, xiv
language-argument from analogy

in reply to species-extinction argument, 190
Laudan, Larry, 294, 298–300, 302
Lewontin, Richard, 111, 143
life cycle epigenesis, 112–13, 115
life cycle epigenesis of DST, 113
life-force as final cause/soul, 31
logical positivism, 294, 296, 302
logical positivists' verifiability criterion of meaning, 294
Lyell, Charles, 56–60, 64, 66
  Principles of Geology, 43
  uniformitarian geologic principles, 56–57
Lyellian gradualism, 69

Main Argument of the *Origin of Species*, 49, 51–52, 55
  inference to the best explanation, 76
malfunctioning
  Mixed Evolutionary Functionalism's accommodation of, 155
  systemic functionalism and, 135–38
Malthus, Thomas
  Essay on the Principles of Population, 43
Malthusian Pressure Cooker, 64–65, 75–76
Marks, Jonathan, 292
"massive modularity of mind" thesis, 258
mate selection, 258, 268–69
material cause, 12–13, 16
Mayr, Ernst, 139–40, 164, 182–84
medicine, 3–4
meiosis, 98
Mendel, Gregor, 93
mental "cheating detectors," 241
mental modules, 256–57, 259, 271
meta-ethical account, 220
meta-ethics, 214–16, 232, 243
Metaphysical Darwinism, 283–84, 286–88
metaphysics, xiv, 168
Metaphysics (Aristotle), 3
methodological naturalism (MN), 288, 299–300, 302
Miller, Kenneth, 290, 292–93

Milliken, Ruth, 140
mixed account of moral theory, 237
mixed causal-empiricist methodology, 5, 68
mixed evolutionary functionalism, xvi, 152
  able to accommodate malfunctioning, 155
  able to distinguish genuine functions from side effects, 153
  endorses dual functionalism, 156
  rejection of conscious design, 155
  special weight to evolution and natural selection, 155
mixed evolutionary functionalism, four-fold criteria (table), 155
mixed species accounts, 173
"the modern synthesis," 85, 93
modularity of mind thesis, 258
monophyletic groups, 194
Moore, G.E., 222
  "is-ought" concern, 222
  open-question argument, 232
Moore's "but is that which is desired good?" query, 243
Moore's Machete, 235, 245
moral compatibilism, 245
"moral facts," 214
moral feelings, 234. *See also* human moral sense faculty
moral goodness, 216, 241
moral principles, 214
moral projects, evolutionary ethics projects, and selection mechanisms (table), 220
moral responsibility, 224–25
moral sense, 233–34, 246
  in aid of survival and reproductive success, 233
  possible evolutionary causes, xvii
moral theory, 220, 232. *See also* ethical theory
morality, 214, 221, 237
morality, three faces of, 219
morphology, 69
Morris, Conway, 228, 230–31

multi-level selection (MLS), 107–10, 116
multi-level selection (MLS) and the unit(s) of selection (table), 109
mutability of species (P2) of Main Argument, 49

natural habitat, 149
natural kind, 174–75
natural kind/essentialism species concepts, 172
natural selection, xviii, 40, 43–45, 48, 60, 128, 241
  argument from analogy to domestic breeding practices, 60–63, 65, 72–75
  favoring genetic survivorship and replication, 87
  gene level, 99
  group level, 86, 92
  highest degree probable claim, 49, 51, 62
  as probable, 49
  responsible for the tree-of-life (possibility), 61
  retention and weeding out for reproductive advantage, 77, 116
  selection of successive slight modification, 67
  as synonymous with principle of preservation, 54
Natural Theology (Paley), 43
naturalism, 283–84
naturalistic account of the nature of moral goodness, 241
naturalistic fallacy, 223–24, 235, 239–40, 243–45
The Naturalistic Fallacy, 221
naturalizing ethics
  objection from agency: Korsgaard's Katana, 224–26
naturalizing human morality, 211, 222
nature is able to maintain itself argument, 19–20
necessary-and-sufficient tactic, 299

"necessary-mechanistic ought," 239
"necessary-practical-ought," 239
nested hierarchy of life, 116, 172
neurological development
  "proliferate and prune" perspective, 259–60
Newtonian blade of grass, 40
Newtonian mechanics, 47
Newtonian Revolution, 40
Newton's Law of Cooling, 39, 46–47
nominalism, 168–69
non-human and human other-regarding behavior toward family members, 218
non-intentional final causation sequential unfolding in, 21
non-intentional-state example, 19
normative guidance account, 246–47
normativity, 221
normativity and biological function, 137, 146

objection from agency, xvii, 220
objection from agency: Korsgaard's Katana, 224–26
objection from progress, 238
objection to evolutionary progress, xvii, 220
objection to progressive evolution: Gould's Guillotine, 226–32
Objections to Naturalizing Morality/Normativity (table), 231
Ockham's Razor, 47, 101–02
Ogle, William, 2
Okasha, Samir, 103–04, 108–09
old-Earth (Darwin argued for), 56, 71
open-area species, 75
open-question argument, 223–24, 232
open-question objection, 220
open-question objection: Moore's Machete, 222–24
Origin of Species (Darwin), xv, 40–45, 54, 60, 73, 78, 91, 130, 278
  argument against divine-designer as cause, xvii

exclusion of divine or intentional context, 138
inferential reasoning, 50
Main Argument, 49, 51–52, 55, 76
refutes God-as-divine-designer belief, 279
tree-of-life conception, 55–56
Orr, H. Allen, 196–97
other-regarding and self-regarding behaviors, 211
other-regarding behavior, 106, 217–18, 236–37. *See also* altruism
   discarded in favor of self-regarding strategic moral sense, 212
other-regarding behavior toward non-kin
   evolutionary sense of, 219
other-regarding community-based intentional psychology, 237
ought, 239–40. *See also* is-ought fallacy
outlaw genes, 98–99

Paley, William, 52–53, 279, 282
   Natural Theology, 43
panpsychism criticism, 20–21
Pargetter, Robert, 141, 146, 148–51
parsimony, 47, 67, 101–02
Parts of Animals (Aristotle), 9
parts of animals have causal priority, 27
parts of individuals as *primary* units of selection
   Gould's argument against, 101–02
parts of organisms
   functional role in the developing organism, 27
Pennock, Robert, 299–300, 302
Pennock's reply to Johnson, 287–88
Perlman, 135–38
Phenetic Species Concept, 172, 179–82
   able to range over all life forms, 181
   nominalist challenge, 181
Phenetic Species Concept and the Four-Field Species Criteria (table), 182
pheneticism, xvii, 165, 173, 195

phenotype (interactor), 100
phenotypes, 101
philosophy of biology, xiv, 15
philosophy of evolution, xiv
philosophy of science, xiii, xiv, 5
phylogenetic analysis, 133
Phylogenetic-Cladistic Species Concept, xvii, 165, 172–73, 192–93
   evolutionary perspective, 198
   monophyletic groups, 194
   nominalist challenge, 198
   reliance on monphyly, 197
   reply to, 195–98
Phylogenetic-Cladistic Species Concept and the Four-Fold Species Criteria (table), 198
Physical-to-Psychological Analogy, 269–73
physics, 5–6
Pinker, Steven, 254–55, 258
Pleistocene adapted mind, 267–68
Pleistocene period, 267, 270–71
pluralism, 103, 204
pneuma (vital heat), 30–31
Politics (Aristotle), 22–23
Popper, Karl, 302
Popper's criterion of falsifiability, 296–97
Popper's falsification standard, 294, 296
population dynamics, xvi, 109
population genetics, 93
population growth beyond available resources (Malthus' pressure cooker), 64–65
population structure, 272
Posterior Analytics (Aristotle), 8
pre-existing final and formal causes, 30
Precision and Longevity Argument, 95–100, 102
predictive or mechanistic ought, 240
preferred male characteristics, 259–60
prescriptive/normative ought, 240
prescriptivity, 221
Preservation Argument, 54–55, 57–58, 60, 65

Preston, Beth, 143–44
primary unit of selection, 88–90, 107
Principles of Geology (Lyell), 43
Prior Analytics (Aristotle), 8
privileging the gene over the individual as *primary* unit of selection, 100
progress, 238. *See also* Argument from Progress
progressive evolution, 230
  objection to, xvii, 220, 226–32
Propensity Evolutionary Functionalism, 148
  circularity charge, 150
  replies to, 149–52
Propensity Evolutionary Functionalism, Four-Fold Criteria (table), 153
Propensity Functionalism, xvi
psychology, xiv
purpose
  Aristotle's views on, 15
  part of causal make-up of natural phenomena, 16

Quine, W.V.O., 297–98
Quine-Duhem thesis, 298
Quinn, Philip, 294

race, xiv
Ramsay, Andrew, 56
Rawls, John, 233
  principle of justice, 233–35
realism/nominalism debate, 184
realists, 168
reasons as causes, 225–26, 236
  ontological status of, 224
reasons distinct from first-order desires, 226
reciprocal altruism, 218–20, 234–35, 247
reductionist objection, 28–33
Relational Biosimilarity Species Concept, xvii, 165, 173, 198–200
  nominalist challenge, 201–03

Relational Biosimilarity Species Concept and the Four-Field Species Criteria (table), 204
replicators, 95
Reznick, David N., 56–57, 66–67
Richards, Robert J., 240
  move from "is" to "ought," 239–40
  non-consequentialist account, 236
Richards' EE2 project and the standard objections, 238–40
Richards' EE2 Project (table), 238
Richards' Kantian/intention based evolutionary ethic, 237
Roth, Martin, 132–35, 137–38, 142
Ruse, Michael, 213, 232–34, 240
  acknowledges free choices, 235–36
  moral-sense-faculty theory, 235
  rejects evolutionary ethic that relies on progress, 236
Ruse's EE1 Project (table), 235
Ruse's Rawlsian account, 236

science and religion relationship, 277–79
science *vs.* non-science, 294–95, 300, 302
science *vs.* non-science (demarcation problem), 294, 296, 299, 302
scientific observation, 9
scientism, 283–84
Sedgwick, Adam, 42
Sedley, David, 23
  use of the Godly world analogy, 24–25
segregation distorter genes, 98
self-correcting (species-specific), 19
self-maintenance, 20, 126–28
self-organizing control, 19
self-regarding strategic advantages of social justice practices, 235
The Selfish Gene (Dawkins), 96
selfish genome genes, 98–99
selfish renegade genes, 98–99
sequential and developmental organization, 18, 21

INDEX

sets, 168
Shackelford, Todd K., 269–70, 273
Skinner, B.F., 20
social cooperation, 234
social Darwinism, 76
soul, 5–6, 23, 27, 31
   as final and formal cause, 18–19
speciation (production of new species), 74–75
species, xiv, 44–45, 166
Species as Essentialist-Natural Kind, 174–79, 186
   Darwinism and, 176–79
   nominalist challenge and, 179
Species as Essentialist-Natural Kind and the four-Fold Species Criteria (table), 179
Species as Individual Lineages, 185–92, 195
   artefact analogy strategy, 189–90
   asexual-groups criticism, 188–89
   nominalist challenge, 191–92
Species as Individual Lineages and the Four-Fold Species Criteria (table), 192
species as individuals, xvii, 165, 173
species as natural kinds, xvii, 165
species ascription, criteria for, 171–72
species debate, xvi, 164–67, 172
   argument from arbitrariness, 170
   argument from causal efficacy, 169
   nominalist challenge, 168–70
species designation, four-fold criteria (table), 171
species-level transformation, 72
species modification, 59
species survival, 76
specific predictions (SP), 117–18
specific predictions (SP) inference, 119
Spencer, Herbert, 222
Stamos, David, 165–67, 184, 190, 199–203
Starratt, Valerie G., 267, 269–70, 272
static condition, 175, 179
Sterelny, Kim, 103, 115, 269

stromatolites, 67
Strong Argument for Natural Selection, 49–55, 65–67, 69, 77–78, 87
Strong Evolutionary Psychology (SEP), xvii, 255, 258–59, 261, 265–69, 271
   evolutionary adaptive traits endorsed by, 271–72
   massive modularity thesis, 260
   principles of, 255–56
   reliance on Physical-to-Psychological Analogy, 273
Succulina, 229
supervenience, 203
surprise/hostility criterion, 62
Symons, David, 257–58, 268
synapomorphies, 194–97
systemic functionalism, xvi, 130–39
systemic functionalism, four-fold criteria and (table), 139

teleological concepts, 15
teleological understanding of biological phenomena, 128
teleology/final causation (Aristotle). *See* Aristotle's final causation (teleology)
Time, Argument from, 267–69, 270
time and improved technology
   production of further knowledge, 292
Tooby, John, 117, 254–55, 258, 265
trait polarity, 272
transitional species, absence of, 69–70
tree-of-life hypothesis, 55–61, 64, 69, 72, 75
"trust-forming" interactions, 247
type III secretory system (TTSS), 290–92

unavoidability condition, 238
uniformitarian geologic principles, 56–57
Uniformity of Degree, 57
Uniformity of Kind, 57
Uniformity of Law, 57
Uniformity of Methodology, 57

unit(s) of selection, xiv
unit(s) of selection debate, xvi, 88–91, 96, 108, 110–11
unit(s) of selection debate (table), 90
universal condition, 175, 178–79
universals, 167–68
"Unmoved Mover," 25
unobserved final causes, 18
utilitarianism, 215–16, 219
utility, 128
utility and goal-directed benefits connection, 128

verification principle, 295–96
Vienna Circle, 294
virtue ethics, 215
visibility argument, 100, 102

Waddington, C.H., 128

waist-to-hip-ratio (WHR), 258, 268–69
"war of nature," 65
Wason Selection Task, 261–63, 265
Weak Conclusion, 49, 67
Weak Evolutionary Psychology (WEP), xvii, 255
Whewell, William, 45–46
Williams, George C., 87, 89
  Adaptation and Natural Selection, 86
Williston's Law, 229
Wilson, D.S., 107–08
Wilson, E.O., 226–28, 231
Wilson, R.A., 110–11, 116
Woolcock, Peter G., 225
working species concepts and four-fold criteria (table), 173
Wynne-Edwards, V.C., 87–88, 104
  Animal Dispersion in Relation to Social Behavior, 86

FROM THE PUBLISHER

A name never says it all, but the word "Broadview" expresses a good deal of the philosophy behind our company. We are open to a broad range of academic approaches and political viewpoints. We pay attention to the broad impact book publishing and book printing has in the wider world; for some years now we have used 100% recycled paper for most titles. Our publishing program is internationally oriented and broad-ranging. Our individual titles often appeal to a broad readership too; many are of interest as much to general readers as to academics and students.

Founded in 1985, Broadview remains a fully independent company owned by its shareholders—not an imprint or subsidiary of a larger multinational.

For the most accurate information on our books (including information on pricing, editions, and formats) please visit our website at www.broadviewpress.com. Our print books and ebooks are also available for sale on our site.

broadview press
www.broadviewpress.com

3 9015 10051 4911

The interior of this book is printed on 100% recycled paper.